Non-SI Units		SI Units*
Multiply	By	To obtain
acre	4047	square meter, m^2
acre	0.405	hectare, ha (10^4 m^2)
acre	4.05×10^{-3}	square kilometer, km^2 (10^6 m^2)
Angstrom unit	10	nanometer, nm (10^{-9} m)
atmosphere	0.101	megapascal, MPa (10^6 Pa)
bar	10^{-1}	megapascal, MPa (10^6 Pa)
British thermal unit	1054	joule, j
calorie	4.19	joule, J
cubic feet	0.028	cubic meter, m^3
cubic feet	28.3	liter, L (10^{-3} m^3)
cubic inch	1.64×10^{-5}	cubic meter, m^3
cubic inch	16.4	cubic centimeter, cm^3 (10^{-6} m^3)
curie	3.7×10^{10}	becquerel, Bq
degrees (angle)	1.75×10^{-2}	radian, rad
dyne	10^{-5}	newton, N
erg	10^{-7}	joule, J
foot	0.305	meter, m
foot-pound	1.356	joule, J
gallon	3.785	liter, L (10^{-3} m^3)
gallon per acre	9.35	liter, per hectare, $L \cdot ha^{-1}$
gram per cubic centimeter	1.00	megagram per cubic meter, $Mg \cdot m^{-3}$
inch	25.4	millimeter, mm (10^{-3} m)
micron	1.00	micrometer, μm (10^{-6} m)
mile	1.61	kilometer, km (10^3 m)
miles per hour	0.477	meter per second, $m \cdot s^{-1}$
millimhos per centimeter	1.00	decisiemen per meter, $dS \cdot m^{-1}$
ounce	28.4	gram, g (10^{-3} kg)
ounce (fluid)	2.96×10^{-2}	liter, L (10^{-3} m^3)
pint (liquid)	0.473	liter, L (10^{-3} m^3)
pound	453.6	gram, g (10^{-3} kg)
pound per acre	1.12	kilogram per hectare, $kg \cdot ha^{-1}$
pound per acre	1.12×10^{-3}	megagram per hectare, $Mg \cdot ha^{-1}$
pounds per cubic foot	16.02	kilogram per cubic meter, $kg \cdot m^{-3}$
pounds per cubic inch	2.77×10^4	kilogram per cubic meter, $kg \cdot m^{-3}$
pounds per square foot	47.88	pascal, Pa
pounds per square inch	6.90×10^3	pascal, Pa
quart (liquid)	0.946	liter, L (10^{-3} m^3)
quintal (metric)	10^2	kilogram, kg
solid angle	12.57	steradian, sr
square centimeter per gram	0.1	square meter per kilogram, $m^2 \cdot kg^{-1}$
square feet	9.29×10^{-2}	square meter, m^2
square inch	645.2	square millimeter, mm^2 (10^{-6} m^2)
square mile	2.59	square kilometer, km^2
square millimeters per gram	10^{-3}	square meter per kilogram, $m^2 \cdot kg^{-1}$
temperature (°F) − 32	0.5555	temperature, C
temperature (°C) + 273	1	temperature, °K
ton (metric)	10^3	kilogram, kg
ton (2000 lb)	907	kilogram, kg
tons (2000 lb) per acre	2.24	megagram per hectare, $Mg \cdot ha^{-1}$

*International System of Units.

FUNDAMENTALS OF SOIL SCIENCE

FUNDAMENTALS OF SOIL SCIENCE

EIGHTH EDITION

HENRY D. FOTH
Michigan State University

JOHN WILEY & SONS
New York · Chichester · Brisbane · Toronto · Singapore

Cover Photo
Soil profile developed from glacio-fluvial sand in a balsam fir-black spruce
forest in the Laurentian Highlands of Quebec, Canada. The soil is classified as
a Spodosol (Orthod) in the United States and as a Humo-Ferric Podzol in
Canada.

Library of Congress Cataloging in Publication Data:

Foth, H. D.
    Fundamentals of soil science / Henry D. Foth.—8th ed.
        p.   cm.
    Includes bibliographical references.
    ISBN 0-471-52279-1
    1. Soil science.   I. Title.
    S591.F68   1990
    631.4—dc20
                                                    90-33890
                                                    CIP

Printed in the United States of America

10 9 8 7 6 5 4

Printed and bound by Malloy Lithographing, Inc.

PREFACE

The eighth edition is a major revision in which there has been careful revision of the topics covered as well as changes in the depth of coverage. Many new figures and tables are included. Summary statements are given at the ends of the more difficult sections within chapters, and a summary appears at the end of each chapter. Many nonagricultural examples are included to emphasize the importance of soil properties when soils are used in engineering and urban settings. The topics relating to environmental quality are found throughout the book to add interest to many chapters. Several examples of computer application are included.

The original Chapter 1, "Concepts of Soil," was split into two chapters. Each chapter emphasizes an important concept of soil—soil as a medium for plant growth and soil as a natural body. Topics covered in Chapter 1 include the factors affecting plant growth, root growth and distribution, nutrient availability (including the roles of root interception, mass flow and diffusion), and soil fertility and productivity. The importance of soils as a source of nutrients and water is stressed in Chapter 1 and elsewhere throughout the book. Chapter 2 covers the basic soil formation processes of humification of organic matter, mineral weathering, leaching, and translocation of colloids. The important theme is soil as a three-dimensional body that is dynamic and ever-changing. The concepts developed in the first two chapters are used repeatedly throughout the book.

The next five chapters relate to soil physical properties and water. The material on tillage and traffic was expanded to reflect the increasing effect of tillage and traffic on soils and plant growth and is considered in Chapter 4. The nature of soil water is presented as a continuum of soil water potentials in Chapter 5. Darcy's law is developed and water flow is discussed as a function of the hydraulic gradient and conductivity. Darcy's Law is used in Chapter 6, "Soil Water Management," in regard to water movement in infiltration, drainage, and irrigation. Chapter 6 also covers disposal of sewage effluent in soils and prescription athletic turf (PAT) as an example of precision control of the water, air, and salt relationships in soils used for plant growth. "Soil Erosion," Chapter 7, has been slightly reorganized with greater emphasis on water and wind erosion processes.

Chapters 8 and 9, "Soil Ecology" and "Soil Organic Matter," are complimentary chapters relating to the biological aspects of soils. The kinds and nature of soil organisms and nutrient cycling remain as the central themes of Chapter 8. An expanded section on the rhizosphere has been included. The distinctions between labile and stable organic matter and the interaction of organic matter with the minerals (especially clays) are central themes of Chapter 9. Also, the concept of cation exchange capacity is minimally developed in the coverage of the nature of soil organic matter in Chapter 9.

Chapter 10, "Soil Mineralogy," and Chapter 11, "Soil Chemistry", are complimentary chapters relating to the mineralogical and chemical properties of soils. The evolution theme included in Chapter 2 is used to develop the concept of changing mineralogical and chemical properties with time. Soils are characterized as being minimally, moderately, and intensively weathered, and these distinctions are used in discussions of soil pH, liming, soil fertility and fertilizer use, soil genesis, and land use.

Chapters 12 through 15 are concerned with the general area of soil fertility and fertilizer use. Chapters 12 and 13 cover the macronutrients and micronutrients plus toxic elements, respectively.

Chapters 14 and 15 cover the nature of fertilizers and the evaluation of soil fertility and the use of fertilizers, respectively. Greater stress has been placed on mass flow and diffusion in regard to nutrient uptake. The interaction of water and soil fertility is developed, and there is expanded coverage of soil fertility evaluation and the methods used to formulate fertilizer recommendations. Recognition is made of the increasing frequency of high soil test results and the implications for fertilizer use and environmental quality. Greater coverage is given to animal manure as both a source of nutrients and a source of energy. Information on land application of sewage sludge and on sustainable agriculture has been added. Throughout these four chapters there is a greater emphasis on the importance of soil fertility and fertilizers and on the environmental aspects of growing crops.

The next four chapters (Chapters 16, 17, 18, and 19) relate to the areas of soil genesis, soil taxonomy, soil geography and land use, and soil survey and land use interpretations. In this edition, the subjects of soil taxonomy (classification) and of soil survey and land use interpretations have received increased coverage in two small chapters. The emphasis in the soil geography and land use chapter is at the suborder level. References to lower categories are few. Color photographs of soil profiles are shown in Color Plates 5 and 6. No reference to Soil Taxonomy (USDA) is made until taxonomy is covered in Chapter 17. This allows a consideration of soil classification after soil properties have been covered. This arrangement also makes the book more desirable for use in two-year agricultural technology programs and overseas, in countries where Soil Taxonomy is not used.

The final chapter, "Land and the World Food Supply," includes a section on the world grain trade and examines the importance of nonagronomic factors in the food-population problem.

Both English and metric units are used in the measurement of crop yields, and for some other parameters. Using both kinds of units should satisfy both United States and foreign readers.

Special thanks to Mary Foth for the artwork and to my late son-in-law, Nate Rufe, for photographic contributions. Over the years, many colleagues have responded to my queries to expand my knowledge and understanding. Others have provided photographs. The reviewers also have provided an invaluable service. To these persons, I am grateful.

Finally, this book is a STORY about soil. The story reflects my love of the soil and my devotion to promoting the learning and understanding of soils for more than 40 years. I hope that all who read this book will find it interesting as well as informative.

Henry D. Foth

East Lansing, Michigan

BRIEF CONTENTS

DETAILED CONTENTS

CHAPTER 1

SOIL AS A MEDIUM FOR PLANT GROWTH

SOIL. Can you think of a substance that has had more meaning for humanity? The close bond that ancient civilizations had with the soil was expressed by the writer of Genesis in these words:

And the Lord God formed Man of dust from the ground.

There has been, and is, a reverence for the ground or soil. Someone has said that "the fabric of human life is woven on earthen looms; everywhere it smells of clay." Even today, most of the world's people are tillers of the soil and use simple tools to produce their food and fiber. Thus, the concept of soil as a medium of plant growth was born in antiquity and remains as one of the most important concepts of soil today (see Figure 1.1).

FACTORS OF PLANT GROWTH

The soil can be viewed as a mixture of mineral and organic particles of varying size and composition in regard to plant growth. The particles occupy about 50 percent of the soil's volume. The remaining soil volume, about 50 percent, is pore space, composed of pores of varying shapes and sizes. The pore spaces contain air and water and serve as channels for the movement of air and water. Pore spaces are used as runways for small animals and are avenues for the extension and growth of roots. Roots anchored in soil suppport plants and roots absorb water and nutrients. For good plant growth, the root-soil environment should be free of inhibitory factors. The three essential things that plants absorb from the soil and use are: (1) *water* that is mainly evaporated from plant leaves, (2) *nutrients* for nutrition, and (3) *oxygen* for root respiration.

Support for Plants

One of the most obvious functions of soil is to provide *support* for plants. Roots anchored in soil enable growing plants to remain upright. Plants grown by hydroponics (in liquid nutrient culture) are commonly supported on a wire framework. Plants growing in water are supported by the buoyancy of the water. Some very sandy soils that are droughty and infertile provide plants with little else than support. Such soils, however, produce high-yielding crops when fertilized and frequently irrigated. There are soils in which the impenetrable nature of the subsoil, or the presence of water-saturated soil close to the soil surface, cause shal-

FIGURE 1.1 Wheat harvest near the India-Nepal border. About one half of the world's people are farmers who are closely tied to the land and make their living producing crops with simple tools.

low rooting. Shallow-rooted trees are easily blown over by wind, resulting in *windthrow*.

Essential Nutrient Elements

Plants need certain *essential nutrient elements* to complete their life cycle. No other element can completely substitute for these elements. At least 16 elements are currently considered essential for the growth of most vascular plants. Carbon, hydrogen, and oxygen are combined in photosynthetic reactions and are obtained from air and water. These three elements compose 90 percent or more of the dry matter of plants. The remaining 13 elements are obtained largely from the soil. Nitrogen (N), phosphorus (P), potassium (K), calcium Ca), magnesium (Mg), and sulfur (S) are required in relatively large amounts and are referred to as the *macronutrients*. Elements required in considerably smaller amount are called the *micronutrients*. They include boron (B), chlorine (Cl), copper (Cu), iron (Fe), manganese (Mn), molybdenum (Mo), and zinc (Zn). Cobalt (Co) is a micronutrient that is needed by only

some plants. Plants deficient in an essential element tend to exhibit symptoms that are unique for that element, as shown in Figure 1.2.

More than 40 other elements have been found

FIGURE 1.2 Manganese deficiency symptoms on kidney beans. The youngest, or upper leaves, have light-green or yellow-colored intervein areas and dark-green veins.

in plants. Some plants accumulate elements that are not essential but increase growth or quality. The absorption of sodium (Na) by celery is an example, and results in an improvement of flavor. Sodium can also be a substitute for part of the potassium requirement of some plants, if potassium is in low supply. Silicon (Si) uptake may increase stem strength, disease resistance, and growth in grasses.

Most of the nutrients in soils exist in the minerals and organic matter. Minerals are inorganic substances occurring naturally in the earth. They have a consistent and distinctive set of physical properties and a chemical composition that can be expressed by a formula. Quartz, a mineral composed of SiO_2, is the principal constituent of ordinary sand. Calcite ($CaCO_3$) is the primary mineral in limestone and chalk and is abundant is many soils. Orthclase-feldspar ($KAlSi_3O_8$) is a very common soil mineral, which contains potassium. Many other minerals exist in soils because soils are derived from rocks or materials containing a wide variety of minerals. Weathering of minerals brings about their decomposition and the production of *ions* that are released into the soil water. Since silicon is not an essential element, the weathering of quartz does not supply an essential nutrient, plants do not depend on these minerals for their oxygen. The weathering of calcite supplies calcium, as Ca^{2+}, and the weathering of orthoclase releases potassium as K^+.

The organic matter in soils consists of the recent remains of plants, microbes, and animals and the resistant organic compounds resulting from the rotting or decomposition processes. Decomposition of soil organic matter releases essential nutrient ions into the soil water where the ions are available for another cycle of plant growth.

Available elements or nutrients are those nutrient ions or compounds that plants and microorganisms can absorb and utilize in their growth. Nutrients are generally absorbed by roots as cations and anions from the water in soils, or the *soil solution*. The ions are electrically charged.

Cations are positively charged ions such as Ca^{2+} and K^+ and anions are negatively charged ions such as NO_3^- (nitrate) and $H_2PO_4^-$ (phosphate). The amount of cations absorbed by a plant is about chemically equal to the amount of anions absorbed. Excess uptake of cations, however, results in excretion of H^+ and excess uptake of anions results in excretion of OH^- or HCO_3^- to maintain electrical neutrality in roots and soil. The essential elements that are commonly absorbed from soils by roots, together with their chemical symbols and the uptake forms, are listed in Table 1.1.

In nature, plants accommodate themselves to the supply of available nutrients. Seldom or rarely is a soil capable of supplying enough of the essential nutrients to produce high crop yields for any reasonable period of time after natural or virgin lands are converted to cropland. Thus, the use of animal manures and other amendments to increase soil fertility (increase the amount of nutrient ions) are ancient soil management practices.

Water Requirement of Plants

A few hundred to a few thousand grams of water are required to produce 1 gram of dry plant material. Approximately one percent of this water becomes an integral part of the plant. The remainder of the water is lost through *transpiration*, the loss of water by evaporation from leaves. Atmospheric conditions, such as relative humidity and temperature, play a major role in determining how quickly water is transpired.

The growth of most economic crops will be curtailed when a shortage of water occurs, even though it may be temporary. Therefore, the soil's ability to hold water over time against gravity is important unless rainfall or irrigation is adequate. Conversely, when soils become water saturated, the water excludes air from the pore spaces and creates an oxygen deficiency. The need for the removal of excess water from soils is related to the need for oxygen.

TABLE 1.1   Chemical Symbols and Common Forms of the Essential Elements Absorbed by Plant Roots from Soils

Nutrient	Chemical Symbol	Forms Commonly Absorbed by Plants
Macronutrients		
Nitrogen	N	NO_3^-, NH_4^+
Phosphorus	P	$H_2PO_4^-, HPO_4^{2-}$
Potassium	K	K^+
Calcium	Ca	Ca^{2+}
Magnesium	Mg	Mg^{2+}
Sulfur	S	SO_4^{2-}
Micronutrients		
Manganese	Mn	Mn^{2+}
Iron	Fe	Fe^{2+}
Boron	B	H_3BO_3
Zinc	Zn	Zn^{2+}
Copper	Cu	Cu^{2+}
Molybdenum	Mo	MoO_4^{2-}
Chlorine	Cl	Cl^-

FIGURE 1.3   The soil in which these tomato plants were growing was saturated with water. The stopper at the bottom of the left container was immediately removed, and excess water quickly drained away. The soil in the right container remained water saturated and the plant became severely wilted within 24 hours because of an oxygen deficiency.

Oxygen Requirement of Plants

Roots have openings that permit gas exchange. Oxygen from the atmosphere diffuses into the soil and is used by root cells for respiration. The carbon dioxide produced by the respiration of roots, and microbes, diffuses through the soil pore space and exits into the atmosphere. Respiration releases energy that plant cells need for synthesis and translocation of the organic compounds needed for growth. Frequently, the concentration of nutrient ions in the soil solution is less than that in roots cells. As a consequence, respiration energy is also used for the active accumulation of nutrient ions against a concentration gradient.

Some plants, such as water lilies and rice, can grow in water-saturated soil because they have morphological structures that permit the diffusion of atmospheric oxygen down to the roots. Successful production of most plants in water culture requires aeration of the solution. Aerobic microorganisms require molecular oxygen (O_2) and use oxygen from the soil atmosphere to decompose organic matter and convert unavailable nutrients in organic matter into ionic forms that plants can reuse (nutrient cycling).

Great differences exist between plants in their ability to tolerate low oxygen levels in soils. Sensi-

FIGURE 1.4   Soil salinity (soluble salt) has seriously affected the growth of sugar beets in the foreground of this irrigated field.

tive plants may be wilted and/or killed as a result of saturating the soil with water for a few hours, as shown in Figure 1.3. The wilting is believed to result from a decrease in the permeability of the roots to water, which is a result of a disturbance of metabolic processes due to an oxygen deficiency.

Freedom from Inhibitory Factors

Abundant plant growth requires a soil environment that is free of inhibitory factors such as toxic substances, disease organisms, impenetrable layers, extremes in temperature and acidity or basicity, or an excessive salt content, as shown in Figure 1.4.

PLANT ROOTS AND SOIL RELATIONS

Plants utilize the plant growth factors in the soil by way of the roots. The density and distribution of roots affect the amount of nutrients and water that roots extract from soils. Perennials, such as oak and alfalfa, do not reestablish a completely new root system each year, which gives them a distinct advantage over annuals such as cotton or wheat. Root growth is also influenced by the soil environment; consequently, root distribution and density are a function of both the kind of plant and the nature of the root environment.

Development of Roots in Soils

A seed is a dormant plant. When placed in moist, warm soil, the seed absorbs water by osmosis and swells. Enzymes activate, and food reserves in the endosperm move to the embryo to be used in germination. As food reserves are exhausted, green leaves develop and photosynthesis begins. The plant now is totally dependent on the sun for energy and on the soil and atmosphere for nutrients and water. In a sense, this is a critical period in the life of a plant because the root system is small. Continued development of the plant requires: (1) the production of food (carbohydrates,

etc.) in the shoot via photosynthesis and translocation of food downward for root growth, and (2) the absorption of water and nutrients by roots and the upward translocation of water and nutrients to the shoot for growth.

After a root emerges from the seed, the root tip elongates by the division and elongation of cells in the meristematic region of the root cap. After the root cap invades the soil, it continues to elongate and permeate the soil by the continued division and elongation of cells. The passage of the root tip through the soil leaves behind sections of root that mature and become "permanent" residents of the soil.

As the plant continues to grow and roots elongate throughout the topsoil, root extension into the subsoil is likely to occur. The subsoil environment will be different in terms of the supply of water, nutrients, oxygen, and in other growth factors. This causes roots at different locations in the soil (topsoil versus subsoil) to perform different functions or the same functions to varying degrees. For example, most of the nitrogen will probably be absorbed by roots from the topsoil because most of the organic matter is concentrated there, and nitrate–nitrogen becomes available by the decomposition of organic matter. By contrast, in soils with acid topsoils and alkaline subsoils, deeply penetrating roots encounter a great abundance of calcium in the subsoil. Under these conditions, roots in an alkaline subsoil may absorb more calcium than roots in an acid topsoil. The topsoil frequently becomes depleted of water in dry periods, whereas an abundance of water still exists in the subsoil. This results in a relatively greater dependence on the subsoil for water and nutrients. Subsequent rains that rewet the topsoil cause a shift to greater dependence on the topsoil for water and nutrients. Thus, the manner in which plants grow is complex and changes continually throughout the growing season. In this regard, the plant may be defined as *an integrator of a complex and ever changing set of environmental conditions*.

The root systems of some common agricultural

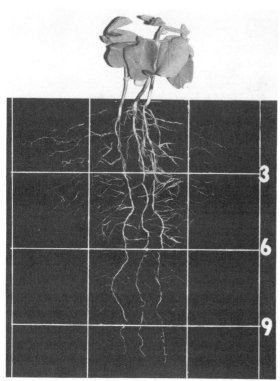

FIGURE 1.5 The tap root systems of two-week old soybean plants. Note the many fine roots branching off the tap roots and ramifying soil. (Scale on the right is in inches.)

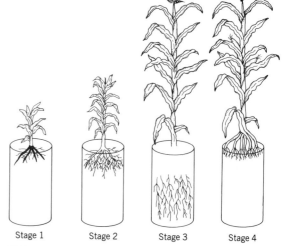

Stage 1   Stage 2   Stage 3   Stage 4

FIGURE 1.6 Four stages for the development of the shoots and roots of corn (Zea maize). Stage one *(left)* shows dominant downward and diagonal root growth, stage two shows "filling" of the upper soil layer with roots, stage three shows rapid elongation of stem and deep root growth, and stage four *(right)* shows development of the ears (grain) and brace root growth.

crops were sampled by using a metal frame to collect a 10-centimeter-thick slab of soil from the wall of a pit. The soil slab was cut into small blocks and the roots were separated from the soil using a stream of running water. Soybean plants were found to have a tap root that grows directly downward after germination, as shown in Figure 1.5. The tap roots of the young soybean plants are several times longer than the tops or shoots. Lateral roots develop along the tap roots and space themselves uniformly throughout the soil occupied by roots. At maturity, soybean taproots will extend about 1 meter deep in permeable soils with roots well distributed throughout the topsoil and subsoil. Alfalfa plants also have tap roots that commonly penetrate 2 to 3 meters deep; some have been known to reach a depth of 7 meters.

Periodic sampling of corn (*Zea maize*) root systems and shoots during the growing season revealed a synchronization between root and shoot growth. Four major stages of development were found. Corn has a fibrous root system, and early root growth is mainly by development of roots from the lower stem in a downward and diagonal direction away from the base of the plant (stage one of Figure 1.6). The second stage of root growth occurs when most of the leaves are developing and lateral roots appear and "fill" or space themselves uniformly in the upper 30 to 40 centimeters of soil. Stage three is characterized by rapid elongation of the stem and extension of roots to depths of 1 to 2 meters in the soil. Finally, during stage four, there is the production of the ear or the grain. Then, brace roots develop from the lower nodes of the stem to provide anchorage of the plant so that they are not blown over by the wind. Brace roots branch profusely upon entering the soil and also function for water and nutrient uptake.

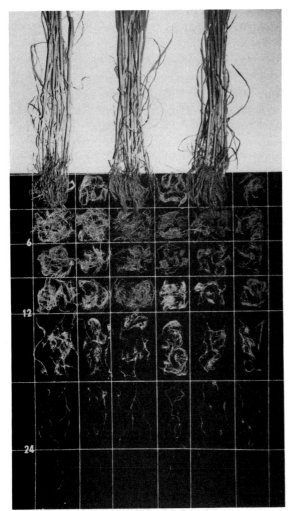

FIGURE 1.7 Roots of mature oat plants grown in rows 7 inches (18 cm) apart. The roots made up 13 percent of the total plant weight. Note the uniform, lateral, distribution of roots between 3 and 24 inches (8 and 60 cm). Scale along the left is in inches.

Extensiveness of Roots in Soil

Roots at plant maturity comprise about 10 percent of the entire mass of cereal plants, such as corn, wheat, and oats. The oat roots shown in Figure 1.7 weighed 1,767 pounds per acre (1,979 kg/ha) and made up 13 percent of the total plant weight. For trees, there is a relatively greater mass of roots as compared with tops, commonly in the range of 15 to 25 percent of the entire tree.

There is considerable uniformity in the lateral distribution of roots in the root zone of many crops (see Figure 1.7). This is explained on the basis of two factors. First, there is a random distribution of pore spaces that are large enough for root extension, because of soil cracks, channels formed by earthworms, or channels left from previous root growth. Second, as roots elongate through soil, they remove water and nutrients, which makes soil adjacent to roots a less favorable environment for future root growth. Then, roots grow preferentially in areas of the soil devoid of roots and where the supply of water and nutrients is more favorable for root growth. This results in a fairly uniform distribution of roots throughout the root zone unless there is some barrier to root extension or roots encounter an unfavorable environment.

Most plant roots do not invade soil that is dry, nutrient deficient, extremely acid, or water saturated and lacking oxygen. The preferential development of yellow birch roots in loam soil, compared with sand soil, because of a more favorable

FIGURE 1.8 Development of the root system of a yellow birch seedling in sand and loam soil. Both soils had adequate supplies of water and oxygen, but, the loam soil was much more fertile. (After Redmond, 1954.)

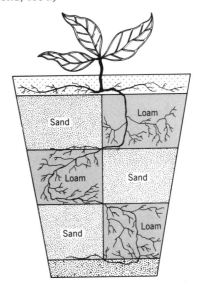

combination of nutrients and water, is shown in Figure 1.8.

Extent of Root and Soil Contact

A rye plant was grown in 1 cubic foot of soil for 4 months at the University of Iowa by Dittmer (this study is listed amoung the references at the end of the chapter). The root system was carefully removed from the soil by using a stream of running water, and the roots were counted and measured for size and length. The plant was found to have hundreds of kilometers (or miles) of roots. Based on an assumed value for the surface area of the soil, it was calculated that 1 percent or less of the soil surface was in direct contact with roots. Through much of the soil in the root zone, the distance between roots is approximately 1 centimeter. Thus, it is necessary for water and nutrient ions to move a short distance to root surfaces for the effective absorption of the water and nutrients. The limited mobility of water and most of the nutrients in moist and well-aerated soil means that only the soil that is invaded by roots can contribute significantly to the growth of plants.

Roles of Root Interception, Mass Flow, and Diffusion

Water and nutrients are absorbed at sites located on or at the surface of roots. Elongating roots directly encounter or intercept water and nutrient ions, which appear at root surfaces in position for absorption. This is *root interception* and accounts for about 1 percent or less of the nutrients absorbed. The amount intercepted is in proportion to the very limited amount of direct root and soil contact.

Continued absorption of water adjacent to the root creates a lower water content in the soil near the root surface than in the soil a short distance away. This difference in the water content between the two points creates a water content gradient, which causes water to move slowly in the direction of the root. Any nutrient ions in the water are carried along by flow of the water to root surfaces where the water and nutrients are both in position for absorption. Such movement of nutrients is called *mass flow*.

The greater the concentration of a nutrient in the soil solution, the greater is the quantity of the nutrient moved to roots by mass flow. The range

TABLE 1.2   Relation Between Concentration of Ions in the Soil Solution and Concentration within the Corn Plant

| | Concentration, parts per million | | | Ratio of Corn Plant Content to Lowest and Highest Soil Solution Contents | |
| | Soil Solution | | Corn Plant[a] | | |
	Low	High	Average	Low	High
Calcium	8	450	2,200	275	4.9
Nitrogen	6	1,700	15,000	2,500	8.8
Magnesium	3	204	1,800	600	8.8
Sulfur	118	655	1,700	155	2.6
Phosphorus	0.03	7.2	2,000	66,666	278.0
Potassium	3	156	20,000	6,666	128.0

Adapted from S. A. Barber, "A Diffusion and Mass Flow Concept of Soil Nutrient Availability, "*Soil Sci.,* **93:**39–49, 1962.
Used by permission of the author and The Williams and Wilkins Co., Baltimore.
[a] Dry weight basis.

of concentration for some nutrients in soil water is given in Table 1.2. The calcium concentration (Table 1.2) ranges from 8 to 450 parts per million (ppm). For a concentration of only 8 ppm in soil water, and 2,200 ppm of calcium in the plant, the plant would have to absorb 275 (2,200/8) times more water than the plant's dry weight to move the calcium needed to the roots via mass flow. Stated in another way, if the transpiration ratio is 275 (grams of water absorbed divided by grams of plant growth) and the concentration of calcium in the soil solution is 8 ppm, enough calcium will be moved to root surfaces to supply the plant need. Because 8 ppm is a very low calcium concentration in soil solutions, and transpiration ratios are usually greater than 275, mass flow generally moves more calcium to root surfaces than plants need. In fact, calcium frequently accumulates along root surfaces because the amount moved to the roots is greater than the amount of calcium that roots absorb.

Other nutrients that tend to have a relatively large concentration in the soil solution, relative to the concentration in the plant, are nitrogen, magnesium, and sulfur (see Table 1.2). This means that mass flow moves large amount of these nutrients to roots relative to plant needs.

Generally, mass flow moves only a small amount of the phosphorus to plant roots. The phosphorus concentration in the soil solution is usually very low. For a soil solution concentration of 0.03 ppm and 2,000 ppm plant concentration, the transpiration ratio would need to be more than 60,000. This illustration and others that could be drawn from the data in Table 1.2, indicate that some other mechanism is needed to account for the movement of some nutrients to root surfaces. This mechanism or process is known as diffusion.

Diffusion is the movement of nutrients in soil water that results from a concentration gradient. Diffusion of ions occurs whether or not water is moving. When an insufficient amount of nutrients is moved to the root surface via mass flow, diffusion plays an important role. Whether or not plants will be supplied a sufficient amount of a nutrient also depends on the amount needed. Calcium is rarely deficient for plant growth, partially because plants' needs are low. As a consequence, these needs are usually amply satisfied by the movement of calcium to roots by mass flow. The same is generally true for magnesium and sulfur. The concentration of nitrogen in the soil solution tends to be higher than that for calcium, but because of the high plant demand for nitrogen, about 20 percent of the nitrogen that plants absorb is moved to root surfaces by diffusion. Diffusion is the most important means by which phosphorus and potassium are transported to root surfaces, because of the combined effects of concentration in the soil solution and plant demands.

Mass flow can move a large amount of nutrients rapidly, whereas diffusion moves a small amount of nutrients very slowly. Mass flow and diffusion have a limited ability to move phosphorus and potassium to roots in order to satisfy the needs of crops, and this limitation partly explains why a large amount of phosphorus and potassium is added to soils in fertilizers. Conversely, the large amounts of calcium and magnesium that are moved to root surfaces, relative to crop plant needs, account for the small amount of calcium and magnesium that is added to soils in fertilizers.

Summary Statement

The available nutrients and available water are the nutrients and water that roots can absorb. The absorption of nutrients and water by roots is dependent on the surface area-density (cm^2/cm^3) of roots. Mathematically:

uptake = availability × surface area-density

SOIL FERTILITY AND SOIL PRODUCTIVITY

Soil fertility is defined as *the ability of a soil to supply essential elements for plant growth without a toxic concentration of any element*. Soil fertility refers to only one aspect of plant growth—the adequacy, toxicity, and balance of plant nutrients. An assessment of soil fertility can be made with a series of chemical tests.

Soil productivity is the soil's capacity to produce a certain yield of crops or other plants with optimum management. For example, the productivity of a soil for cotton production is commonly expressed as kilos, or bales of cotton per acre, or hectare, when using an optimum management system. The optimum managment system specifies such factors as planting date, fertilization, irrigation schedule, tillage, cropping sequence, and pest control. Soil scientists determine soil productivity ratings of soils for various crops by measuring yields (including tree growth or timber production) over a period of time for those production uses that are currently relevant. Included in the measurement of soil productivity are the influence of weather and the nature and aspect of slope, which greatly affects water runoff and erosion. Thus, soil productivity is an expression of all the factors, soil and nonsoil, that influence crop yields.

For a soil to produce high yields, it must be fertile for the crops grown. It does not follow, however, that a fertile soil will produce high yields. High yields or high soil productivity depends on optimum managment systems. Many fertile soils exist in arid regions but, within management systems that do not include irrigation, these soils are unproductive for corn or rice.

SUMMARY

The concept of soil as a medium for plant growth is an ancient concept and dates back to at least the beginning of agriculture. The concept emphasizes the soil's role in the growth of plants. Important aspects of the soil as a medium for plant growth are: (1) the role of the soil in supplying plants with growth factors, (2) the development and distribution of roots in soils, and (3) the movement of nutrients, water, and air to root surfaces for absorption. Soils are productive in terms of their ability to produce plants.

The concept of soil as a medium for plant growth views the soil as a material of fairly uniform composition. This is entirely satisfactory when plants are grown in containers that contain a soil mix. Plants found in fields and forests, however, are growing in soils that are not uniform. Differences in the properties between topsoil and subsoil layers affect water and nutrient absorption. It is natural for soils in fields and forests to be composed of horizontal layers that have different properties, so it is also important that agriculturists and foresters consider soils as *natural bodies*. This concept is also useful for persons involved in the building of engineering structures, solving environment problems such as nitrate pollution of groundwater, and using the soil for waste disposal. The soil as a natural body is considered in the next chapter.

REFERENCES

Barber, S. A. 1962. "A Diffusion and Mass Flow Concept of Soil Nutrient Availability." *Soil Sci.* **93**:39–49.

Dittmer, H. J. 1937. "A Quantitative Study of the Roots and Root Hairs of a Winter Rye Plant." *Am. Jour. Bot.* **24**:417–420.

Foth, H. D. 1962. "Root and Top Growth of Corn." *Agron. Jour.* **54**:49–52.

Foth, H. D., L. S. Robertson, and H. M. Brown. 1964. "Effect of Row Spacing Distance on Oat Performance." *Agron. Jour.* **56**:70-73.

Foth, H. D. and B. G. Ellis. 1988. *Soil Fertility.* John Wiley, New York.

Redmond, D. R. 1954. "Variations in Development of Yellow Birch Roots in Two Soil Types." *Forestry Chronicle.* **30**:401–406.

Simonson, R. W. 1968. "Concept of Soil." *Adv. in Agron.* **20**:1-47. Academic Press, New York.

Wadleigh, C. H. 1957. "Growth of Plants," in *Soil,* USDA Yearbook of Agriculture. Washington, D.C.

CHAPTER 2

SOIL AS A NATURAL BODY

One day a colleague asked me why the alfalfa plants on some research plots were growing so poorly. A pit was dug in the field and a vertical section of the soil was sampled by using a metal frame. The sample of soil that was collected was 5 centimeters thick, 15 centimeters wide, and 75 centimeters long. The soil was glued to a board and a vacuum cleaner was used to remove loose soil debris and expose the natural soil layers and roots. Careful inspection revealed four soil layers as shown in Figure 2.1.

The upper layer, 9 inches (22 cm) thick, is the plow layer. It has a dark color and an organic matter content larger than any of the other layers. Layer two, at the depth of 9 to 14 inches (22 to 35 cm) differs from layer one by having a light-gray color and a lower organic matter content. Both layers are porous and permeable for the movement of air and water and the elongation of roots. In layer three, at a depth of 14 to 23 inches (35 to 58 cm) many of the soil particles were arranged into blocklike aggregrates. When moist soil from layer three was pressed between the fingers, more stickiness was observed than in layers one and two, which meant that layer three had a greater clay content than the two upper layers. The roots penetrated this layer with no difficulty, however. Below layer three, the alfalfa tap root encountered a layer (layer four) that was impenetrable (too compact), with the root growing above it in a lateral direction. From these observations it was concluded that the alfalfa grew poorly because the soil material below a depth of 58 centimeters: (1) created a barrier to deep root penetration, which resulted in a less than normal supply of water for plant growth during the summer, and (2) created a water-saturated zone above the fourth layer that was deficient in oxygen during wet periods in the spring. The fact that the soil occurred naturally in a field raises such questions as: What kinds of layers do soils have naturally? How do the layers form? What are their properties? How do these layers affect how soils are used? The answers to these questions require an understanding that landscapes consist of three-dimensional bodies composed of unique horizontal layers. These naturally occurring bodies are *soils*. A recognition of the kinds of soil layers and their properties is required in order to use soils effectively for many different purposes.

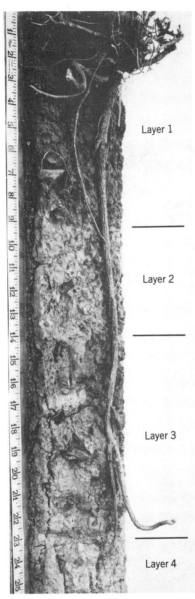

FIGURE 2.1 This alfalfa taproot grew vertically downward through the upper three layers. At a depth of 23 inches (58 cm), the taproot encountered an impenetrable layer (layer 4) and grew in a lateral direction above the layer.

THE PARENT MATERIAL OF SOILS

Soil formation, or the development of soils that are natural bodies, includes two broad processes.

First is the formation of a *parent material* from which the soil evolves and, second, the evolution of soil layers, as shown in Figure 2.1. Approximately 99 percent of the world's soils develop in mineral parent material that was or is derived from the weathering of bedrock, and the rest develop in organic materials derived from plant growth and consisting of muck or peat.

Bedrock Weathering and Formation of Parent Material

Bedrock is not considered soil parent material because soil layers do not form in it. Rather, the unconsolidated debris produced from the weathering of bedrock is soil parent material. When bedrock occurs at or near the land surface, the weathering of bedrock and the formation of parent material may occur simultaneously with the evolution of soil layers. This is shown in Figure 2.2, where a single soil horizon, the topsoil layer, overlies the R layer, or bedrock. The topsoil layer is about 12 inches (30 cm) thick and has evolved slowly at a rate controlled by the rate of rock weathering. The formation of a centimeter of soil in hundreds of years is accurate for this example of soil formation.

Rates of parent material formation from the direct weathering of bedrock are highly variable. A weakly cemented sandstone in a humid environment might disintegrate at the rate of a centimeter in 10 years and leave 1 centimeter of soil. Con-

FIGURE 2.2 Rock weathering and the formation of the topsoil layer are occurring simultaneously. Scale is in feet.

versely, quartzite (metamorphosed sandstone) nearby might weather so slowly that any weathered material might be removed by water or wind erosion. Soluble materials are removed during limestone weathering, leaving a residue of insoluble materials. Estimates indicate that it takes 100,000 years to form a foot of residue from the weathering of limestone in a humid region. Where soils are underlain at shallow depths by bedrock, loss of the soil by erosion produces serious consequences for the future management of the land.

Sediment Parent Materials

Weathering and erosion are two companion and opposing processes. Much of the material lost from a soil by erosion is transported downslope and deposited onto existing soils or is added to some sediment at a lower elevation in the landscape. This may include alluvial sediments along streams and rivers or marine sediments along ocean shorelines. Glaciation produced extensive sediments in the northern part of the northern hemisphere.

Four constrasting parent material–soil environments are shown in Figure 2.3. Bare rock is exposed on the steep slopes near the mountaintops. Here, any weathered material is lost by erosion and no parent material or soil accumulates. Very

thick alluvial sediments occur in the valley. Very thick glacial deposits occur on the tree-covered lateral moraine that is adjacent to the valley floor along the left side. An intermediate thickness of parent material occurs where trees are growing below the bare mountaintops and above the thick alluvial and moraine sediments. Most of the world's soils have formed in sediments consisting of material that was produced by the weathering of bedrock at one place and was transported and deposited at another location. In thick sediments or parent materials, the formation of soil layers is not limited by the rate of rock weathering, and several soil layers may form simultaneously.

SOIL FORMATION

Soil layers are approximately parallel to the land surface and several layers may evolve simultaneously over a period of time. The layers in a soil are genetically related; however, the layers differ from each other in their physical, chemical, and biological properties. In soil terminology, the layers are called *horizons*. Because soils as natural bodies are characterized by genetically developed horizons, soil formation consists of the evolution of soil horizons. A vertical exposure of a soil consisting of the horizons is a *soil profile*.

FIGURE 2.3 Four distinct soil-forming environments are depicted in this landscape in the Rocky Mountains, United States. On the highest and steepest slopes, rock is exposed because any weathered material is removed by erosion as fast as it forms. Thick alluvial sediments occur on the valley floor and on the forested lateral moraine adjacent to the valley floor along the left side. Glacial deposits of varying thickness overlying rock occur on the forested mountain slopes at intermediate elevations.

Soil-Forming Processes

Horizonation (the formation of soil horizons) results from the differential *gains, losses, transformations*, and *translocations* that occur over time within various parts of a vertical section of the parent material. Examples of the major kinds of changes that occur to produce horizons are: (1) addition of organic matter from plant growth, mainly to the topsoil; (2) transformation represented by the weathering of rocks and minerals and the decomposition of organic matter; (3) loss of soluble components by water moving downward through soil carrying out soluble salts; and, (4) translocation represented by the movement of suspended mineral and organic particles from the topsoil to the subsoil.

Formation of A and C Horizons

Many events, such as the deposition of volcanic ash, formation of spoil banks during railroad construction, melting of glaciers and formation of glacial sediments, or catastrophic flooding and formation of sediments have been dated quite accurately. By studying soils of varying age, soil scientists have reconstructed the kinds and the sequence of changes that occurred to produce soils.

Glacial sediments produced by continental and alpine glaciation are widespread in the northern hemisphere, and the approximate dates of the formation of glacial parent materials are known. After sediments have been produced near a retreating ice front, the temperature may become favorable for the invasion of plants. Their growth results in the addition of organic matter, especially the addition of organic matter at or near the soil surface. Animals, bacteria, and fungi feed on the organic materials produced by the plants, resulting in the loss of much carbon as carbon dioxide. During digestion or decomposition of fresh organic matter, however, a residual organic fraction is produced that is resistant to further alteration and accumulates in the soil. The resistant organic matter is called *humus* and the process is *humification*. The microorganisms and animals feeding on the organic debris eventually die and thus contribute to the formation of humus. Humus has a black or dark-brown color, which greatly affects the color of A horizons. In areas in which there is abundant plant growth, only a few decades are required for a surface layer to acquire a dark color, due to the humification and accumulation of organic matter, forming an *A horizon*.

The uppermost horizons shown in Figures 2.1 and 2.2 are A horizons. The A horizon in Figure 2.1 was converted into a plow layer by frequent plowing and tillage. Such A horizons are called *Ap horizons*, the p indicating plowing or other disturbance of the surface layer by cultivation, pasturing, or similar uses. For practical purposes, the topsoil in agricultural fields and gardens is synonymous with Ap horizon.

At this stage in soil evolution, it is likely that the upper part of the underlying parent material has been slightly altered. This slightly altered upper part of the parent material is the *C horizon*. The soil at this stage of evolution has two horizons—the A horizon and the underlying C horizon. Such soils are AC soils; the evolution of an AC soil is illustrated in Figure 2.4.

Formation of B Horizons

The subsoil in an AC soil consists of the C horizon and, perhaps, the upper part of the parent material. Under favorable conditions, this subsoil layer

FIGURE 2.4 Sequential evolution of some soil horizons in a sediment parent material.

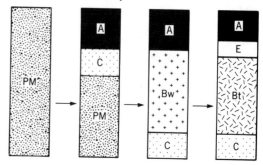

A Horizon—zone of organic matter accumulation

B Horizon—zone of colloid accumulation

C Horizon—zone of minimal weathering

FIGURE 2.5 A soil scientist observing soil properties near the boundary between the A and B horizons in a soil with A, B, and C horizons. As roots grow downward, or as water percolates downward, they encounter a different environment in the A, B, and C horizons. (Photograph USDA.)

eventually develops a distinctive color and some other properties that distinguish it from the A horizon and underlying parent material, commonly at a depth of about 60 to 75 centimeters. This altered subsoil zone becomes a *B horizon* and develops as a layer sandwiched between the A and a new deeper C horizon. At this point in soil evolution, insufficient time has elapsed for the B horizon to have been significantly enriched with fine-sized (colloidal) particles, which have been translocated downward from the A horizon by percolating water. Such a weakly developed B horizon is given the symbol w (as in Bw), to indicate its weakly developed character. A Bw horizon can be distinguished from A and C horizons primarily by color, arrangement of soil particles, and an intermediate content of organic matter. A soil with A, B, and C horizons is shown in Figure 2.5.

During the early phases of soil evolution, the soil formation processes progressively transform parent material into soil, and the soil increases in thickness. The evolution of a thin AC soil into a thick ABwC soil is illustrated in Figure 2.4.

The Bt Horizon Soil parent materials frequently contain calcium carbonate ($CaCO_3$), or lime, and are alkaline. In the case of glacial parent materials, lime was incorporated into the ice when glaciers overrode limestone rocks. The subsequent melting of the ice left a sediment that contains limestone particles. In humid regions, the lime dissolves in percolating water and is removed from the soil, a process called *leaching*. Leaching effects are progressive from the surface downward. The surface soil first becomes acid, and subsequently leaching produces an acid subsoil.

An acid soil environment greatly stimulates mineral weathering or the dissolution of minerals with the formation of many ions. The reaction of orthoclase feldspar ($KAlSiO_3$) with water and H^+ is as follows:

$$2 KAlSiO_3 + 9H_2O + 2H^+$$
(orthoclase)

$$= H_4Al_2Si_2O_9 + 2K^+ + 4H_4SiO_4$$
(kaolinite) (silicic acid)

The weathering reaction illustrates three important results of mineral weathering. First, clay particles (fine-sized mineral particles) are formed—in the example, kaolinite. In effect, soils are "clay factories." Second, ions are released into the soil solution, including nutrient ions such as K^+. Third, other compounds (silicic acid) of varying solubility are formed and are subject to leaching and removal from the soil.

Clay formation results mainly from chemical weathering. Time estimates for the formation of 1 percent clay in rock parent material range from 500 to 10,000 years. Some weathered rocks with small areas in which minerals are being converted into clay are shown in Figure 2.6.

Many soil parent materials commonly contain some clay. Some of this clay, together with clay produced by weathering during soil formation, tends to be slowly translocated downward from the A horizon to the B horizon by percolating water. When a significant increase in the clay content of a Bw horizon occurs due to clay translocation, a Bw horizon becomes a *Bt horizon*.

FIGURE 2.6 Weathering releases mineral grains in rocks and results in the formation of very fine-sized particles of clay, in this case, kaolinite.

Thin layers or films of clay can usually be observed along cracks and in pore spaces with a 10-power hand lens. The process of accumulation of soil material into a horizon by movement out of some other horizon is *illuviation*. The t (as in Bt) refers to an illuvial accumulation of clay. The Bt horizon may be encountered when digging holes for posts or trenching for laying underground pipes.

Alternating periods of wetting and drying seem necessary for clay translocation. Some clay parti-

cles are believed to disperse when dry soil is wetted at the end of a dry season and the clay particles migrate downward in percolating water during the wet season. When the downward percolating water encounters dry soil, water is withdrawn into the surrounding dry soil, resulting in the deposition of clay on the walls of pore spaces. Repeated cycles of wetting and drying build up layers of oriented clay particles, which are called *clay skins*.

Many studies of clay illuviation have been made. The studies provide evidence that thousands of years are needed to produce a significant increase in the content of clay in B horizons. An example is the study of soils on the alluvial floodplain and adjacent alluvial fans in the Central Valley of California. Here, increasing elevation of land surfaces is associated with increasing age. The soils studied varied in age from 1,000 to more than 100,000 years.

The results of the study are presented in Figure 2.7. The Hanford soil developed on the floodplain is 1,000 years old; it shows no obvious evidence of illuviation of clay. The 10,000-year-old Greenfield soil has about 1.4 times more clay in the subsoil (Bt horizon) than in the A horizon. Snelling soils are 100,000 years old and contain 2.5 times more clay in the Bt horizon than in the A horizon. The

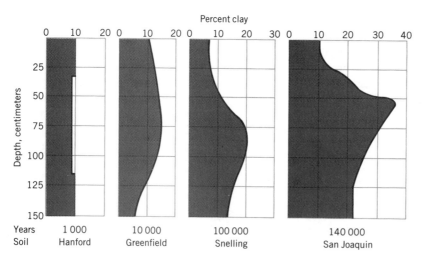

FIGURE 2.7 Clay distribution as a function of time in soils developed from granitic parent materials in the Central Valley of California. The Hanford soil, only 1,000 years old, does not have a Bt horizon. The other three soils have Bt horizons. The Bt horizon of the San Joaquin is a claypan that inhibits roots and the downward percolation of water. (After Arkley, 1964.)

San Joaquin soil is 140,000 years old, and has 3.4 times more clay in the horizon of maximum clay accumulation as compared to the A horizon.

The three youngest soils (Hanford, Greenfield, and Snelling) are best suited for agriculture because the subsoil horizons are permeable to water and air, and plant roots penetrate through the B horizons and into the C horizons. Conversely, the impermeable subsoil horizon in San Joaquin soil causes shallow rooting. The root zone above the impermeable horizon becomes water saturated in the wet seasons. The soil is dry and droughty in the dry season.

Water aquifers underlie soils and varying thicknesses of parent materials and rocks. Part of the precipitation in humid regions migrates completely through the soil and recharges underlying aquifers. The development of water-impermeable claypans over an extensive region results in less water recharge and greater water runoff. This has occurred near Stuttgart, Arkansas, where wells used for the irrigation of rice have run dry because of the limited recharge of the aquifer.

The Bhs Horizon Many sand parent materials contain very little clay, and almost no clay forms in them via weathering. As a consequence, clay illuviation is insignificant and Bt horizons do not evolve. Humus, however, reacts with oxides of aluminum and/or iron to form complexes in the upper part of the soil. Where much water for leaching (percolation) is present, as in humid regions, these complexes are translocated downward in percolating water to form illuvial accumulations in the B horizon. The illuvial accumulation of humus and oxides of aluminum and/or iron in the B horizon produces *Bhs horizons*. The h indicates the presence of an illuvial accumulation of humus and the s indicates the presence of illuvial oxides of aluminum and/or iron. The symbol s is derived from sesquioxides (such as Fe_2O_3 and Al_2O_3). Bhs horizons are common in very sandy soils that are found in the forested areas of the eastern United States from Maine to Florida.

The high content of sand results in soils with low fertility and low water-retention capacity (droughtiness).

Formation of E Horizons

The downward translocation of colloids from the A horizon may result in the concentration of sand and silt-sized particles (particles larger than clay size) of quartz and other resistant minerals in the upper part of many soils. In soils with thin A horizons, a light-colored horizon may develop at the boundary of the A and B horizons (see Figure 2.4). This horizon, commonly grayish in color, is the *E horizon*. The symbol E is derived from *eluviation*, meaning, "washed-out."Both the A and E horizons are eluvial in a given soil. The main feature of the A horizon, however, is the presence of organic matter and a dark color, whereas that of the E horizon is a light-gray color and having low organic matter content and a concentration of silt and sand-sized particles of quartz and other resistant minerals.

The development of E horizons occurs more readily in forest soils than in grassland soils, because there is usually more eluviation in forest soils, and the A horizon is typically much thinner. The development of E horizons occurs readily in soils with Bhs horizons, and the E horizons may have a white color (see soil on book cover).

A soil with A, E, Bt, and C horizons is shown in Figure 2.8. At this building site, the suitability of the soil for the successful operation of a septic effluent disposal system depends on the rate at which water can move through the least permeable horizon, in this example the Bt horizon. Thus, the value of rural land for home construction beyond the limits of municipal sewage systems depends on the nature of the subsoil horizons and their ability to allow for the downward migration and disposal of sewage effluent. Suitable sites for construction can be identified by making a percolation test of those horizons through which effluent will be disposed.

FIGURE 2.8 A soil with A, E, Bt, and C horizons that formed in 10,000 years under forest vegetation. The amount of clay in the Bt horizon and permeability of the Bt horizon to water determine the suitability of this site for home construction in a rural area where the septic tank effluent must be disposed by percolating through the soil.

Formation of O Horizons

Vegetation produced in the shallow waters of lakes and ponds may accumulate as sediments of peat and muck because of a lack of oxygen in the water for their decomposition. These sediments are the parent material for *organic soils*. Organic soils have *O horizons*; the O refers to soil layers dominated by organic material. In some cases, extreme wetness and acidity at the surface of the soil produce conditions unfavorable for decomposition of organic matter. The result is the formation of O horizons on the top of mineral soil horizons. Although a very small proportion of the world's soils have O horizons, these soils are widely scattered throughout the world.

SOILS AS NATURAL BODIES

Various factors contribute to making soils what they are. One of the most obvious is parent material. Soil formation, however, may result in many different kinds of soils from a given parent material. Parent material and the other factors that are responsible for the development of soil are the *soil-forming factors*.

The Soil-Forming Factors

Five soil-forming factors are generally recognized: *parent material, organisms, climate, topography,* and *time*. It has been shown that Bt and Bhs horizon development is related to the clay and sand content within the parent material and/or the amount of clay that is formed during soil evolution.

Grass vegetation contributes to soils with thick A horizons because of the profuse growth of fine roots in the upper 30 to 40 centimeters of soil. In forests, organic matter is added to soils mainly by leaves and wood that fall onto the soil surface. Small-animal activities contribute to some mixing of organic matter into and within the soil. As a result, organic matter in forest soils tends to be incorporated into only a thin layer of soil, resulting in thin A horizons.

The climate contributes to soil formation through its temperature and precipitation components. If parent materials are permanently frozen or dry, soils do not develop. Water is needed for plant growth, for weathering, leaching, and translocation of clay, and so on. A warm, humid climate promotes soil formation, whereas dry and/or cold climates inhibit it.

The topography refers to the general nature of the land surface. On slopes, the loss of water by runoff and the removal of soil by erosion retard soil formation. Areas that receive runoff water may have greater plant growth and organic matter content, and more water may percolate through the soil.

The extent to which these factors operate is a function of the amount of time that has been available for their operation. Thus, soil may be defined as:

unconsolidated material on the surface of the earth that has been subjected to and influenced by the genetic and environmental factors of parent material, climate, organisms, and topography, all acting over a period of time.

FIGURE 2.10 Soil scientists studying soils as natural bodies in the field. Soils are exposed in the pit and soils data are displayed on charts for observation and discussion.

Soil Bodies as Parts of Landscapes

At any given location on the landscape, there is a particular soil with a unique set of properties, including kinds and nature of the horizons. Soil properties may remain fairly constant from that location in all directions for some distance. The area in which soil properties remain reasonably constant is a *soil body*. Eventually, a significant change will occur in one or more of the soil-forming factors and a different soil or soil body will be encountered.

Locally, changes in parent material and/or

slope (topography) account for the existence of different soil bodies in a given field, as shown in Figure 2.9. The dark-colored soil in the foreground receives runoff water from the adjacent slopes. The light-colored soil on the slopes developed where water runoff and erosion occurred. Distinctly different management practices are required to use effectively the poorly drained soil in the foreground and the eroded soil on the slope.

The boundary between the two different soils soils is easily seen. In many instances the boundaries between soils require an inspection of the soil, which is done by digging a pit or using a soil auger.

How Scientists Study Soils as Natural Bodies

A particular soil, or soil body, occupies a particular part of a landscape. To learn about such a soil, a pit is usually dug and the soil horizons are described and sampled. Each horizon is described in terms of its thickness, color, arrangement of particles, clay content, abundance of roots, presence or absence of lime, pH, and so on. Samples from each horizon are taken to the laboratory and are analyzed for their chemical, physi-

FIGURE 2.9 A field or landscape containing black-colored soil in the lowest part of the landscape. This soil receives runoff water and eroded sediment from the light-colored soil on the sloping areas.

FIGURE 2.11 Muck (organic) soil layer being replaced with sand to increase the stability of the roadbed.

cal, and biological properties. These data are presented in graphic form to show how various soil properties remain the same or change from one horizon to another (shown for clay content in Figure 2.7). Figure 2.10 shows soil scientists studying a soil in the field. Pertinent data are presented by a researcher, using charts, and the properties and genesis of the soil are discussed by the group participitants.

Importance of Concept of Soils as Natural Bodies

The nature and properties of the horizons in a soil determine the soil's suitability for various uses. To use soils prudently, an inventory of the soil's properties is needed to serve as the basis for making predictions of soil behavior in various situations. Soil maps, which show the location of the soil bodies in an area, and written reports about soil properties and predictions of soil behavior for various uses began in the United States in 1896. By the 1920s, soil maps were being used to plan the location and construction of highways in Michigan. Soil materials that are unstable must be re-

moved and replaced with material that can withstand the pressures of vehicular traffic. The scene in Figure 2.11 is of section of road that was built without removing a muck soil layer. The road became unstable and the muck layer was eventually removed and replaced with sand.

SUMMARY

The original source of all mineral soil parent material is rock weathering. Some soils have formed directly in the products of rock weathering at their present location. In these instances, horizon formation may be limited by the rate of rock weathering, and soil formation may be very slow. Most soils, however, have formed in sediments resulting from the erosion, movement, and deposition of material by glaciers, water, wind, and gravity. Soils that have formed in organic sediments are organic soils.

The major soil-forming processes include: (1) humification and accumulation of organic matter, (2) rock and mineral weathering, (3) leaching of soluble materials, and (4) the eluviation and illuviation of colloidal particles. The operation of the soil formation processes over time produces soil horizons as a result of differential changes in one soil layer, as compared to another.

The master soil horizons or layers include the O, A, E, B, C, and R horizons.

Different kinds of soil occur as a result of the interaction of the soil-forming factors: parent material, organisms, climate, topography, and time.

Landscapes are composed of three-dimensional bodies that have naturally (genetically) developed horizons. These bodies are called soils.

Prudent use of soils depends on a recognition of soil properties and predictions of soil behavior under various conditions.

REFERENCES

Arkley, R. J. 1964. *Soil Survey of the Eastern Stanislaus Area, California*. U.S.D.A. and Cal. Agr. Exp. Sta.

Barshad, I. 1959. "Factors Affecting Clay Formation." *Clays and Clay Minerals*. E. Ingerson (ed.), *Earth Sciences Monograph 2*. Pergamon, New York.

Jenny, H. 1941. *Factors of Soil Formation*. McGraw-Hill, New York.

Jenny, H. 1980. *The Soil Resource*. Springer-Verlag, New York.

Simonson, R. W. 1957. "What Soils Are." USDA Agricultural Yearbook, Washington, D. C.

Simmonson, R. W. 1959. "Outline of Generalized Theory of Soil Genesis." *Soil Sci. Soc. Am. Proc.* **23**:161–164.

Twenhofel, W, H. 1939. "The Cost of Soil in Rock and Time." *Am J. Sci.* **237**:771–780.

CHAPTER 3

SOIL PHYSICAL PROPERTIES

Physically, soils are composed of mineral and organic particles of varying size. The particles are arranged in a matrix that results in about 50 percent pore space, which is occupied by water and air. This produces a three-phase system of solids, liquids, and gases. Essentially, all uses of soils are greatly affected by certain physical properties. The physical properties considered in this chapter include: texture, structure, consistence, porosity, density, color, and temperature.

SOIL TEXTURE

The physical and chemical weathering of rocks and minerals results in a wide range in size of particles from stones, to gravel, to sand, to silt, and to very small clay particles. The particle-size distribution determines the soil's coarseness or fineness, or the soil's *texture*. Specifically, texture is *the relative proportions of sand, silt, and clay in a soil*.

The Soil Separates

Soil *separates* are the size groups of mineral particles less than 2 millimeters (mm) in diameter or the size groups that are smaller than gravel. The diameter and the number and surface area per gram of the separates are given in Table 3.1.

Sand is the 2.0 to 0.05 millimeter fraction and, according to the United States Department of Agriculture (USDA) system, the sand fraction is subdivided into very fine, fine, medium, coarse, and very coarse sand separates. Silt is the 0.05 to 0.002 millimeter (2 microns) fraction. At the 0.05 millimeter particle size separation, between sand and silt, it is difficult to distinguish by feel the individual particles. In general, if particles feel coarse or abrasive when rubbed between the fingers, the particles are larger than silt size. Silt particles feel smooth like powder. Neither sand nor silt is sticky when wet. Sand and silt differ from each other on the basis of size and may be composed of the same minerals. For example, if sand particles are smashed with a hammer and particles less than 0.05 millimeters are formed, the sand has been converted into silt.

The sand and silt separates of many soils are dominated by quartz. There is usually a significant amount of weatherable minerals, such as feldspar and mica, that weather slowly and re-

TABLE 3.1 Some Characteristics of Soil Separates

Separate	Diameter, mm[a]	Diameter, mm[b]	Number of Particles per Gram	Surface Area in 1 Gram, cm^2
Very coarse sand	2.00–1.00	—	90	11
Coarse sand	1.00–0.50	2.00–0.20	720	23
Medium sand	0.50–0.25	—	5,700	45
Fine sand	0.25–0.10	0.20–0.02	46,000	91
Very fine sand	0.10–0.05	—	722,000	227
Silt	0.05–0.002	0.02–0.002	5,776,000	454
Clay	Below 0.002	Below 0.002	90,260,853,000	8,000,000[c]

[a] United States Department of Agriculture System.
[b] International Soil Science Society System.
[c] The surface area of platy-shaped montmorillonite clay particles determined by the glycol retention method by Sor and Kemper. (See Soil Science Society of America Proceedings, Vol. 23, p. 106, 1959.) The number of particles per gram and surface area of silt and the other separates are based on the assumption that particles are spheres and the largest particle size permissible for the separate.

lease ions that supply plant needs and recombine to form secondary minerals, such as clay. The greater specific surface (surface area per gram) of silt results in more rapid weathering of silt, compared with sand, and greater release of nutrient ions. The result is a generally greater fertility in soil having a high silt content than in soils high in sand content.

Clay particles have an effective diameter less than 0.002 millimeters (less than 2 microns). They tend to be plate-shaped, rather than spherical, and very small in size with a large surface area per gram, as shown in Table 3.1. Because the specific surface of clay, is many times greater than that of sand or silt, a gram of clay adsorbs much more water than a gram of silt or sand, because water adsorption is a function of surface area. The clay particles shown in Figure 3.1 are magnified about 30,000 times; their plate-shaped nature contributes to very large specific surface. Films of water between plate-shaped clay particles act as a lubricant to give clay its plasticity when wet. Conversely, when soils high in clay are dried, there is an enormous area of contact between plate-shaped soil particles and great tendency for very hard soil clods to form. Although the preceding

statements apply to most clay in soils, some soil clays have little tendency to show stickiness and expand when wetted. The clay fraction usually has a net negative charge. The negative charge adsorbs nutrient cations, including Ca^{2+}, Mg^{2+}, and K^+, and retains them in available form for use by roots and microbes.

FIGURE 3.1 An electron micrograph of clay particles magnified 35,000 times. The platy shape, or flatness, of the particles results in very high specific surface. (Photograph courtesy Mineral Industries Experiment Station, Pennsylvania State University.)

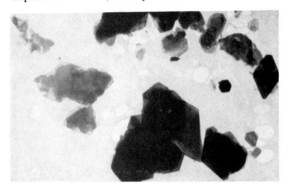

Particle Size Analysis

Sieves can be used to separate and determine the content of the relatively large particles of the sand and silt separates. Sieves, however, are unsatisfactory for the separation of the clay particles from the silt and sand. The hydrometer method is an empirical method that was devised for rapidly determining the content of sand, silt, and clay in a soil.

In the hydrometer method a sample (usually 50 grams) of air-dry soil is mixed with a dispersing agent (such as a sodium pyrophosphate solution) for about 12 hours to promote dispersion. Then, the soil-water suspension is placed in a metal cup with baffles on the inside, and stirred on a mixer for several minutes to bring about separation of the sand, silt, and clay particles. The suspension

is poured into a specially designed cylinder, and distilled water is added to bring the contents up to volume.

The soil particles settle in the water at a speed directly related to the square of their diameter and inversely related to the viscosity of the water. A hand stirrer is used to suspend the soil particles thoroughly and the time is immediately noted. A specially designed hydrometer is carefully inserted into the suspension and two hydrometer readings are made. The sand settles in about 40 seconds and a hydrometer reading taken at 40 seconds determines the *grams* of silt and clay remaining in suspension. Subtraction of the 40-second reading from the sample weight gives the *grams* of sand. After about 8 hours, most of the silt has settled, and a hydrometer reading taken at 8 hours determines the grams of *clay* in the sample (see Figure 3.2). The silt is calculated by difference: add the percentage of sand to the percentage of clay and subtract from 100 percent.

FIGURE 3.2 Inserting hydrometer for the 8-hour reading. The sand and silt have settled. The 8-hour reading measures grams of clay in suspension and is used to calculate the percentage of clay.

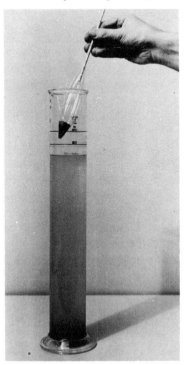

PROBLEM: Calculate the percentage of sand, clay, and silt when the 40-second and 8-hour hydrometer readings are 30 and 12, respectively; assume a 50 gram soil sample is used:

$$\frac{\text{sample weight} - \text{40-second reading}}{\text{sample weight}} \times 100 = \% \text{ sand}$$

$$\frac{50 \text{ g} - 30 \text{ g}}{50 \text{ g}} \times 100 = 40\% \text{ sand}$$

$$\frac{\text{8-hour reading}}{\text{sample weight}} \times 100 = \% \text{ clay}$$

$$\frac{12 \text{ g}}{50 \text{ g}} \times 100 = 24\% \text{ clay}$$

$$100\% - (40\% + 24\%) = 36\% \text{ silt}$$

After the hydrometer readings have been obtained, the soil-water mixture can be poured over a screen to recover the entire sand fraction. After it is dried, the sand can be sieved to obtain the various sand separates listed in Table 3.1.

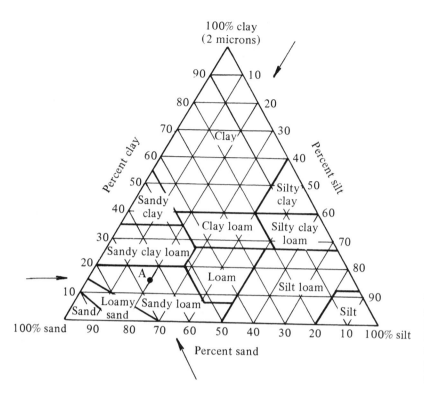

FIGURE 3.3 The textural triangle shows the limits of sand, silt, and clay contents of the various texture classes. Point A represents a soil that contains 65 percent sand, 15 percent clay, and 20 percent silt, and is a sandy loam.

Soil Textural Classes

Once the percentages of sand, silt, and clay have been determined, the soil can be placed in one of 12 major textural classes. For a soil that contains 15 percent clay, 65 percent sand, and 20 percent silt, the logical question is: What is the texture or textural class of the soil?

The texture of a soil is expressed with the use of *class* names, as shown in Figure 3.3. The sum of the percentages of sand, silt, and clay at any point in the triangle is 100. Point A represents a soil containing 15 percent clay, 65 percent sand, and 20 percent silt, resulting in a textural class name of sandy loam. A soil containing equal amounts of sand, silt, and clay is a clay loam. The area outlined by the bold lines in the triangle defines a given class. For example, a loam soil contains 7 to 27 percent clay, 28 to 50 percent silt, and between 22 and 52 percent sand. Soils in the loam class are influenced almost equally by all three sep-

arates—sand, silt, and clay. For sandy soils (sand and loamy sand), the properties and use of the soil are influenced mainly by the sand content of the soil. For clays (sandy clay, clay, silty clay), the properties and use of the soil are influenced mainly by the high clay content.

Determining Texture by the Field Method

When soil scientists make a soil map, they use the field method to determine the texture of the soil horizons to distinguish between different soils. When investigating a land-use problem, the ability to estimate soil texture on location is useful in diagnosing the problem and in formulating a solution.

A small quantity of soil is moistened with water and kneaded to the consistency of putty to determine how well the soil forms casts or ribbons (plasticity). The kind of cast or ribbon formed is related to the clay content and is used to categorize soils as loams, clay loams, and clays. This is shown in Figure 3.4.

If a soil is a loam, and feels very gritty or sandy, it is a sandy loam. Smooth-feeling loams are high in silt content and are called silt loams. If the sample is intermediate, it is called a loam. The same applies to the clay loams and clays. Sands are loose and incoherent and do not form ribbons. More detailed specifications are given in Appendix 1.

Influence of Coarse Fragments on Class Names

Some soils contain significant amounts of gravel and stones and other coarse fragments that are larger than the size of sand grains. An appropriate adjective is added to the class name in these cases. For example, a sandy loam in which 20 to 50 percent of the volume is made up of gravel is a gravelly sandy loam. Cobbly and stony are used for fragments 7.5 to 25 centimeters and more than 25 centimeters in diameter, respectively. Rockiness expresses the amount of the land surface that is bedrock.

Texture and the Use of Soils

Plasticity, water permeability, ease of tillage, droughtiness, fertility, and productivity are all

FIGURE 3.5 Foundation soil with high clay content expanded and shrunk with wetting and drying, cracking the wall. (Photograph courtesy USDA.)

closely related to soil texture. Many clay soils expand and shrink with wetting and drying, causing cracks in walls and foundations of buildings, as shown in Figure 3.5. Conversely, many red-colored tropical soils have clay particles composed mainly of kaolinite and oxides of iron and alumimum. These particles have little capacity to develop stickiness and to expand and contract on wetting and drying. Such soils can have a high clay content and show little evidence of stickiness when wet, or expansion and contraction with wetting and drying.

Many useful plant growth and soil texture relationships have been established for certain geographic areas. Maximum productivity for red pine (*Pinus resinosa*) occurs on coarser-textured soil than for corn (*Zea maize*) in Michigan. The greatest growth of red pine occurs on soils with sandy

FIGURE 3.4 Determination of texture by the field method. Loam soil on left forms a good cast when moist. Clay loam in center forms a ribbon that breaks somewhat easily. Clay on the right forms a long fexible ribbon.

loam texture where the integrated effects of the nutrients, water, and aeration are the most desirable. Loam soils have the greatest productivity for corn. When soils are irrigated and highly fertilized, however, the highest yields of corn occur on the sand soils. The world record corn yield, set in 1977, occurred on a sandy soil that was fertilized and irrigated.

The recommendations for reforestation in Table 3.2 assume a uniform soil texture exists with increasing soil depth, but this is often not true in the field. Species that are more demanding of water and nutrients can be planted if the underlying soil horizons have a finer texture. In some instances an underlying soil layer inhibits root penetration, thus creating a shallow soil with limited water and nutrient supplies.

Young soils tend to have a similar texture in each horizon. As soils evolve and clay is translocated to the B horizon, a decrease in permeability of the subsoil occurs. Up to a certain point, however, an increase in the amount of clay in the subsoil is desirable, because the amount of water and nutrients stored in that zone can also be increased. By slightly reducing the rate of water movement through the soil, fewer nutrients will be lost by leaching. The accumulation of clay in time may become sufficient to restrict severely the movement of air and water and the penetration of roots in subsoil Bt horizons. Water saturation of the soil above the Bt horizon may occur in wet seasons. Shallow rooting may result in a deficiency of water in dry seasons. A Bt horizon with a high clay content and very low water permeability is called a *claypan*. A claypan is a dense, compact layer in the subsoil, having a much higher clay content than the overlying material from which it is separated by a sharply defined boundary (see Figure 3.6).

Claypan soils are well suited for flooded rice production. The claypan makes it possible to maintain flooded soil conditions during the growing season with a minimum amount of water. Much of the rice produced in the United States is produced on claypan soils located on nearly level land in Arkansas, Louisiana, and Texas. The other major kind of rice soil in the United States has very limited water permeability because there is a clay texture in all horizons.

SOIL STRUCTURE

Texture is used in reference to the size of soil particles, whereas *structure* is used in reference to the arrangement of the soil particles. Sand, silt, and clay particles are typically arranged into secondary particles called *peds*, or *aggregates*. The

TABLE 3.2   Recommendations for Reforestation on Upland Soils in Wisconsin in Relation to Soil Texture[a]

Percent Silt Plus Clay		Species Recommended
Less than 5		(Only for wind erosion control)
5–10		Jack pine, Red cedar
10–15		Red pine, Scotch pine, Jack pine
15–25	Increasing	All pines
25–35	demand for	White pine, European larch, Yellow birch, White elm, Red oak, Shagbark hickory, Black locust
Over 35	water and nutrients ↓	White spruce, Norway spruce, White cedar, White ash, Basswood, Hard maple, White oak, Black walnut

[a] Wilde, S. A., "The Significance of Soil Texture in Forestry, and Its Determination by a Rapid Field Method," *Journal of Forestry,* **33**:503–508, 1935. By permission *Journal of Forestry.*

shape and size of the peds determine the soil's structure.

Importance of Structure

Structure modifies the influence of texture with regard to water and air relationships and the ease of root penetration. The macroscopic size of most peds results in the existence of interped spaces that are much larger than the spaces existing between adjacent sand, silt, and clay particles. Note the relatively large cracks between the structural peds in the claypan (Bt horizon) shown in Figure 3.6. The peds are large and in the shape of blocks or prisms, as a result of drying and shrinkage upon exposure. When cracks are open, they are avenues for water movement. When this soil wets, however, the clay expands and the cracks between the peds close, causing the claypan horizon to become impermeable. Root penetration and the downward movement of water, however, are not inhibited in the B horizon of most soils.

Genesis and Types of Structure

Soil peds are classified on the basis of shape. The four basic structural types are spheroid, platelike, blocklike, and prismlike. These shapes give rise to granular, platy, blocky, and prismatic types of structure. Columar structure is prismatic-shaped peds with rounded caps.

Soil structure generally develops from material that is without structure. There are two structure-less conditions, the first of which is sands that remain loose and incoherent. They are referred to as *single grained*. Second, materials with a significant clay content tend to be *massive* if they do not have a developed structure. Massive soil has no observable aggregation or no definite and orderly arrangement of natural lines of weakness.

FIGURE 3.6 A soil with a strongly developed Bt horizon; a claypan. On drying, cracks form between structural units. On wetting, the soil expands, the cracks close, and the claypan becomes impermeable to water and roots. (Photograph courtesy Soil Conservation Service, USDA.)

Claypan

Massive soil breaks up into random shaped clods or chunks. Structure formation is illustrated in Figure 3.7.

Note in Figure 3.8 that each horizon of the soil has a different structure. This is caused by different conditions for structure formation in each horizon. The A horizon has the most abundant root and small-animal activity and is subject to frequent cycles of wetting and drying. The structure tends to be granular (see Figure 3.8). The Bt horizon has more clay, less biotic activity, and is under constant pressure because of the weight of the overlying soil. This horizon is more likely to crack markedly on drying and to develop either a blocky or prismatic structure. The development of structure in the E horizon is frequently weak and the structure hard to observe. In some soils the E horizon has a weakly developed platy structure (see Figure 3.8).

Grade and Class

Complete descriptions of soil structure include: (1) the type, which notes the shape and arrangement of peds; (2) the class, which indicates ped size; and (3) the grade, which indicates the distinctiveness of the peds. The class categories depend on the type. A table of the types and classes of soil structure is given in Appendix 2. Terms for grade of structure are as follows:

0. Structureless—no observable aggregation or no definite and orderly arrangement of natural lines of weakness. Massive, if coherent, as in clays. Single grained, if noncoherent, as in sands.
1. Weak—poorly formed indistinctive peds, barely observable in place.
2. Moderate—well-formed distinctive peds, moderately durable and evident, but not distinct in undisturbed soil.
3. Strong—durable peds that are quite evident in undisturbed soil, adhere weakly to one another, and become separated when the soil is disturbed.

The sequence followed in combining the three terms to form compound names is first the grade, then the class, and finally the type. An example of a soil structural description is "strong fine granular."

Managing Soil Structure

Practical management of soil structure is generally restricted to the topsoil or plow layer in regard to use of soils for plant growth. A stable structure at the soil surface promotes more rapid infiltration of water during storms and results in reduced water runoff and soil erosion and greater water storage within the soil. The permanence of peds depends on their ability to retain their shape

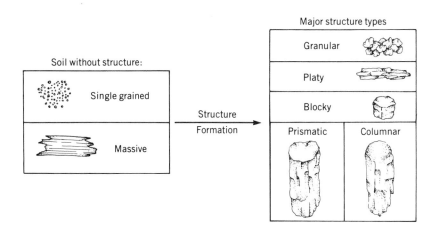

FIGURE 3.7 Illustration of the formation of various types of soil structure from single-grained and massive soil conditions. Structure types are based on the shape of aggregates.

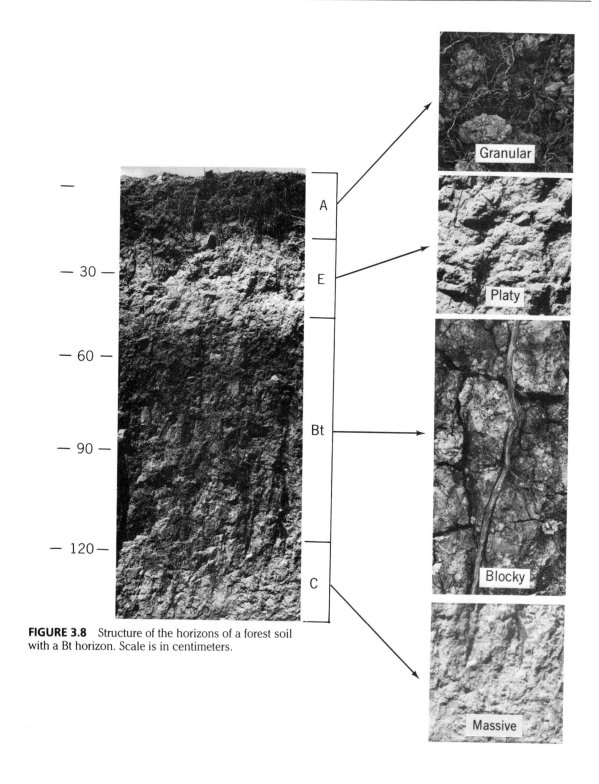

FIGURE 3.8   Structure of the horizons of a forest soil with a Bt horizon. Scale is in centimeters.

after being subjected to the disruptive effects of raindrops and tillage. In addition, for peds to be permanent, the particles in the peds must hold together when dry peds are wetted.

When dry soil is suddenly wetted, water moves into the peds from all directions and compresses the air in the pore spaces. Peds unable to withstand the pressure exerted by the entrapped air are disrupted and fall apart. Unstable peds at the soil surface in cultivated fields are also broken apart by the force of falling raindrops. Disruption of peds produces smaller chunks and some individual particles, which float over the soil surface and seal off the surface to additional infiltrating water. On drying, the immediate surface soil may acquire a massive or structureless condition.

Researchers at the University of Wisconsin measured the development of peds in four soils in which they also determined the content of agents known to bind particles together and give peds stability. These binding agents were microbial gum, organic carbon, iron oxide, and clay. The data in Table 3.3 show that microbial gum was the most important in promoting ped formation. Iron oxide, formed during weathering from minerals that contain iron, was most important in one soil and second most important when the four soils were grouped together. Soil management practices that result in frequent additions of organic matter to the soil in the form of plant residue, or manure, will tend to produce more microbial gum and increase ped formation and

stability. Practices that result in keeping a vegetative cover also reduce raindrop impact and thus reduce the disruption of peds.

SOIL CONSISTENCE

Consistence is the resistance of the soil to deformation or rupture. It is determined by the cohesive and adhesive properties of the entire soil mass. Whereas structure deals with the shape, size, and distinctiveness of natural soil aggregates, consistence deals with the strength and nature of the forces between the sand, silt, and clay particles. Consistence is important for tillage and traffic considerations. Dune sand exhibits minimal cohesive and adhesive properties, and because sand is easily deformed, vehicles can easily get stuck in it. Clay soils can become sticky when wet, and thus make hoeing or plowing difficult.

Soil Consistence Terms

Consistence is described for three moisture levels: wet, moist, and dry. A given soil may be sticky when wet, firm when moist, and hard when dry. A partial list of terms used to describe consistence includes:

1. Wet soil—nonsticky, sticky, nonplastic, plastic
2. Moist soil—loose, friable, firm
3. Dry soil—loose, soft, hard

TABLE 3.3 Order of Importance of Soil Constituents in Formation of Peds over 0.5 mm in Diameter in the A Horizons of Four Wisconsin Soils

Soil	Order of Importance of Soil Constituents
Parr silt loam	Microbial gum > clay > iron oxide > organic carbon
Almena silt loam	Microbial gum > iron oxide
Miami silt loam	Microbial gum > iron oxide > organic carbon
Kewaunee silt loam	Iron oxide > clay > microbial gum
All soils	Microbial gum > iron oxide > organic carbon > clay

Adapted from Chesters, G., O. J., Attoe, and O. N., Allen, "Soil Aggregation in Relation to Various Soil Constituents," *Soil Science Society of America Proceedings* Vol. 21, 1957, p. 276, by permission of the Soil Science Society of America.

Plastic soil is capable of being molded or deformed continuously and permanently, by relatively moderate pressure, into various shapes when wet. Friable soils readily break apart and are not sticky when moist.

Two additional consistence terms for special situations are *cemented* and *indurated*. Cementation is caused by cementing agents such as calcium carbonate, silica, and oxides of iron and aluminum. Cemented horizons are not affected by water content and limit root penetration. When a cemented horizon is so hard that a sharp blow of a hammer is required to break the soil apart, the soil is considered to be indurated.

A silica cemented horizon is called a *duripan*. Large rippers pulled by tractors are used to break up duripans to increase rooting depth in soils. Sometimes a layer with an accumulation of carbonates (a Ck horizon, for example) accumulates so much calcium carbonate as to become cemented or indurated, and is transformed into a *petrocalcic* horizon. Petrocalcic horizons are also called *caliche*. Cemented and indurated horizons tend to occur in soils of great age.

DENSITY AND WEIGHT RELATIONSHIPS

Two terms are used to express soil density. *Particle density* is the average density of the soil particles, and *bulk density* is the density of the bulk soil in its natural state, including both the particles and pore space.

Particle Density and Bulk Density

A soil is composed of mineral and organic particles of varying composition and density. Feldspar minerals (including orthoclase) are the most common minerals in rocks of the earth's crust and very common in soils. They have densities ranging from 2.56 g/cm^3 to 2.76 g/cm^3. Quartz is also a common soil mineral; it has a density of 2.65 g/cm^3. Most mineral soils consist mainly of a wide variety of minerals and a small amount of organic matter. The average particle density for mineral soils is usually given as 2.65 g/cm^3.

Bulk samples of soil horizons are routinely collected in the field and are used for chemical analyses and for some physical analyses, such as particle-size distribution. Soil cores are collected to determine soil bulk density. The soil core samples are collected to obtain a sample of soil that has the undisturbed density and pore space representative of field conditions. Coring machines are used to remove soil cores with a diameter of about 10 centimeters to a depth of a meter or more. The technique for hand sampling of individual soil horizons is shown in Figure 3.9. Care

FIGURE 3.9 Technique for obtaining soil cores that are used to determine bulk density. The light-colored metal core fits into the core sampler just to the left. This unit is then driven into the soil in pile-driver fashion, using the handle and weight unit.

.nust be exercised in the collection of soil cores so that the natural structure is preserved. Any change in structure, or compaction of soil, will alter the amount of pore space and, therefore, will alter the bulk density. The bulk density is the mass per unit volume of oven dry soil, calculated as follows:

$$\text{bulk density} = \frac{\text{mass oven dry}}{\text{volume}}$$

PROBLEM: If the soil in the core of Figure 3.9 weighs 600 grams oven dry and the core has a volume of 400 cm³, calculate the bulk density.

$$\text{bulk density} = \frac{600 \text{ g}}{400 \text{ cm}^3} = 1.5 \text{ g/cm}^3$$

The bulk density of a soil is inversely related to porosity. Soils without structure, such as single-grained and massive soil, have a bulk density of about 1.6 to 1.7 g/cm³. Development of structure results in the formation of pore spaces between peds (interped spaces), resulting in an increase in the porosity and a decrease in the bulk density. A value of 1.3 g/cm³ is considered typical of loam surface soils that have granular structure. As the clay content of surface soils increases, there tends to be an increase in structural development and a decrease in the bulk density. As clay accumulates in B horizons, however, the clay fills existing pore space, resulting in a decrease in pore space volume. As a result, the formation of Bt horizons is associated with an increase in bulk density.

The bulk densities of the various horizons of a soil are given in Figure 3.10. In this soil the C horizon has the greatest bulk density,— 1.7 g/cm³. The C horizon is massive, or structureless, and this is associated with high bulk density and low porosity. Note that the Bt horizon is more dense than the A horizon and that there is an inverse relationship between bulk density and total porosity (macropore space plus micropore space).

Organic soils have very low bulk density compared with mineral soils. Considerable varia-

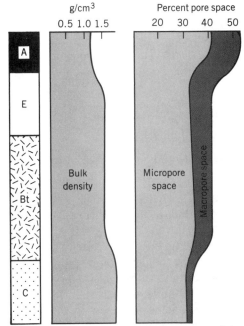

FIGURE 3.10 Bulk density and pore space of the horizons of a soil with A, E, Bt, and C horizons. (Data for Miami loam from Wascher, 1960.)

tion exists, depending on the nature of the organic matter and the moisture content at the time of sampling to determine bulk density. Bulk densities for organic soils commonly range from 0.1 to 0.6 g/cm³.

Changes in soil porosity due to compaction are commonly evaluated in terms of changes in bulk density.

Weight of a Furrow-Slice of Soil

Soil erosion losses are commonly expressed as tons of soil per acre. An average annual soil loss of 10 tons per acre is quite meaningless unless the weight of an inch of soil over 1 acre is known. In this instance, an erosion rate of 10 tons annually would result in the loss of a layer about 7 inches thick, every 100 years.

The acre-furrow-slice has customarily been considered to be a layer of soil about 7 inches thick over 1 acre. An acre has 43,560 ft². The volume of soil in an acre-furrow-slice, therefore, is 25,410 ft^3 to a depth of 7 inches. If the soil has a bulk density of 1.3 g/cm^3, the soil has a density 1.3 times that of water. Because a cubic foot of water weighs 62.4 pounds, a cubic foot of the soil would weigh 81.1 (62.4 × 1.3) pounds. The weight of a 7-inch-thick acre-furrow-slice of a soil with a bulk density of 1.3 g/cm^3 (81 lb/ft^3) is calculated as follows:

$$[7/12 \text{ ft} \times 43\,560 \text{ ft}^2] \times [81.1 \text{ lb/ft}^3]$$
$$= 2{,}061{,}259 \text{ lb}$$

This value—2,061,259—is usually rounded off as 2,000,000 pounds as the weight of an acre-furrow-slice of typical oven-dry loamy soil.

Soil samples to assess the fertility of a soil are usually taken to the depth of the plow layer. The results are commonly reported as pounds of potassium, for example, per acre. Because it is usually assumed that the plow layer of an acre weighs 2 million pounds, the test results are sometimes reported as parts per 2 million (pp2m). For example, a soil test result for available potassium of 100 pounds per acre (acre-furrow-slice) is sometimes given as 100 pp2m. In recent years there has been a tendency to plow deeper and, in many cases, the furrow slice is considered to be 9 or 10 inches thick and to weigh proportionally more.

PROBLEM: Calculate the weight of a 10-inch-thick acre-furrow-slice that has a bulk density of 1.2 g/cm^3.

$$[10/12 \text{ ft} \times 43\,560 \text{ ft}^2] \times [74.9 \text{ lb/ft}^3]$$
$$= 2{,}718{,}870 \text{ lb}$$

Lime recommendations for increasing soil pH are based on assumptions in regard to the weight of the soil volume that must be limed.

Soil Weight on a Hectare Basis

A hectare is an area 100 meters square and contains 10,000 square meters (m^2) or 100,000,000 cm^2. A 15 centimeter thick layer of soil over a hectare that has a bulk density of 1.3 g/cm^3 would weigh

$$100{,}000{,}000 \text{ cm}^2 \times 15 \text{ cm} \times 1.3 \text{ g/cm}^3$$
$$= 1{,}950{,}000{,}000 \text{ g}$$

The weight of a soil layer for a hectare that is 15 centimeters thick is, therefore, 1,950,000 kg/ha.

SOIL PORE SPACE AND POROSITY

The fact that mineral soils have a particle density and bulk density of about 2.65 and 1.3 g/cm^3, respectively, means that soils have about 50 percent total pore space or porosity. In simple terms, rocks without pore space are broken down by weathering to form mineral soils that have about 50 percent porosity. The pore spaces vary in size, and the size of pore spaces themselves can be as important as the total amount of pore space.

Determination of Porosity

Oven dry soil cores, which have been used to determine bulk density, can be placed in a pan of water, allowed to saturate or satiate, and then reweighed to obtain the data needed to calculate soil porosity. When the soil is satiated, the pore space is filled with water, except for a small amount of entrapped air. The volume of water in a satiated soil core approximates the amount of pore space and is used to calculate the soil's approximate porosity. For example, if the soil in a 400 cm^2 core weighed 600 grams at oven dry, and 800 grams at water satiation, the satiated soil would contain 200 grams of water that occupies 200 cubic centimeters of space (1 g of water has a volume of 1 cm^3). Porosity is calculated as follows:

$$\% \text{ porosity} = \frac{\text{cm}^3 \text{ pore space}}{\text{cm}^3 \text{ soil volume}} \times 100$$

$$= \frac{200 \text{ cm}^3}{400 \text{ cm}^3} \times 100 = 50\%$$

PROBLEM: A 500 cm^3 oven dry core has a bulk density of 1.1 g/cm^3. The soil core is placed in a pan of water and becomes water saturated. The oven dry soil and water at saturation weigh 825 grams. Calculate the total soil porosity.

$$\text{weight of oven dry soil} = 500 \text{ cm}^3 \times 1.1 \text{ g/cm}^3$$
$$= 550 \text{ g}$$

$$\text{weight of water in saturated core} =$$
$$825 \text{ g} - 550 \text{ g} = 275 \text{ g}$$

$$\frac{275 \text{ cm}^3 \text{ pore space}}{500 \text{ cm}^3 \text{ soil volume}} \times 100 = 55\%$$

Effects of Texture and Structure on Porosity

Spheres in closest packing result in a porosity of 26 percent, and spheres in open packing have a porosity of 48 percent. Stated in another way, a ball that just fits in a box occupies 52 percent of the volume of the box, whereas 48 percent of the volume is empty space. These facts are true regardless of size of the spheres or balls. Single-grained sands have a porosity of about 40 percent. This suggests that the sand particles are not perfect spheres and the sand particles are not in a perfect close packing arrangement. The low porosity of single-grained sands is related to the absence of structure (peds) and, therefore, an absence of interped spaces.

Fine-textured A horizons, or surface soils, have a wide range of particle sizes and shapes, and the particles are usually arranged into peds. This results in pore spaces within and between peds. These A horizons with well-developed granular structure may have as much as 60 percent porosity and bulk density values as low as 1.0 g/cm^3. Fine-textured Bt horizons have a different structural condition and tend to have less porosity and, consequently, a greater bulk density than fine-textured A horizons. This is consistent with the filling of pore space by translocated clay and the effects of the weight of the overlying soil, which applies a pressure on the Bt horizon. The pore spaces within the peds will generally be smaller than the pore spaces between the peds, resulting in a wide range in pore sizes.

It has been pointed out that sand surface soils have less porosity than clayey surface soils. Yet, our everyday experiences tell us that water moves much more rapidly through the sandy soil. The explanation of this apparent paradox lies in the pore size differences in the two soils. Sands contain mostly *macropores* that normally cannot retain water against gravity and are usually filled with air. As a consequence, macropores have also been called aeration pores. Since the porosity of sands is composed mainly of macropores, sands transmit water rapidly. Pores that are small enough to retain water against gravity will remain water filled after soil wetting by rain or irrigation, and are called *capillary* or *micropores*.

Because sands have little micropore space they are unable to retain much water. Fine-textured soils tend to contain mainly micropores and thus are able to retain a lot of water but have little ability to transmit water rapidly. An example of the distribution of micropore and macropore space in the various horizons of a soil profile is given in Figure 3.10. Note that the amounts of both total porosity and macropore space are inversely related to the bulk density.

Porosity and Soil Aeration

The atmosphere contains by volume nearly 79 percent nitrogen, 21 percent oxygen, and 0.03 percent carbon dioxide. Respiration of roots, and other organisms, consumes oxygen and produces carbon dioxide. As a result, soil air commonly contains 10 to 100 times more carbon dioxide and slightly less oxygen than does the atmosphere (nitrogen remains about constant). Differences in the pressures of the two gases are created beween the soil and atmosphere. This causes carbon dioxide to diffuse out of the soil and oxygen to diffuse into the soil. Normally, this diffusion is sufficiently rapid to prevent an oxygen deficiency or a carbon dioxide toxicity for roots.

Although water movement through a uniformly porous medium is greatly dependent on pore size, the movement and diffusion of gas are closely correlated with *total* porosity. Gaseous diffusion in soil, however, is also dependent on pore space continuity. When oxygen is diffusing through a macropore and encounters a micropore that is filled with water, the water-filled micropore acts as a barrier to further gas movement. Diffusion of oxygen through the water barrier is essentially zero because oxygen diffusion through water is about 10,000 times slower than through air. Clayey soils are particularly susceptible to poor soil aeration when wet, because most of the pore space consists of micropores that may be filled with water. Sands tend to have good aeration or gaseous diffusion because most of the porosity is composed of macropores. In general, a desirable soil for plant growth has a total porosity of 50 percent, which is one half macropore porosity and one half micropore porosity. Such a soil has a good balance between the retention of water for plant use and an oxygen supply for root respiration.

Oxygen deficiencies are created when soils become water saturated or satiated. The wilting of tomato, due to water-saturated soil, is shown in Figure 1.3. Plants vary in their susceptibilty to oxygen deficiency. Tomatoes and peas are very susceptible to oxygen deficiency and may be killed when soils are water saturated for a few hours (less than a day). Two yew plants that were planted along the side of a house at the same time are shown in Figure 3.11. The plant on the right side died from oxygen stress caused by water-saturated soil as a result of flooding from the downspout water. The soil had a high clay content. The use of pipes to carry water away from building foundations is also important in preventing building foundation damage, where soils with high clay content expand and contract because of cycles of wetting and drying.

When changes in soil aeration occur slowly, plants are able to make some adjustment. All plant roots must have oxygen for respiration, and plants growing on flooded soil or in swamps have special mechanisms for obtaining oxygen.

SOIL COLOR

Color is about the most obvious and easily determined soil property. Soil color is important because it is an indirect measure of other important characteristics such as water drainage, aeration,

FIGURE 3.11 Both of these yews were planted at the same time along the side of a new house. The failure to install a pipe on the right, to carry away the downspout water, caused water-saturated soil and killed the plant.

and the organic matter content. Thus, color is used with other characteristics to make many important inferences regarding soil formation and land use.

Determination of Soil Color

Soil colors are determined by matching the color of a soil sample with color chips in a Munsell soil-color book. The book consists of pages, each having color chips arranged systematically according to their *hue*, *value*, and *chroma*, the three variables that combine to give colors. Hue refers to the dominant wavelength, or color of the light. Value, sometimes color brillance, refers to the quantity of light. It increases from dark to light colors. Chroma is the relative purity of the dominant wavelength of the light. The three properties are always given in the order of hue, value, and chroma. In the notation, 10YR 6/4, 10YR is the hue, 6 is the value, and 4 is the chroma. This color is a light-yellowish brown. This color system enables a person to communicate accurately the color of a soil to anyone in the world.

Factors Affecting Soil Color

Organic matter is a major coloring agent that affects soil color, depending on its nature, amount, and distribution in the soil profile. Raw peat is usually brown; well-decomposed organic matter, such as humus, is black or nearly so. Many organic soils have a black color.

In most mineral soils, the organic matter content is usually greatest in the surface soil horizons and the color becomes darker as the organic matter content increases. Even so, many A horizons do not have a black color. Black-colored A horizons, however, are common in soils that developed under tall grass on the pampa of Argentina and the prairies of the United States. An interesting color phenomenon occurs in the clayey soils of the Texas Blacklands. The A horizon has a black color; however, the black color may extend to the depth of a meter even though there is a considerable decrease in organic matter content with increasing soil depth. In some soils, finely divided manganese oxides contribute to black color.

The major coloring agents of most subsoil horizons are iron compounds in various states of oxidation and hydration. The rusting of iron is an oxidation process that produces a rusty or reddish-colored iron oxide. The bright red color of many tropical soils is due to dehydrated and oxidized iron oxide, hematite (Fe_2O_3). Organic matter accumulation in the A horizons of these soils results in a brownish-red or mahogany color. Hydrated and oxidized iron, goethite (FeOOH), has a yellow or yellowish-brown color, and reduced and hydrated iron oxide has a gray color. Various combinations of iron oxides result in brown and yellow-brown colors. Thus, the color of the iron oxides is related to aeration and hydration conditions as controlled by the absence or presence of water. Soils on slopes that never saturate with water have subsoils indicative of well-drained and aerated soil—subsoils with reddish and brownish colors. Soils in depressions that collect water, and poorly drained locations in which soils are water saturated much of the time, will tend to have gray-colored B horizons. Soils in intermediate situations will tend to have yellowish-colored B horizons. If a soil has a gray-colored B horizon, the soil is likely to be water saturated at least part of the time unless it is artificially drained (see Figure 3.12).

The light and grayish colors of E horizons are related to the illuviation of iron oxides and low organic matter content. Some soil horizons may have a white color because of soluble salt accumulation at the soil surface or calcium carbonate accumulation in subsoils. Horizons in young soils may be strongly influenced by the color of the soil parent material.

Significance of Soil Color

Many people have a tendency to equate black-colored soils with fertile and productive soils.

FIGURE 3.12 Mineral soils that develop under the influence of much soil wetness have a dark-colored A horizon and a gray-colored subsoil. The A horizon is about 40 centimeters thick.

Broad generalizations between soil color and soil fertility are not always valid. Within a local region, increases in organic matter content of surface soils may be related to increases in soil fertility, because organic matter is an important reservoir of nitrogen.

Subsoil colors are very useful in predicting the likelihood of subsoil saturation with water and poor aeration. Gray subsoil color indicates a fairly constant water-saturated condition. Such soils are poor building sites because basements tend to be wet and septic tank filter fields do not operate properly in water-saturated soil. Installation of a drainage system is necessary to use the soil as a building site successfully. Subsoils that have bright brown and red colors are indicative of good aeration and drainage. These sites are good locations for buildings and for the production of tree fruits. Many landscape plants have specific needs and tolerances for water, and aeration and soil color are a useful guide in selection of plant species.

Alternating water saturation and drying of the subsoil, which may occur because of alternating wet and dry seasons, may produce an intermediate color situation. During the wet season, iron is hydrated and reduced, and during the dry season the iron is dehydrated and oxidized. This causes a mixed pattern of soil colors called *mottling*. Mottled-colored B horizons are indicative of soils that are intermediate between frequent water saturation and soils with continuous well-drained conditions.

SOIL TEMPERATURE

Below freezing there is extremely limited biological activity. Water does not move through the soil as a liquid and, unless there is frost heaving, time stands still for the soil. A soil horizon as cold as 5° C acts as a deterrent to the elongation of roots. The chemical processes and activities of microorganisms are temperature dependent. The alternate freezing and thawing of soils results in the alternate expansion and contraction of soils. This affects rock weathering, structure formation, and the heaving of plant roots. Thus, temperature is an important soil property.

Heat Balance of Soils

The heat balance of a soil consists of the gains and losses of heat energy. Solar radiation received at the soil surface is partly reflected back into the atmosphere and partly absorbed by the soil surface. A dark-colored soil and a light-colored quartz sand may absorb about 80 and 30 percent of the incoming solar radiation, respectively. Of the total solar radiation available for the earth, about 34 percent is reflected back into space, 19 percent is absorbed by the atmosphere, and 47 percent is absorbed by the land.

Heat is lost from the soil by: (1) evaporation of water, (2) radiation back into the atmosphere,

(3) heating of the air above the soil, and (4) heating of the soil. For the most part the gains and losses balance each other. But during the daytime or in the summer, the gains exceed the losses, whereas the reverse is true for nights and winters.

The amount of heat needed to increase the temperature of soil is strongly related to water content. It takes only 0.2 calories of heat energy to increase the temperature of 1 gram of dry soil 1° C; compared with 1.0 calorie per gram per degree for water. This is important in the temperate regions where soils become very cold in winter and planting dates in the spring depend on a large rise in soil temperature. In general, sandy soils warm more quickly and allow earlier planting than do fine-textured soils, because sands retain less water and heat up faster.

Location and Temperature

In the northern hemisphere, soils located on southern slopes have a higher temperature than soils on north-facing slopes. Soils on south-facing slopes are more perpendicular to the sun's rays and absorb more heat energy per unit area than do soils on northern slopes. This is very obvious in tundra regions where soils on north-facing slopes may have permanently frozen subsoil layers within the normal rooting depth of trees, whereas soils on southern slopes do not. Trees tend to grow only on the south-facing slopes without permanently frozen subsoils. Considerable variation in the microenvironment for plants can exist around a building. Southern exposures are warmer, drier, and have more light than northern exposures.

Large bodies of water act as heat sinks and buffer temperature changes nearby. In the spring and fall the temperature changes more slowly and gradually in the area adjacent to the Great Lakes than at locations further inland. Near these lakes in the spring, the temperatures of both air and soil increase slowly, which delays the blossoms on fruit trees and thereby reduces the hazard of late spring frosts. Killing frosts in the fall are delayed, resulting in an extension of the growing season. As a consequence, production of vine and tree fruits is concentrated in areas adjacent to the Great Lakes. The effect of distance from Lake Erie on the dates of last killing frost in spring and first killing frost in the fall is shown in Figure 3.13.

Control of Soil Temperature

Two practical steps can be taken to change soil temperature. Wet soils can be drained to remove

FIGURE 3.13 Dates for last killing frost in spring and first killing frost in the fall at left and right, respectively, for Erie County, Pennsylvania. The longer growing season along Lake Erie favors land use for fruits and vegetables. (Data from Taylor, 1960.)

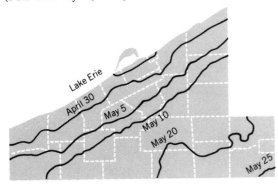

water, and mulches can be applied on the surface of the soil to alter the energy relationships. When used for agriculture, wet soils are drained to create an aerated root zone. The removal of water also causes the soil to warm more quickly in the spring.

A light-colored organic matter mulch, such as straw, will tend to lower soil temperature because (1) more solar radiation will be reflected and less absorbed, and (2) the water content of the soil will tend to be greater because more water will infiltrate. The mulch will, however, tend to increase soil temperature because the mulch tends to: (1) reduce loss of heat by radiation, and (2) reduce water evaporation from the soil surface, which requires energy. The net effect, however, is to reduce soil temperature in the spring. In the northern corn belt, crop residues left on the soil surface promote cooler soils in the spring and tend to result in slightly lower corn yields. A black plastic mulch increases soil temperature because it increases absorption of solar radiation, reduces heat loss by radiation, and reduces evaporation of water from the soil surface. Black mulches have been used to increase soil temperatures in order to produce vegetables and melons for an earlier market.

Permafrost

When the mean annual soil temperature is below 0° C, the depth of freezing in winter may exceed the depth of thawing in summer. As a consequence, a layer of permanently frozen soil, called *permafrost*, may develop. Permafrost ranges from material that is essentially all ice to frozen soil, which appears ordinary except that it is frozen and hard. The surface layer that thaws in summer and freezes in winter is the *active* layer. The active layer is used by plants. Permafrost is widespread in tundra regions and where it occurs there is, generally, an absence of trees. Low-growing shrubs, herbs, and grasses grow on the soils with permafrost if the temperature is above freezing in the summer.

FIGURE 3.14   Soil with permafrost having a layer of water-saturated soil overlying the permafrost. Even in summer, with little precipitation, soils are wet. Frost heaving causes an irregular soil surface with rounded hummocks.

The base of the active layer is the upper surface of the permafrost, or the *permafrost table*. The melting of frozen soil ice in summer results in wet soil conditions, even if there is little or no rainfall, because the permafrost inhibits the downward movement of water from the soil. Alternate freezing and thawing of wet soil above the permafrost produces *cryoturbation* (movement and mix-

FIGURE 3.15   The utilidor system for servicing buildings in the Northwest Territories, Canada. Utility lines (water, sewage) in above-ground utilidor are protected from breakage resulting from soil movement associated with freezing and thawing in soils with permafrost. The utilidor also heats the water and sewage pipes to prevent freezing.

ing of soil due to freezing and thawing). Soil wetness from melting of ice and the irregular soil surface (hummocks) due to cryoturbation are shown in Figure 3.14. Excessive soil movement necessitates the distribution of utilities to buildings above ground with utilidors, as shown in Figure 3.15.

When the vegetative cover is disturbed by traffic or building construction, more heat enters the soil and more ice melts. In some cases enough ice melts to cause soil subsidence. Buildings are built on stilts to shade the soil in summer, to promote minimum heat transfer from buildings to soil, and to allow maximum cooling and freezing of soil in winter.

SUMMARY

Soil physical properties affect virtually every use made of the soil.

Texture relates to the amount of sand, silt, and clay in the soil, and structure relates to the arrangement of the sand, silt, and clay into peds. Texture and structure greatly affect plant growth by influencing water and air relationships. Soils that expand and shrink with wetting and drying affect the stability of building foundations.

About one half of the volume of mineral soils is pore space. Such soils have a bulk density of about 1.3 g/cm^3 and 50 percent porosity. In soils with favorable conditions for water retention and aeration, about one half of the porosity is macropore space and one half is micropore space.

Soil color is used as an indicator of organic matter content, drainage, and aeration.

Soil temperature affects plant growth. Soil temperature is greatly affected by soil color, water content, and the presence or absence of surface materials, such as mulches. Permafrost occurs in soils with average temperature below freezing.

REFERENCES

Chesters, G. O., O. J. Attoe, and O. N. Allen. 1957. "Soil Aggregation in Relation to Various Soil Constituents." *Soil Sci. Soc. Am. Proc.* **21:**272–277.

Day, P. R. 1953. "Experimental Confirmation of Hydrometer Theory."*Soil Sci.* **75:**181-186.

Foth, H. D. 1983. "Permafrost in Soils." *J. Agron. Educ.* **12:**77-79.

Soil Survey Staff. 1951. *Soil Survey Manual.* U.S.D.A. Handbook 18. Washington, D.C.

Taylor, D. C. 1960. *Soil Survey of Erie County, Pennsylvania.* U.S.D.A. and Pennsylvania State University

Tedrow, J. C. F. 1977. *Soils of Polar Landscapes.* Rutgers University Press, New Brunswick, N.J.

Wascher, H. L. 1960. "Characteristics of Soils Associated with Glacial Tills in Northeastern Illinois." *Univ. Illinois Agr. Epp. Sta. Bul. 665.*

Wilde, S. A. 1935. "The Significance of Texture in Forestry and Its Determination by a Rapid Field Method." *Jour. For.* **33:**503-508.

CHAPTER 4

TILLAGE AND TRAFFIC

The beginning of agriculture marks the beginning of soil tillage. Crude sticks were probably the first tillage tools used to establish crops by planting the vegetative parts of plants, including stem sections, roots, and tubers. Paintings on the walls of ancient Egyptian tombs, dating back 5,000 years, depict oxen yoked together pulling a plow made from a forked tree. By Roman times, tillage tools and techniques had advanced to the point that thorough tillage was a recommended practice for crop production. Childe (1951) considered the discovery and development of the plow to be one of the 19 most important discoveries or applications of science in the development of civilization. Now, paradoxically, the use of the plow for primary tillage is being challenged.

EFFECTS OF TILLAGE ON SOILS AND PLANT GROWTH

Tillage is the *mechanical manipulation of soil* for any reason. In agriculture and forestry, tillage is usually restricted to the modification of soil conditions for plant growth. There are three com-monly accepted purposes of tillage: (1) to manage crop residues, (2) to kill weeds, and (3) to alter soil structure, especially preparation of the soil for planting seeds or seedlings.

Management of Crop Residues

Crops are generally grown on land that contains the plant residues of a previous crop. The moldboard plow is widely used for burying these crop residues in humid regions. Fields free of trash permit precise placement of seed and fertilizer at planting and permit easy cultivation of the crop during the growing season (see Figure 4.1).

In subhumid and semiarid regions, by contrast, the need for wind erosion control and conservation of moisture have led to the development of tillage practices that successfully establish crops without plowing. The plant residues remaining on the soil surface provide some protection from water and wind erosion. Plant residues left on the land over winter may trap snow and increase the water content of soils when the trapped snow melts.

FIGURE 4.1 Tillage (disking) in the spring to prepare the land for corn planting where the land was plowed the previous fall and virtually all of the previous crop residues were buried. This type of tillage allows planting and cultivating without interference from plant residues, but leaves the soil bare and susceptible to water runoff and soil erosion. Such tillage is representative of conventional tillage systems.

Tillage and Weed Control

Weeds compete with crop plants for nutrients, water, and light. Weeds, however, can be controlled with herbicides. If weeds are eliminated without tillage, can cultivation of row crops be eliminated? Data from many experiments support the conclusion that the major benefit of cultivating corn is weed control. Cultivation late in the growing season may be detrimental, because roots may be pruned and yields reduced. Cultivation to control weeds is important in subhumid regions to increase and conserve soil water when crops are not growing during a fallow period.

Effects of Tillage on Structure and Porosity

All tillage operations change the structure of the soil. The lifting, twisting, and turning action of a moldboard plow leaves the soil in a more aggregated condition. Cultivation of a field to kill weeds may also have the immediate effect of loosening the soil and increasing water infiltration and aeration. The resistance of peds to

disintegration or breakdown remains unchanged. Consequently, tillage with cultivators, disks, and packers crushes some of the soil peds and tends to reduce soil porosity. Exposed cultivated land suffers from disruption of peds by raindrop impact in the absence of a vegetative cover. In addition, tillage hastens organic matter decomposition. Therefore, the long-term effect of plowing and cultivation is a more compacted soil as a result of the crushing of peds and subsequent settling of soil.

When forest or grassland soils are converted to use for crop production, there is a decline in soil aggregation and soils become more compact because of the long-term effects of tillage. Compaction results in a decrease in the total pore space and an increase in the bulk density. Pushing particles together as a result of tillage results in a decrease in the average pore size. Some of the macropores are reduced to micropores, and thus there is an increase in the volume of micropore space and a decrease in the amount of macropore space. The overall decrease in total porosity results from a greater decrease in macropore space than the increase in micropore space. An increase in micropore space, or water-filled space, because of the compaction of clayey soils, is detrimental because of reduced aeration and water movement. By contrast, the compaction of a sand soil could be desirable because the soil could then retain more water and still be sufficiently aerated.

A study was made of the effects of growing cotton for 90 years on Houston Black Clay soils in the U.S. Texas Blacklands. Soil samples from cultivated fields were compared with samples obtained from an adjacent area that had remained uncultivated in grass. Long-time cultivation caused a significant decrease in soil aggregation and total porosity and an increase in bulk density. Changes in total porosity inversely paralleled the changes in bulk density. These changes are shown in Figure 4.2. The most striking and important change was the decrease in macropore space, which was reduced to about one half of

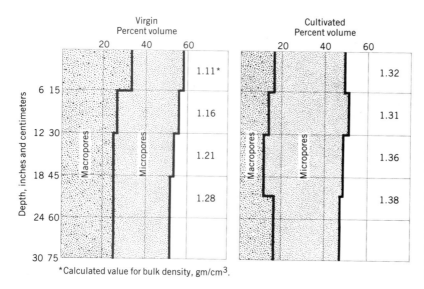

FIGURE 4.2 Effects of 90 years of cultivation on pore space and bulk density (gm/cm³) of Houston Black Clay soils. Macropore and total pore space decreased and the micropore space increased. Decreases in total porosity resulted in increases in bulk density. (Data from Laws and Evans, 1949.)

*Calculated value for bulk density, gm/cm³.

that of the uncultivated soil. Note that these changes occurred to a depth of at least 75 centimeters.

Surface Soil Crusts

During crop production, the surface of the soil is exposed and left bare for varying periods of time without the protection of vegetation or crop residues. Soil peds at the soil surface are broken apart by raindrops, and the primary particles (sand, silt, clay) are dispersed. The splashing raindrops and the presence of water cause the smallest particles to move into the spaces between the larger particles. A thin, dense-surface crust may form and reduce the water infiltration rate by a thousandfold or more. Thus, surface crust formation increases water runoff and erosion on sloping land. When the surface crust dries, the crust may become very hard and massive, inhibiting seedling emergence. Crust hardness increases with increased clay content of the soil and the extent of soil drying.

Minimum and Zero Tillage Concepts

It is obvious that plants in natural areas grow without tillage of the soil. Even though plants grow well in experiments without tillage, the production of most crops will generally require some tillage. Sugar beet yields were lowest when the land was not worked or tilled at all and when the land was worked four or more times, between plowing and planting. The highest yields occurred where the land was worked or tilled only once or twice between plowing and planting, as shown in Table 4.1. The data in Table 4.1 show that excessive tillage is detrimental. Some tillage is necessary for the practical production of some crops; however, the total amount of tillage used is declining. The result has been considerable pro-

TABLE 4.1 Sugar Beet Yields as Affected by the Number of Times a Field Was Worked Prior to Planting

Times Worked	Tons per Acre	Metric Tons per Hectare
None	14.0	31.4
One	16.8	37.6
Two	16.7	37.4
Three	15.2	34.0
Four	14.8	33.2
Five	14.2	31.8
Six	14.3	32.0

Adapted from Cook, et al., 1959.

gress in the development of *minimum tillage* and *no-till tillage* systems.

Minimum tillage is the minimum soil manipulation necessary for crop production under the existing soil and climatic conditions. Some of the minimum tillage equipment prepares the land for planting, plants the seeds, and applies fertilizer and herbicide in one trip over the field. Under favorable conditions, no further tillage will be required during the growing season. Press wheels behind the planter shoes pack soil only where the seed is placed, leaving most of the soil surface in a state conducive to water infiltration. Experimental evidence shows that minimum tillage reduces water erosion on sloping land. It also makes weed control easier because planting immediately follows plowing. Weed seeds left in loose soil are at a disadvantage because the seeds have minimum contact with soil for absorbing water. Crop yields are similar to those where more tillage is used, but the use of minimum tillage can reduce energy use and overall production costs. Some of the savings, conversely, may be used to pay for herbicide costs.

No-till is a procedure whereby a crop is planted directly into the soil with no preparatory tillage since the harvest of the previous crop. The planting of wheat in the wheat stubble from the previous crop without any prior tillage (such as plowing) is shown in Figure 4.3.

FIGURE 4.3 No-till planting of wheat in the plant residues of the previous year's wheat crop. (Photograph courtesy Soil Conservation Service, USDA.)

In no-till systems for row crop production, a narrow slot is made in the untilled soil so that seed can be planted where moisture is adequate for germination. Weed control is by herbicides. The residues of the previous crop remain on the soil surface, which results in greatly reduced water runoff and soil erosion. There was 4,000 pounds per acre of wheat residue on the field when the soybeans were planted in the field shown in Figure 4.4. The coverage of the soil surface with crop residues with various tillage systems is shown in Table 4.2. Tillage systems with reduced tillage and those where crop residues are left on the soil surface minimize water runoff and soil erosion; this is called *conservation tillage*.

From this discussion of tillage, it should be obvious that tillage is not a requirement of plant growth. Generally, gardens do not need to be plowed. The only tillage necessary is that needed for the establishment of the plants and the control of weeds. Mulches in gardens, like crop residues in fields, protect the soil from raindrop impact and promote desirable soil structure. A major benefit is a greater water-infiltration rate and a shorter amount of time needed to irrigate.

Tilth and Tillage

Tilth is the physical condition of the soil as related to the ease of tillage, fitness of soil as a seedbed, and desirability for seedling emergence and root penetration. Tilth is related to structural conditions, the presence or absence of impermeable layers, and the moisture and air content of the soil. The effect of tillage on tilth is strongly related to soil water content at the time of tillage. This is especially true when tillage of wet clay soils creates large chunks or clods that are difficult to break down into a good seedbed when the chunks dry. As shown earlier, cropping usually causes aggregate deterioration and reduced porosity. Farm managers need to exercise considerable judgment concerning both the kind of tillage and the timing of tillage operations to minimize the

FIGURE 4.4 Soybeans planted no-till in the residues of a previous wheat crop. The previous crop left about 4,000 pounds of wheat residue per acre. (Photograph courtesy Soil Conservation Service, USDA.)

detrimental effects and to maximize the beneficial effects of tillage on soil tilth.

TRAFFIC AND SOIL COMPACTION

The changes in soil porosity, owing to compaction, are caused by the long-term effects of cropping and tillage and by the pressure of tractor tires, animal hooves, and shoes. As tractors and machinery get larger, there is greater potential for

TABLE 4.2 Effect of Tillage System on Amount of Residue Cover in Continuous Corn Immediately After Planting

Tillage System	Soil Covered
	%
Conventional, fall, moldboard plow	1
Till-plant	8
Disk twice	13
Chisel plow	19
Strip rotary	62
No-tillage	76

Griffith, Mannering, and Moldenhauer, 1977.

soil compaction. These changes in porosity are important enough to warrant their summarization as an aid to a further discussion of the effects of traffic on soils. Soil compaction results in: (1) a decrease in total pore space, (2) a decrease in macropore space, and (3) an increase in micropore space.

Compaction Layers

Compaction at the bottom of a plow furrow frequently occurs during plowing because of tractor wheels running in a furrow that was made by the previous trip over the field. Subsequent tillage is usually too shallow to break up the compacted soil, and compaction gradually increases. This type of compaction produces a layer with high bulk density and reduced porosity at the bottom of the plow layer, and it is appropriately termed a *plow pan*, or *pressure pan*. Pressure pans are a problem on sandy soils that have insufficient clay content to cause enough swelling and shrinking, via wetting and drying, to naturally break up the compacted soil. Bulk densities as high as

1.9 g/cm^3 have been found in plow pans of sandy coastal plain soils in the southeastern United States. Cotton root extension was inhibited in these soils when the bulk density exceeded 1.6 g/cm^3. The effect of a compact soil zone on the growth of bean roots is shown in Figure 4.5.

Effects of Wheel Traffic on Soils and Crops

As much as 75 percent of the entire area of an alfalfa field may be run over by machinery wheels in a single harvest operation. The potential for plant injury and soil compaction is great, because there are 10 to 12 harvests annually in areas where alfalfa is grown year-round with irrigation. Wheel traffic damages plant crowns, causing plants to become weakened and more susceptible to disease infection. Root development is restricted. Alfalfa stands (plant density) and yields have been reduced by wheel traffic.

FIGURE 4.5 Bean root growth in compacted soil at 4-inch depth on left and no soil compaction on the right.

Potatoes are traditionally grown on sandy soils that permit easy development of well-shaped tubers. Resistance to tuber enlargement in compacted soils not only reduces tuber yield but increases the amount of deformed tubers, which have greatly reduced market value. In studies where bulk density was used as a measure of soil compaction, potato tuber yield was negatively correlated and the amount of deformed tubers was positively correlated with bulk density.

Effects of Recreational Traffic

Maintenance of a plant cover in recreational areas is necessary to preserve the natural beauty and to prevent the undesirable consequences of water runoff and soil erosion. Three campsites in the Montane zone of Rocky Mountain National Park in Colorado were studied to determine the effects of camping activities on soil properties. Soils in areas where tents were pitched, and where fireplaces and tables were located, had a bulk density of 1.60 g/cm^3 compared with 1.03 g/cm^3 in areas with little use. The results show that camping activities can compact soil as much as tractors and other heavy machinery.

Snowmobile traffic on alfalfa fields significantly reduced forage growth at one out of four locations in a Wisconsin study. It appears that injury to alfalfa plants and reduced yields are related to snow depth. There is less plant injury as snow depth increases. The large surface area of the snowmobile treads makes it likely that soil compaction will be minimal or nonexistent. Moderate traffic on fields with a good snow cover appears to have little effect on soil properties or dormant plants buried in the snow.

Human traffic compacts soil on golf courses and lawns. Aeration and water infiltration can be increased by using coring aerators. Numerous small cores about 2 centimeters in diameter and 10 centimeters long are removed by a revolving drum and left lying on the ground, as shown in Figure 4.6.

One of the most conspicuous recent changes in

some of the arid landscapes of the southwestern United States has been caused by use of all-terrain vehicles. In the Mojave Desert one pass of a motorcycle increases bulk density of loamy sand soils from 1.52 to 1.60 g/cm^3, and 10 passes increase the bulk density to 1.68 g/cm^3, according to one study. Parallel changes in porosity occurred with the greatest reduction in the largest pores. Water infiltration was decreased and water runoff and soil erosion were greatly accelerated. Recovery of natural vegetation and restoration of soils occur very slowly in deserts. Estimates indicate that about 100 years will be required to restore bulk density, porosity, and infiltration capacity to their original values based on extrapolation of 51 years of data from an abandoned town in southern Nevada.

Effects of Logging Traffic on Soils and Tree Growth

The use of tractors to skid logs has increased because of their maneuverability, speed, and economy. About 20 to 30 percent of the forest land may be affected by logging. Tractor traffic can disturb or break shallow tree roots that are growing just under the surface soil layer.

The ease with which water can flow through the soil was found to be 65 and 8 percent as great on cutover areas and on logging roads, respectively, in comparison with undisturbed areas in southwestern Washington. The reductions in permeability increase water runoff and soil erosion, resulting in less water being available for the growth of tree seedlings or any remaining trees. Reduced growth and survival of chlorotic Douglas-fir seedlings on tractor roads in western Oregon was considered to be caused by poor soil aeration and low nitrogen supply.

Changes in bulk density and pore space of forest soils, because of logging, are similar to changes produced by longtime cropping. After logging, and in the absence of further traffic, however, forest soils are naturally restored to their former condition. Extrapolation of 5 years' data suggests that restoration of logging trails takes 8 years and wheel ruts require 12 years in northern Mississippi. In Oregon, however, soil compaction in tractor skid trails was still readily observable after 16 years.

FIGURE 4.6 Soil aerator used to increase soil aeration on a college campus in California. The insert shows detail of removal of small cores that are left lying on the surface of the lawn.

Controlled Traffic

The pervasive detrimental effects of soil compaction have caused great interest in tillage research (see Figure 4.7). The extensive coverage of the soil surface by machinery wheels, producing soil compaction, has resulted in the development of *controlled traffic* tillage systems. Controlled traffic systems have all tillage operations performed in fixed paths so that recompaction of soil by traffic (wheels) does not occur outside the selected paths. In fields with controlled traffic, only the small amount of soil in the wheel tracks is compacted, and this compacted soil has almost no effect on crop yields. One additional benefit of controlled traffic is that the machines run on hard and firm paths that permit the planting of crops in soils with much greater water content than normal. Tractors do not get stuck and crops can be planted at an earlier date.

An experiment was conducted in Minnesota to study the effects of machinery wheels on soil compaction in two tillage systems. The 9-year experiment showed there was no difference in soil bulk density between a regular tillage system using a moldboard plow and a conservation (reduced) tillage system with normal trafficking of machinery wheels. When the tillage was carried out with the machinery wheels running on the same path, no-wheel or controlled traffic, compaction caused by wheel traffic was eliminated and bulk densities were lower in both moldboard plow and conservation tillage than for wheel traffic. In the no-wheel traffic experiments, the conservation tillage treatment eventually showed lower bulk densities (less compaction) than the moldboard plow treatment, as shown in Figure 4.8.

A controlled traffic experiment on sandy loam soil in California showed that alfalfa yields with normal traffic were 10 percent less as compared to controlled traffic. The lower yields were also associated with more compact soils having greater bulk density. An experimental controlled traffic machine for planting small grain is shown in Figure 4.9.

In instances where repeated traffic is necessary,

FIGURE 4.7 Tillage research facilities at the USDA National Soil Dynamics Laboratory in Auburn, Alabama. (Photograph courtesy of James H. Taylor.)

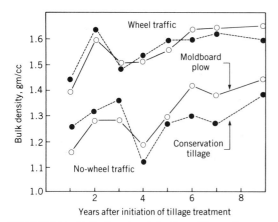

FIGURE 4.8 Bulk density of the 0- to 15-centimeter soil layer as affected by tillage and wheel traffic over time. (Reproduced from *Soil Science Society America Journal,* Vol. 48, No. 1, January–February 1984, p. 152–156 by permission of the Soil Science Society of America.)

it is generally best to use the same tracks because most of the compaction occurs with the first pass over the soil.

FLOODING AND PUDDLING OF SOIL

Rice or paddy (wetland rice) is one of the world's three most important food crops: more than 90

FIGURE 4.9 Incorporation of wheat seed at a depth of 5 centimeters in a wet soil in a controlled traffic system where machinery wheels constantly run in the same tracks. (Photograph courtesy National Soil Dynamics Laboratory, USDA.)

percent of the rice is grown in Asia, mostly in small fields or paddies that have been leveled and bunded (enclosed with a ridge to retain water). The paddy fields are flooded and tilled before rice transplanting. A major reason for the tillage is the destruction of soil structure and formation of soil that is dense and impermeable to water. The practice permits much land that is normally permeable to water to become impermeable and suited for wetland rice production. In the United States, naturally water-impermeable soils are used for rice production and the soils are not puddled. A small amount of upland rice is grown, like wheat, on unflooded soils.

Effects of Flooding

Flooding of dry soil causes water to enter peds and to compress the air in the pores, resulting in small explosions that break the peds apart. The anaerobic conditions (low oxygen supply) from prolonged flooding cause the reduction and dissolution of iron and manganese compounds and the decomposition of organic structural-binding materials. Ped or aggregate stability is greatly reduced and the aggregates are easily crushed. The disruption of soil peds and the clogging of pores with microbial wastes reduce soil permeability.

Effects of Puddling

Puddling is the tillage of water-saturated soil when water is standing on the field, as shown in Figure 4.10. Peds that are already weakened by flooding are worked into a uniform mud, which becomes a two-phase system of solids and liquids. Human or animal foot traffic is effective in forming a compact layer, or *pressure pan*, about 5 to 10 centimeters thick at the base of the puddled layer. The development of a pressure pan allows for the conversion of soils with a wide range in texture and water permeability to become impermeable enough for rice production. Puddling increases bulk density, eliminates large pores, and increases capillary porosity in the soil above the

FIGURE 4.10 Puddling soil to prepare land for paddy (rice) transplanting.

pressure pan. Soil stratification may occur in soils when sand settles before the silt and clay. This results in the formation of a thin surface layer enriched with silt and clay, also having low permeability. These changes are desirable for rice production because soil permeability is reduced by a factor of about 1,000 and the amount of water needed to keep the field flooded is greatly reduced. Much of the rice is grown without a dependable source of irrigation water, and rainfall can be erratic. Low soil permeability is, therefore, of great importance in maintaining standing water on the paddy during the growing season. Standing water promotes rice growth (because of no water stress and an increased supply of nutrients), and the nearly zero level of soil oxygen inhibits growth of many weeds.

Paddies are drained and allowed to dry before harvest. Where winters are dry and no irrigation water is available, a dryland crop is frequently planted. Drying of puddled soil promotes ped formation; however, large clods often form that are difficult to work into a good seedbed. Pressure pans are retained from year to year and create a shallow root zone that severely limits the supply of water and nutrients for dry season crops.

Oxygen Relationships in Flooded Soils

The water standing on a flooded soil has an oxygen content that tends toward equilibrium with the oxygen in the atmosphere. This oxygenated water layer supplies oxygen via diffusion to a thin upper layer of soil about a centimeter thick, as shown in Figure 4.11. This thin upper layer retains the brownish or reddish color of oxidized soil. The soil below the thin surface layer and above the pressure pan is oxygen deficient, and reducing conditions exist. Colors indicative of reduced iron and manganese occur, that is, gray and black colors. The soil in the immediate vicinity of rice roots is oxidized because of the transport of oxygen into the root zone from the atmosphere. This produces a thin layer of yellowish soil around the roots where reduced (ferrous) iron has been oxidized to ferric iron and precipitated at the root surfaces.

The soil below the pressure pan can be either reduced or oxidized. If the soil occurs in a depression and is naturally poorly drained, a gray subsoil that is indicative of reduced soil is likely. Many rice paddies are formed high on the landscape, and they naturally have well-aerated subsoils below pressure pans that may retain their

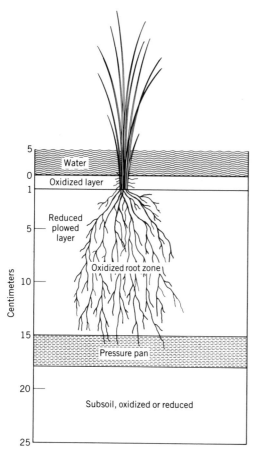

FIGURE 4.11   Oxidized and reduced zones in flooded paddy soil.

All soil compaction results in higher bulk density and lower total porosity. The volume of macropores is greatly reduced with an accompanying increase in micropores. Soil layers with a bulk density greater than about 1.6 to 1.8 g/cm^3 are barriers for root penetration.

Pressure pans, or plow soles, tend to develop at the bottom of plow furrows, especially in sandy soils with minimal capacity for expansion and contraction due to wetting and drying.

Minimum and no-till tillage systems have been developed to minimize the effects of tillage and traffic on soil compaction. Only the minimum amount of tillage necessary for the crop and climatic conditions is used in the first method. In no-till, crops are planted directly in the residues of the previous crop with no prior tillage. For row crops, a slit is made in the soil in which the seed is planted. Minimum and no-till tillage leave more residues on the soil surface than conventional tillage, resulting in reduced water runoff and soil erosion, thus they are considered conservation tillage.

In Asia, many soils are flooded and puddled to create a soil with minimal water permeability for flooded rice production. A pressure pan is created below a puddled soil layer (root zone).

bright oxidized colors after paddies are established.

SUMMARY

Tillage is the mechanical manipulation of soil to modify soil conditions for plant growth. Tillage is beneficial for preparing land for planting, controlling weeds, and managing crop residues. Tillage involves mechanical manipulation of soil and vehicular or animal traffic, which results in soil compaction.

REFERENCES

Childe, V. E. 1951. *Man Makes Himself*. Mentor, New York.

Cook, R. L., J. F. Davis, and M. G. Frakes. 1959. "An Analysis of Production Practices of Sugar Beet Farmers in Michigan." *Mi. Agr. Exp. Sta. Quart. Bul.* **42**:401–420.

Dickerson, B. P. 1976. "Soil Changes Resulting from Tree-Length Skidding." *Soil Sci. Soc. Am. Proc.* **40**:965–966.

Dotzenko, A. D., N. T. Papamichos, and D. S. Romine. 1967. "Effect of Recreational Use on Soil and Moisture Conditions in Rocky Mountain National Park." *J. Soil and Water Con.* **22**:196–197.

Froehlich, H. A. 1979. "Soil Compaction from Logging

Equipment: Effects on Growth of Young Ponderosa Pine." *J. Soil and Water Con.* **34**:276–278.

Griffith, D. R., J. V. Mannering, and W. C. Moldenhauer. 1977. "Conservation Tillage in the Eastern Corn Belt." *J. Soil and Water Con.* **32**:20–28.

Grimes, D. W. and J. C. Bishop. 1971. "The Influence of Some Soil Physical Properties on Potato Yields and Grade Distribution." *Am. Potato J.* **48**:414–422.

Iverson, R. M., B. S. Hinckley, and R. M. Webb. 1981. "Physical Effects of Vehicular Disturbances on Arid Landscapes." *Science.* **212**:915–916.

Laws, W. D. and D. D. Evans. 1949. "The Effects of Long-Time Cutltivation on Some Physical and Chemical Properties of Two Rendzina Soils." *Soil Sci. Soc. Am. Proc.* **14**:15–19.

Meek, B. D., E. A. Rechel, L. M. Carter, and W. R. DeTar. 1988. "Soil Compaction and its Effect on Alfalfa in Zone Production Systems." *Soil Sci. Soc. Am. J.* **52**:232–236.

Moormann, F. R. and N. van Breemen. 1978. *Rice: Soil, Water and Land.* Int. Rice Res. Inst., Manila.

Phillips, R. E., R. L. Blevins, G. W. Thomas, W. R. Frye, and S. H. Phillips. 1980. "No-Tillage Agriculture." *Science.* **208**:1108–1113.

Steinbrenner, E. C. and S. P. Gessel. 1955. "The Effect of Tractor Logging on Physical Properties of Some Forest Soils in Southwestern Washington." *Soil Sci. Soc. A. Proc.* **19**:372–376.

Taylor, J. H. 1986. "Controlled Traffic: A Soil Compaction Management Concept." *SAE Tech. Paper Series 860731.* Soc. Auto. Eng. Inc., Warrendale, Pa.

Voorhees, W. B. and M. J. Lindstrom. 1984. "Long-Term Effects of Tillage Method on Soil Tilth Independent of Wheel Traffic Compaction." *Soil Sci. Soc. Am. J.* **48**:152–156.

Walejko, R. N., J. W. Pendleton, W. H. Paulson, R. E. Rand, G. H. Tenpas, and D. A. Schlough. 1973. "Effect of Snowmobile Traffic on Alfalfa." *J. Soil and Water Con.* **28**:272–273.

Youngberg, C. T. 1959. "The Influence of Soil Conditions Following Tractor Logging on the Growth of Planted Douglas-Fir Seedlings." *Soil Sci. Soc. Am. Proc.* **23**:76–78.

CHAPTER 5

SOIL WATER

Water is the most common substance on the earth; it is necessary for all life. The supply of fresh water on a sustained basis is equal to the annual precipitation, which averages 66 centimeters for the world's land surface. The soil, located at the atmosphere-lithosphere interface, plays an important role in determining the amount of precipitation that runs off the land and the amount that enters the soil for storage and future use. Approximately 70 percent of the precipitation in the United States is evaporated from plants and soils and returned to the atmosphere as vapor, with the soil playing a key role in water retention and storage. The remaining 30 percent of the precipitation represents the longtime annual supply of fresh water for use in homes, industry, and irrigated agriculture. This chapter contains important concepts and principles that are essential for gaining an understanding of the soil's role in the hydrologic cycle and the intelligent management of water resources.

SOIL WATER ENERGY CONTINUUM

As water cascades over a dam, the potential energy (ability of the water to do work) decreases. If water that has gone over a dam is returned to the reservoir, work will be required to lift the water back up into the reservoir, and the energy content of the water will be restored. Therefore, in a cascading waterfall, the water at the top has the greatest energy and water at the bottom has the lowest energy. As water cascades down a falls, the continuous decrease in energy results in an energy continuum from the top to the bottom of the falls. As an analogy, when wet soil dries, there is a continuous decrease in the energy content of the remaining water. When dry soil gets wetter, there is a continuous increase in the energy of soil water. The continuous nature of the changes in the amount of soil water, and corresponding changes in energy, produce the *soil water energy continuum*.

Adhesion Water

Water molecules (H_2O) are electrically neutral; however, the electrical charge within the molecule is asymmetrically distributed. As a result, water molecules are strongly polar and attract each other through H bonding. Soil particles have sites that are both electrically negative and positive. For example, many oxygen atoms are exposed on the faces of soil particles. These oxygen atoms are sites of negativity that attract the positive poles of water molecules. The mutual attraction between water molecules and the surfaces of soil particles results in strong adhesive forces. If a drop of water falls onto some oven dry soil, the water molecules encountering soil particles will be strongly adsorbed and the molecules will spread themselves over the surfaces of the soil particles to form a thin film of water. This layer, or film of water, which will be several water molecules thick, is called *adhesion water* (see Figure 5.1).

Adhesion water is so strongly adsorbed that it moves little, if at all, and some scientists believe that the innermost layer of water molecules exists in a crystalline state similar to the structure of ice. Adhesion water is always present in field soils and on dust particles in the air, but the water can be removed by drying the soil in an oven. Adhesion water has the lowest energy level, is the most immobile water in the soil, and is generally unavailable for use by plant roots and microorganisms.

Cohesion Water

The attractive forces of water molecules at the surface of soil particles decrease inversely and logarithmically with distance from the particle surface. Thus, the attraction for molecules in the second layer of water molecules is much less than for molecules in the innermost layer. Beyond the sphere of strong attraction of soil particles for water molecules of the adhesive water layer, co-

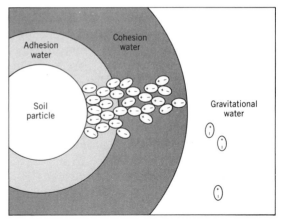

FIGURE 5.1 Schematic drawing of relationship between various forms of soil water. Adhesion water is strongly absorbed and very immobile and unavailable to plants. The gravitational water, by contrast, is beyond the sphere of cohesive forces, and gravity causes gravitational water to flow rapidly down and out of the soil (unless downward flow is inhibited). The cohesion water is intermediate in properties and is the most important for plant growth.

hesive forces operate between water molecules because of hydrogen bonding. This causes water molecules to attach themselves to the adhesion water, resulting in an increase in the thickness of the water film. Water that is retained in soils because of cohesive forces is called *cohesion water,* and is shown in Figure 5.1.

About 15 to 20 molecular layers is the maximum amount of water that is normally adsorbed to soil particle surfaces. With increasing distance from the soil particle surface toward the outer edge of the water film, the strength of the forces that holds the water decreases and the energy and mobility of the water increase. Adhesion water is so strongly held that it is essentially immobile, and thus is unavailable to plants. The cohesion water is slightly mobile and generally available to plants. There is a water-energy gradient across the water films, and the outermost water molecules have the greatest energy, the greatest tendency to move, and the greatest availability for plant roots. As roots absorb water over time, the soil water content decreases, and water films become thinner; roots encounter water with decreasing energy content or mobility. At some point, water may move so slowly to roots that plants wilt because of a lack of water. If wilted indicator plants are placed in a humid chamber and fail to recover, the soil is at the *permanent wilting point* (see Figure 5.2). This occurs when most of the cohesion water has been absorbed.

When soil particles are close to, or touch, adjacent particles, water films overlap each other. Small interstices and pores become water filled. In general, the pores in field soils that are small enough to remain water filled, after soil wetting and downward drainage of excess water has occurred, are the capillary or micropores. At this point, the soil retains all possible water against gravity (adhesion plus cohesion water), and the soil is considered to be at *field capacity*. The aeration or macropores at field capacity are air filled. A typical loamy surface soil at field capacity has a volume composition of about 50 percent solids, 25 percent water, and 25 percent air.

FIGURE 5.2 Wilted coleus plant on the right. If the plant fails to revive when placed in a humid chamber, the soil is at the permanent wilting point in regard to this plant.

Gravitational Water

In soils with claypans, a very slowly permeable subsoil layer permits little water, in excess of field capacity, to move downward and out of the soil. After a long period of rainfall, water accumulates above the claypan in these soils. As the rainy period continues, the soil above the claypan saturates from the bottom upward. The entire A horizon or root zone may become water saturated. Soil water that exists in aeration pores, and that is normally removed by drainage because of the force of gravity, is *gravitational* water. When soils are saturated with water, the soil volume composition is about 50 percent solids and 50 percent water. Gravitational water in soil is detrimental when it creates oxygen deficiency. Gravitational water is not considered available to plants because it normally drains out of soils within a day or two after a soil becomes very wet (see Figure 5.1).

The soil in the depression, of Figure 5.3 is water saturated because underlying soil is too slowly permeable. Note the very poor growth of corn in the soil that remains water saturated after rainfall. Along the right side, the soil is not water saturated. The gravitational water has drained out, the aeration pores are air filled, and the soil is at field capacity. Here, the growth of corn is good.

An energy gradient exists for the water within a macropore, including the gravitational water. When water filled macropores are allowed to drain, the water in the center of the macropore has the most energy and is the most mobile. This water leaves the pore first and at the fastest speed. This is followed by other water molecules closer to the pore edges until all of the gravitational water has drained, leaving the cohesion and adhesion water as a film around the periphery of the macropore.

Summary Statement

1. The forces that affect the energy level of soil water, and the mobility and availability of wa-

FIGURE 5.3 Water-saturated soil occurs in the depression where the corn growth is poor. The surrounding soil at higher elevation is at field capacity because of recent rain, and corn growth is good owing to high availability of both water and oxygen.

ter to plants, are adhesion, cohesion, and gravity.

2. A water-saturated soil contains water with widely varying energy content, resulting in an energy continuum between the oven dry state and water saturation.

3. The mobility and availability of water for plant use parallel the energy continuum. As dry soil wets over time, the soil water increases in energy, mobility, and availability for plants. As wet soil dries over time, the energy, mobility, and plant availability of the remaining water decreases.

4. Plants wilt when soil water becomes so immobile that water movement to plant roots is too slow to meet the transpiration (evaporation from leaves) needs of plants.

ENERGY AND PRESSURE RELATIONSHIPS

There is a relationship between the energy of water and the pressure of water. Because it is much easier to determine the pressure of water, the energy content of water is usually categorized on the basis of pressure.

The hull of a submarine must be sturdily built to withstand the great pressure encountered in a dive far below the ocean's surface. The column (head) of water above the submarine exerts pressure on the hull of the submarine. If the water pressure exerted against the submarine was directed against the blades of a turbine, the energy

in the water could be used to generate electricity. The greater the water pressure, the greater the tendency of water to move and do work, and the greater the energy of the water.

The next consideration will be that of water pressure relationships in water-saturated soils, which is analogous to the water pressure relationships in a beaker of water or in any body of water.

Pressure Relationships in Saturated Soil

The beaker shown in Figure 5.4 has a bottom area assumed to be 100 cm^2. The water depth in the beaker is assumed to be 20 centimeters. The volume of water, then, is 2,000 cm^3 and it weighs 2,000 grams. The pressure, P, of the water at the bottom of the beaker is equal to

$$P = \frac{\text{force}}{\text{area}} = \frac{2,000 \text{ g}}{100 \text{ cm}^2} = 20 \text{ g/cm}^2$$

The water pressure at the bottom of the beaker could also be expressed simply as equal to a column of water 20 centimeters high. This is analogous to saying that the atmospheric pressure at sea level is equal to 1 atmosphere, is equivalent to a column of water of about 33 feet, or to a column of mercury of about 76 centimeters. These are also equal to a pressure of 14.7 lb/in.^{2}

At the 10-centimeter depth, the water pressure in the beaker is one half of that at the 20- centimeter depth and, therefore, is 10 g/cm^2. The water pressure decreases with distance toward the surface and becomes 0 g/cm^2, at the free

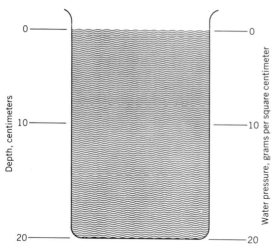

FIGURE 5.4 A beaker with a cross-sectional area of 100 square centimeters contains 2,000 grams of water when the water is 20 centimeters deep. At the water surface, water pressure is 0; the water pressure increases with depth to 20 grams per square centimeter at the bottom.

water surface, as shown in Figure 5.4. In water-saturated soil, the water pressure is zero at the surface of the water table and the water pressure increases with increasing soil depth.

Pressure Relationships in Unsaturated Soil

If the tip of a small diameter glass or capillary tube is inserted into the water in a beaker, adhesion causes water molecules to migrate up the interior wall of the capillary tube. The cohesive forces between water molecules cause other water molecules to be drawn up to a level above the water in the beaker. Now, we need to ask: "What is the pressure of the water in the capillary tube?" and "How can this knowledge be applied to unsaturated soils?"

As shown previously, the water pressure decreases from the bottom to the top of a beaker filled with water and, at the top of the water surface, the water pressure is zero. Beginning at the

open water surface and moving upward into a capillary tube, water pressure continues to decrease. Thus, the water pressure in the capillary tube is less than zero, or is negative. As shown in Figure 5.5, the water pressure decreases in the capillary tube with increasing height above the open water surface. At a height of 20 centimeters above the water surface in the beaker, the water pressure in the capillary is equal to -20 g/cm^2. Figure 5.5 also shows that at this same height above the water surface, the water pressure in unsaturated soil is the same at the 20-centimeter height. The two pressures are the same at any one height, and there is no net flow of water from the soil or capillary tube after an equilibrium condition is established.

Applying the consideration of water in a capillary tube to an unsaturated soil, allows the following statements:

FIGURE 5.5 Water pressure in a capillary tube decreases with increasing distance above the water in the beaker. It is -20 grams per square centimeter at a height 20 centimeters above the water surface. Since the water column of the capillary tube is continuous through the beaker and up into the soil column, the water pressure in the soil 20 centimeters above the water level in the beaker is also -20 grams per square centimeter.

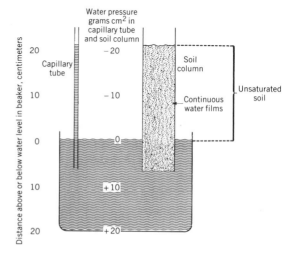

1. Water in unsaturated soil has a negative pressure or is under tension.
2. The water pressure in unsaturated soil decreases with increasing distance above a free water surface or water table.

THE SOIL WATER POTENTIAL

Most of the concerns about soil water involve movement: water movement into the soil surface, water movement from the soil surface downward through soil, and movement of water from the soil to, and into, roots, microorganisms, and seeds. The pressure of soil water, which is an indication of tendency for soil water to move, is expressed by the *soil water potential*. Technically, the soil water potential is defined as the amount of work that must be done per unit quantity of water to transport or move reversibly (without energy loss due to friction) and isothermally (without energy loss due to temperature change) an infinitesimal quantity of water from a pool of pure water at specified elevation and at atmospheric pressure to the soil water at the point under consideration. It is important to note that the water potential is the *amount of work needed to move water from a reference pool to another point*.

The symbol for the water potential is psi, ψ. The total water potential, ψ_t, is made up of several subpotentials, including the gravitational, matric, and osmotic.

The Gravitational Potential

The gravitational potential, ψ_g, is due to the position of water in a gravitational field. The gravitational potential is very important in water-saturated soils. It accounts primarily for the movement of water through saturated soils or the movement of water from high to low elevations. Customarily, the soil water being considered is higher in elevation than the reference pool of pure water. Since the gravitational potential represents the work that must be done to move water from a reference pool (which is at a lower elevation) to soil that is at a higher elevation, the sign is positive. The higher the water above the reference point, the greater the gravitational water potential.

Visualize a water-saturated soil that is underlain at the 1-meter depth with plastic tubing containing small holes. Normally, the plastic tubing allows gravitational water to drain out of the soil. Suppose, however, that the tube outlet is closed and the soil remains water saturated. The gravitational water potential at the top of the soil that is saturated, relative to the plastic tubing, is equal to a 1-meter water column. This is the same as saying that the water pressure is equal to a 1-meter column of water in reference to the point of water drainage through the tubing.

If the drainage tube outlet is opened, gravitational water will leave the water-saturated soil and drain away through the plastic tubing. The soil will desaturate from the surface downward. The water pressure or gravitational potential at the top of water-saturated soil will decrease as the water level drops. When the gravitational water has been removed, and drainage stops, the top of the new level of water-saturated soil will be at the same elevation as the drainage tubing (the reference point). At this elevation, the gravitational potential will be zero.

The Matric Potential

When a soil is unsaturated and contains no gravitational water, the major movement of water is laterally from soil to plant roots. The important forces affecting water movement are adhesion and cohesion. Adhesion and cohesion effects are intimately affected by the size and nature of primary soil particles and peds. The resulting physical arrangement of surfaces and spaces, owing to texture and structure, is the soil *matrix*. The interaction of the soil matrix with the water produces the *matric* water potential. The matric potential is ψ_m.

When rain falls on dry soil, it is analogous to moving water molecules from a pool of pure water to the water films on soil particles. No energy input is required for the movement of water in raindrops to water films around dry soil particles. In fact, a release of energy occurs, resulting in a decrease in the potential of the water. Adsorbed water has lower potential to do work than water in raindrops. In other words, negative work is required to transfer water from raindrops to the films on soil particles. This causes the matric potential to have a negative sign.

The matric potential varies with the content of soil water. The drier a soil is, the greater is the tendency of the soil to wet and the greater is the release of energy when it becomes wetted. In fact, when oven dry clay is wetted, a measureable increase in temperature occurs due to the release of energy as heat. This release of heat is called the *heat of wetting*. Generally, the drier a soil is, the lower is the matric potential. The number preceded by a negative sign becomes larger (becomes more negative) when a soil dries from field capacity to wilt point. In reference to the previous situation where the gravitational water had just exited a soil with drainage tubing at a depth of 1 meter, the matric water potential in unsaturated soil at the soil surface is equal to a -1-meter column of water.

The Osmotic Potential

The osmotic potential, ψ_o, is caused by the forces involved in the adsorption of water molecules by ions (hydration) from the dissolution of soluble salt. Since the adsorption of water molecules by ions releases heat (energy of hydration), the sign of the osmotic potential is also negative.

The osmotic potential is a measure of the work that is required to pull water molecules away from hydrated ions. Normally, the salt content of the soil solution is low and the osmotic potential has little significance. In soils containing a large amount of soluble salt (saline soils), however, the osmotic potential has the effect of reducing water uptake by roots, seeds, and microorganisms. The

osmotic potential has little, if any, effect on water movement within the soil, since diffusion of the hydrated ions creates a uniform distribution and the hydrated ions do not contribute to a differential in matric potential in one region of the soil as compared to another.

Measurement and Expression of Water Potentials

Since the gravitational potential is the distance between the reference point and the surface of the water-saturated soil, or the water table, it can be measured with a ruler. Reference has been made to a gravitational potential equal to a meter-long water column. Water potentials are frequently expressed as bars that are approximately equal to atmospheres. A bar is equivalent to a 1,020 centimeter column of water.

The matric potential can be measured in several ways. Vacuum gauge potentiometers consist of a rigid plastic tube having a porous fired clay cup on one end and a vacuum gauge on the other. The potentiometer is filled with pure water, and at this point the vacuum gauge reads zero. The potentiometer is then buried in the soil so that the porous cup has good contact with the surrounding soil (see Figure 5.6). Since the potential of pure water in the potentiometer is greater than water in unsaturated soil, water will move from the potentiometer into the soil. At equilibrium, the vacuum gauge will record the matric potential. Vacuum gauge potentiometers work in the range 0 to -0.8 bars, which is a biologically important range for plant growth. They are used in irrigated agriculture to determine when to irrigate.

A pressure chamber is used to measure matric potentials less than -0.8 bars. Wet soil is placed on a porous ceramic plate with very fine pores, which is confined in a pressure chamber. Air pressure is applied, and the air forces water through the fine pores in the ceramic plate. At equilibrium, when water is no longer leaving the soil, the air pressure applied is equated to the matric poten-

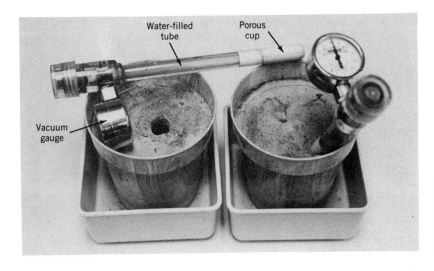

FIGURE 5.6 Soil water potentiometers for measuring the matric potential. The one on the left has been removed from the soil to show the porous clay cup.

tial. At equilibrium, an air pressure of 15 bars is equal to a matric potential of −15 bars. This technique is used to determine the matric potential in the range of −0.3 to −15 bars, which is the range between field capacity and the permanent wilting point, respectively.

Many books and publications express the water potential in bar units. However, *kilopascals and megapascals* are being used more frequently to express the water potential. A bar is equivalent to 100 kilopascals (kPa). Using kilopascals, the water potential at field capacity is −30 kPa and at the permanent wilting point is −1,500 kPa.

SOIL WATER MOVEMENT

Water movement in soils, and from soils to plant roots, like water cascading over a dam, is from regions of higher-energy water to regions of lower-energy water. Thus, water runs downhill.

The driving force for water movement is the difference in water potentials between two points. The *water potential gradient (f)* is the water potential difference between two points divided by the distance between the two points. The rate of water flow is directly related to the water potential difference and inversely related to the flow distance.

The velocity of water flow is also affected by the soil's ability to transmit water or the *hydraulic conductivity (k)*. Water flow through large pores is faster than through small pores; flow is faster when the conductivity is greater. Therefore, the velocity of flow, V, is equal to the water potential gradient times the hydraulic conductivity, k. Mathematically,

$$V = kf$$

The hydraulic conductivity, or the ability of the soil to transmit water, is determined by the nature and size of the spaces and pores through which the water moves. The conductivity of a soil is analogous to the size of the door to a room. As the door to a room becomes larger, the velocity or speed at which people can go in or out of the room increases. As long as the door size remains constant, the speed with which people can move into or out of the room remains constant. As long as the physical properties of a soil (including water content) remain constant, k is constant.

Conductivity (k) is closely related to pore size.

Water flow in a pipe is directly related to the fourth power of the radius. Water can flow through a large pore, having a radius 10 times that of a small pore, 10,000 times faster than through the small pore. In water-saturated soils, water moves much more rapidly through sands than clays because sands have larger pores. The larger pores give the sandy soils a greater ability to transmit water or a greater conductivity when saturated as compared to clayey soils.

The size of pores through which water moves through soil, however, is related to the water content of the soil. In saturated soil, all soil pores are water filled and water moves rapidly through the largest macropores. At the same time, water movement in the smaller pores is slow or nonexistent. As the water content of soil decreases, water moves through pores with decreasing size, because the largest pores are emptied first, followed by the emptying of pores of decreasing size. Therefore, k (hydraulic conductivity of the soil) is closely related to soil water content; k decreases with decreasing soil water content.

Water Movement in Saturated Soil

As shown previously, the potential gradient (f) is the difference in water potentials between the two points of flow divided by the distance between the points of flow. The difference in water potentials between two points in water-saturated soil is the head (h). The gradient, f, is inversely related to the distance of flow through the soil, d. Therefore,

$$f = \frac{h}{d}$$

We can now write

$$V = k \times \frac{h}{d}$$

The quantity of water, Q_w, that will flow through a soil (or a pipe) depends on the cross sectional area, A, and the time of flow, t. Therefore,

$$Q_w = k \frac{h}{d} At$$

This equation is named after Darcy, who studied the flow of water through a porous medium with a setup like that shown in Figure 5.7. The setup includes a column in which a 12-centimeter-thick layer of water rests on top of a 15-centimeter-thick layer of water-saturated soil. In this case, the potential difference between the two points is equal to 12 centimeters (point A to point B) plus 15 centimeters (point B to point C) or 27 centimeters; h equals 27 centimeters. The distance of flow through soil, d, is 15 centimeters. If the amount of water that flowed through the soil in 1 hour was 360 cm³, and the cross-sectional area of the soil column is 100 cm², k can be calculated as follows:

$$360 \text{ cm}^3 = k \times \frac{27 \text{ cm}}{15 \text{ cm}} \times 100 \text{ cm}^2 \times 1 \text{ hr}$$

Then, $k = 2$ cm/hr

The value for k of 2 cm/hr is in the range of soils with moderate saturated hydraulic conductivity; in the range of 0.5 cm/hr to 12.5 cm/hr. Soils with slow saturated hydraulic conductivity range from less than 0.125 cm/hr to 0.5 cm/hr. Soils with rapid saturated hydraulic conductivity have a conductivity in the range 12.5 cm/hr to more than

FIGURE 5.7 A setup similar to that which Darcy used to study water movement in porous materials.

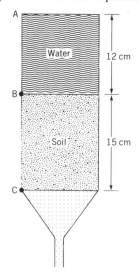

25 cm/hr. Thus, a sand soil could have a saturated hydraulic conductivity that is several hundred times greater than that of a clay soil.

According to Darcy's equation, if the distance of water movement through soil is doubled, the quantity of water flow is halved. This is an important consideration in soil drainage, because the further soil is from the drain lines, the longer it will take for water to exit the soil. Therefore, the distance between drainage lines is dependent on the hydraulic conductivity. Another important decision based on the saturated hydraulic conductivity includes determining the size of seepage beds for septic tank effluent disposal.

Water Movement in Unsaturated Soil

If water is allowed to drain from a saturated soil, water leaves the soil first and most rapidly via the largest pores. As drainage proceeds over time, water movement shifts to smaller and smaller pores. The result is a rapid decrease in both hydraulic conductivity and in the rate of drainage, or movement of water from the soil. Field soils with an average saturated hydraulic conductivity will, essentially, stop draining within a day or two after a thorough wetting, meaning the gravitational water has exited the soil and the soil is retaining the maximum amount of water against the force of gravity. Such soil in the field is at field capacity and will have a matric potential varying from about -10 to -30 kPa (-0.1 to -0.3 bars). The data in Table 5.1 show that the conductivity of the Geary silt loam decreased from 9.5 cm/day at water-saturation (matric potential 0) to only 1.1×10^{-1} near field capacity (matric potential equal to -33 kPa or -0.3 bars). Thus, the hydraulic conductivity at field capacity was about 1 percent as great as at saturation.

Gravity is always operating. After field capacity is attained, however, matric forces control water movement. Additional movement is very slow and over short distances. The flow tends to be horizontally from soil to roots, except for some vertical movement at the soil surface, resulting from loss of water by evaporation. As a consequence, gravity plays a minimum role in the movement of water in water-unsaturated soil.

Darcy's law also applies to the movement of water in unsaturated soil. The value for h, the hydraulic head, becomes the difference in the matric potentials between the points of flow. For water movement from soil to roots, h will commonly be a few hundred kPa (a few bars); for example, the difference between -100 and -300 kPa (-1 to -3 bars).

A most important consideration for unsaturated flow is the fact that unsaturated hydraulic conductivity can vary by a factor of about 1 million within the range that is biologically important. Unsaturated hydraulic conductivity decreases rapidly when plants absorb water and soils become drier than field capacity. For the Geary soil (see Table 5.1), k near field capacity at -33 kPa is 1.1×10^{-1} compared with 6.4×10^{-5} cm/day for a potential of -768 kPa. This means that the conductivity in the soil at field capacity is more than 1,700 times greater than in soil midway between field capacity and the permanent wilting point. The water in unsaturated soil is very immobile and significant movement occurs only over very short distances. The distance between roots must be small to recover this water. It is not surprising that Dittmer (1937) found that a rye plant, growing in a cubic foot of soil, increased root length about 3 miles per day when consideration was made for root hair growth. When plants wilt for a lack of water, the wilting is due more to the slow transmission rate of water from soil to roots rather than a low water content of the soil per se.

Water Movement in Stratified Soil

Stratified soils have layers, or horizons, with different physical properties, resulting in differences in hydraulic conductivity. As a consequence, the downward movement of water may be altered when the downward moving water front encounters a layer with a different texture.

When rain or irrigation water is added to the

TABLE 5.1 Volume Water Content, Hydraulic Conductivities, and Matric Potentials of a Loam and Silt Loam Soil

Volume Water Content	Sarpy Loam		Geary Silt Loam	
	Hydraulic Conductivity (cm/day)	Matric Potential kPa	Hydraulic Conductivity (cm/day)	Matric Potential kPa
0.05	4.5×10^{-5}	-697	—	—
0.06	6.7×10^{-5}	-336	—	—
0.08	4.1×10^{-4}	-125	—	—
0.10	4.8×10^{-3}	$- 44$	—	—
0.12	2.6×10^{-2}	$- 33$	—	—
0.14	5.2×10^{-2}	$- 25$	—	—
0.16	7.8×10^{-2}	$- 20$	—	—
0.18	1.1×10^{-1}	$- 16$	6.4×10^{-5}	-768
0.20	2.7×10^{-1}	$- 13$	4.1×10^{-4}	-402
0.22	7.4×10^{-1}	$- 10$	2.0×10^{-3}	-267
0.24	1.6	$- 8$	3.6×10^{-3}	-167
0.26	3.6	$- 6$	1.6×10^{-2}	$- 81$
0.28	4.7	$- 5$	4.5×10^{-2}	$- 52$
0.30	7.4	$- 4$	1.1×10^{-1}	$- 33$
0.32	1.1×10	$- 3$	2.8×10^{-1}	$- 21$
0.34	1.9×10	$- 3$	5.4×10^{-1}	$- 14$
0.36	3.4×10	$- 2$	1.0	$- 9$
0.38	6.9×10	$- 1$	2.1	$- 6$
0.40	1.1×10^{-2}	$- .3$	4.1	$- 4$
0.41	1.2×10^{-2}	0	5.1	$- 3$
0.42	—	—	6.2	$- 2$
0.44	—	—	7.3	$- 1$
0.46	—	—	9.5	0

Data from Hanks, 1965.

surface of dry, unsaturated soils, the immediate soil surface becomes water saturated. Infiltration, or water movement through the soil surface, is generally limited by the small size of the pores in the soil surface. Water does not rush into the underlying drier unsaturated soil, but moves slowly as unsaturated flow. The downward flow of water below the soil surface is in the micropores, while the macropores remain air-filled. The water moves slowly through the soil as a wetting front, creating a rather sharp boundary between moist and dry soil. Similarly, water moves slowly from an irrigation furrow in all directions. Matric forces (cohesion and adhesion) control water movement; water tends to move upward almost as rapidly as downward.

If the water is moving through loam soil and the wetting front encounters a dry sand layer, the wetting front stops its downward movement. The wetting front then continues to move laterally above the sand layer, as shown in Figure 5.8, because the unsaturated hydraulic conductivity in the dry sand is less than that of the wet silt loam soil above. Water in the wetting front is moving as a liquid and, in dry sand, there are essentially no continuous water films between sand grains. That is, the path of water movement is interrupted because the water films in many adjacent sand grains do not touch each other. The abrupt stoppage of the water front, shown in Figure 5.8, is analogous to a vehicle moving at a high speed and abruptly stopping when it reaches a bridge that

FIGURE 5.8 Photographs illustrating water movement in stratified soil where water is moving as unsaturated flow through a silt loam into a sand layer. In the upper photograph, water movement into the sand is inhibited by the low unsaturated hydraulic conductivity of dry sand. The lower photograph shows that after an elapsed time of 1.5 to 5 hours, the water content in the lower part of the silt loam soil became high enough to cause water drainage downward into the sand. (Photographs courtesy Dr. W. H. Gardner, Washington State University.)

has collapsed. There is a lack of road continuity where the bridge has collapsed. The distance between particles in clay soils is small, and there is greater likelihood of water films bridging between the particles in dry clay soil as compared to dry sand. This gives dry clay soils greater hydraulic conductivity than dry sands.

As the soil under the irrigation furrow, and above the sand layer, becomes wetter over time, the water potential gradient between the moist soil and dry soil increases. In time, the water breaks through the silt loam soil and enters the sand layer, as shown in the lower part of Figure 5.8. The low unsaturated hydraulic conductivity of

any sand and gravel layers underlying loamy soils, enables the loam soil layer to retain more water than if the the soil had a loam texture throughout. A gravel layer under golf greens results in greater water storage in the overlying soil. Conversely, during periods of excessive rainfall, drainage will occur and prevent complete water saturation of the soil on the green.

If a clay layer underlies a loamy soil, the wetting front will immediately enter the clay. Now, however, the clay's capacity to transmit water is very low by comparison, and water will continue to move downwward in the loam soil at a faster rate than through the clay layer. Under these conditions, the soil saturates above the clay layer, as it does when a claypan exists in a soil. From this discussion it should be apparent that well-drained soils tend never to saturate unless some underlying layer interrupts the downward movement of a wetting front and soil saturates from the bottom upward.

At the junction of soil layers or horizons of different texture, downward moving water tends to build-up and accumulate. This causes a temporary increase in the availability of water and of nutrients, unless very wet and saturated soil is produced, and results in increased root growth and uptake of nutrients and water near or at these junctions.

Water Vapor Movement

When soils are unsaturated, some of the pore space is air filled and water moves through the pore space as vapor. The driving force is the difference in water potentials as expressed by differences in vapor pressure. Generally, vapor pressure is high in warm and moist soil and low in cold and dry soil. Vapor flow, then, occurs from warm and moist soil to and into cold and dry soil. In the summer, vapor flow travels from the warmer upper soil horizons to the deeper and colder soil horizons. During the fall, with transition to winter, vapor flow tends to be upward. It is encouraged by a low air temperature and upward water vapor movement, and condensation onto the surface of the soil commonly creates a slippery smear in the late night and early morning hours.

Soil conductivity for vapor flow is little affected by pore size, but it increases with an increase in both total porosity and pore space continuity. Water vapor movement is minor in most soils, but it becomes important when soils crack at the surface in dry weather and water is lost from deep within the soil along the surfaces of the cracks. Tillage that closes the cracks, or mulches that cover the cracks, will reduce water loss by evaporation.

PLANT AND SOIL WATER RELATIONS

Plants use large quantities of water, and this section considers the ability of plants to satisfy their water needs by absorbing water from the soil and the effect of the soil water potential on nutrient uptake and plant growth.

Available Water Supplying Power of Soils

The water-supplying power of soils is related to the amount of available water a soil can hold. The available water is the difference in the amount of water at field capacity (-30 kPa or -0.3 bar) and the amount of water at the permanent wilting point ($-1,500$ kPa or -15 bars).

The amount of available water is related to both texture and structure, because it is dependent on the nature of the surfaces and pores, or soil matrix. Soils high in silt (silt loams) tend to have the most optimum combination of surfaces and pores. They have the largest available water-holding capacity, as shown in Figure 5.9. These soils contain about 16 centimeters of plant-available water per 100 centimeters of depth, or 16 percent by volume. This is about 2 inches of water per foot of soil depth. When such soil is at the permanent wilting point, a 2-inch rain or irrigation, which infiltrates completely, will bring the soil to field capacity to a depth of about 1 foot.

From Figure 5.9, it can be noted that sands have the smallest available water-holding capacity. This tends to make sands droughty soils. Another reason why sandy soils tend to be droughty is that most of the available water is held at relatively high water potential, compared with loams and clays. It should be noted in Figure 5.10 that field capacity for sand is more likely nearer -10 kPa (-0.1 bar) than -30 kPa (-0.3 bar) as in loamy and clayey soils. This means that much of the plant available water in sandy soils can move rapidly to roots when water near roots is depleted, because of the relatively high hydraulic conductivity at these relatively high potentials. Plants growing on clay soils, by contrast, may not absorb

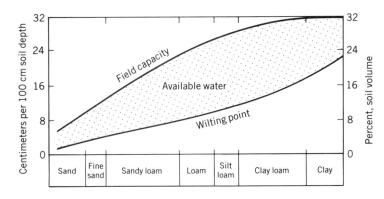

FIGURE 5.9 Relationship of soil texture to available water-holding capacity of soils. The difference between the water content at field capacity and the water content at the permanent wilting point is the available water content.

as much water as they could use during the day because of lower hydraulic conductivity and slower replacement of water near roots when water is depleted. This results in a tendency to conserve the available water in fine-textured soils. In essence, the water retained at field capacity in sands is more plant available than the water in fine-textured soils at field capacity. Therefore, sands tend to be more droughty than clays because they retain less water at field capacity, and the water retained is consumed more rapidly.

Water Uptake from Soils by Roots

Plants play a rather passive role in their use of water. Water loss from leaves by transpiration is

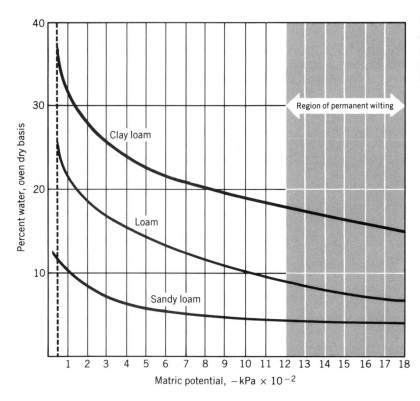

FIGURE 5.10 Soil water characteristic curves for three soils. Field capacity of the sandy loam is about -0.1 kPa $\times 10^{-2}$ (-10 kPa) and for the finer-textured soils is about -0.3 kPa $\times 10^{-2}$ (-30 kPa).

mainly dependent on the surrounding environment. Water in the atmosphere has a much lower water potential than water in a leaf, causing the atmosphere to be a sink for the loss of water from the plant via transpiration. The water-conducting tissue of the leaves (xylem) is connected to the xylem of the stem, and a water potential gradient develops between leaves and stems due to water loss by transpiration. The xylem of the stem is connected to the xylem of the roots, and a water potential gradient is established between the stem and roots. The water potential gradient established between roots of transpiring plants and soil causes water to move from the soil into roots. This water potential gradient, or continuum, is called the soil-plant-atmosphere-continuum, or SPAC. Some realistic water potentials in the continuum during midday when plants are actively absorbing water are −50,000 kPa (−500 bars) in the atmosphere, − 2,500 kPa (−25 bars) in the leaf, − 800 kPa (−8 bars) in the root, and −700 kPa (−7 bars) in the soil. There is considerable resistance to upward water movement in the xylem, so that a considerable water potential gradient can be produced between the leaf and the root.

Diurnal Pattern of Water Uptake

Suppose that it is a hot summer morning, and there has been a soaking rain or irrigation during the night. The soil is about at field capacity and water availability is high. The demand for water is nil, however, because the water-potential differences in the leaves, roots, and soil are small. There is no significant water gradient and no water uptake. After the sun rises and the temperature increases, thus decreasing relative humidity, the water potential gradient increases between the atmosphere and the leaves. The leaves begin to transpire. A water potential gradient is established between the leaves, roots, and soil. Roots begin water uptake. The low water conductivity in (the plant causes the water potential difference between leaves and roots, and between roots and soil, to increase to a maximum at about 2 P.M., as shown in Figure 5.11. This is a period of greater water loss than water uptake, which creates a water deficit in the leaves. It is normal for these leaves to be less than fully turgid and appear slightly wilted.

After 2 P.M., decreasing temperature and increasing relative humidity tend to reduce water

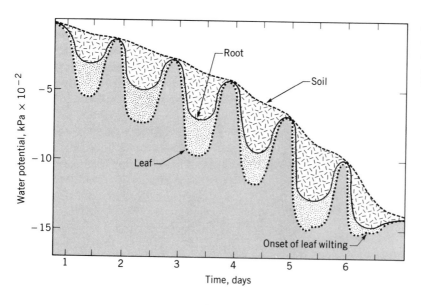

FIGURE 5.11 Diurnal changes in water potential of soil, roots, and leaves during a period of soil drying. (After Slatyer, 1967.)

loss from leaves. With continued water uptake, the plant absorbs water faster than it is lost, and the water potential gradient between leaves and soil decreases. The water deficit in the plant is decreased, and any symptoms of wilting disappear. By the next morning, the water potential gradient between the leaves and soil is eliminated (see Figure 5.11).

As this pattern is repeated, the soil dries and the hydraulic conductivity decreases rapidly and the distance water must move to roots increases. In essence, the availability of soil water decreases over time. As a consequence, each day the potential for water uptake decreases and the plant increasingly develops more internal water deficit, or water stress, during the afternoon. The accumulated effects of these diurnal changes, in the absence of rain or irrigation or the elongation of roots into underlying moist soil, are shown in Figure 5.11. Eventually, leaves may wilt in midday and not recover their turgidity at night. Permanent wilting occurs, the soil is at the permanent wilting point, and the soil has a water potential of about $-1,500$ kPa (-15 bars).

The sequence of events just described is typical where roots are distributed throughout the soil from which water uptake occurs. For the sequence of events early in the growing season of an annual plant, continual extension of roots into underlying moist soil may occur to delay the time when the plant wilts in the absence of rain or irrigation.

The permanent wilting point is not actually a point but a range. Experiments have shown that some common plants could recover from wilting when the potential decreased below $-1,500$ kPa (-15 bars). High temperature and strong winds, on the other hand, can bring about permanent wilting at potentials greater than $-1,500$ kPa. If a plant can survive and, perhaps, make slow growth with a small water demand, the soil water potential may be much less than $-1,500$ kPa. For most soils and plant situations, the permanent wilting range appears to be about $-1,000$ to $-6,000$ kPa (-10 bars to -60 bars), which represents a rather narrow range in differences in soil water content. Desert plants can thrive with very low water potentials and, typically, have some special feature that enables them to withstand a limited supply of soil water. Some plants shed their leaves during periods of extreme water stress.

Pattern of Water Removal from Soil

With each passing summer day, water moves more slowly from soil to roots as soils dry, and the availability of the remaining water decreases. This encourages root extension into soil devoid of roots where the water potential is higher and water uptake is more rapid. For young plants with a small root system, there is a continual increase in rooting depth to contact additional supplies of available water in the absence of rain or irrigation during the summer. In this way the root zone is progressively depleted of water as shown in Figure 5.12. When the upper soil layers are rewetted by rain or irrigation, water absorption shifts back toward the surface soil layer near the base of the plant. This pattern of water removal results in: (1) more deeply penetrating roots in dry years than in wet years, and (2) greater absorption of water from the uppermost soil layers, as compared to the subsoil layers, during the growing season. Crops that have a long growing season and a deep root system, like alfalfa, absorb a greater proportion of their water below 30 centimeters (see Figure 5.13).

Soil Water Potential Versus Plant Growth

Most plants cannot tolerate the low oxygen levels of water-saturated soils. In fact, water saturation of soil kills many kinds of plants. These plants experience an increase in plant growth from saturation to near field capacity. As soils dry beyond field capacity, increased temporary wilting during the daytime causes a reduction in photosynthesis.

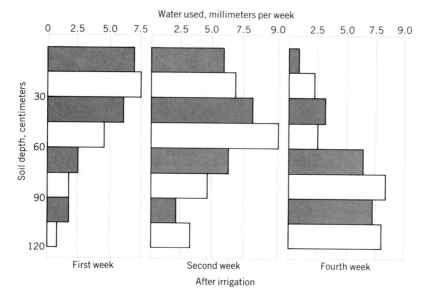

FIGURE 5.12 Pattern of water used from soil by sugar beets during a 4-week period following irrigation. Water in the upper soil layers was used first. Then, water was removed from increasing soil depths with time since irrigation. (Data from Taylor, 1957.)

FIGURE 5.13 Percentage of water used from each 30-centimeter layer of soil for crops produced with irrigation in Arizona. Onions are a shallow-rooted annual crop that is grown in winter and used a total of 44 centimeters of water. Alflalfa, by contrast, grew the entire year and used a total of 186 centimeters of water, much of it from deep in the soil. (Data from Arizona Agr. Exp. Sta. Tech. Bull. 169, 1965.)

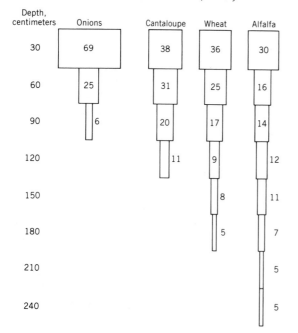

Thus, plant growth tends to be at a maximum when growing in soil near field capacity, where the integrated supply of both oxygen and water is the most favorable.

During periods of water stress, the damage to plants is related to the stage of development. The data in Table 5.2 show that four days of visible wilting during the early vegetative stage caused a 5 to 10 percent reduction in corn yield compared with a 40 to 50 percent reduction during the time of silk emergence and pollination. Injury from water stress at the dough stage was 20 to 30 percent.

TABLE 5.2 Effect of Four Days of Visible Wilting on Corn Yield

Stage of Development	Percent Yield Reduction
Early vegetative	5–10
Tassel emergence	10–25
Silk emergence, pollen shedding	40–50
Blister stage	30–40
Dough stage	20–30

Claassen and Shaw, 1970.

Role of Water Uptake for Nutrient Uptake

Mass flow carries ions to the root surfaces, where they are in a position to be absorbed by roots. An increased flow of water increases the movement of nutrient ions to roots, where they are available for uptake. This results in a positive interaction between water uptake and nutrient uptake. Droughts reduce both water and nutrient uptake. Fertilizers increase drought resistance of plants when their use results in greater root growth and rooting depth (see Figure 5.14).

FIGURE 5.14 Fertilizer greatly increased the growth of both tops and roots of wheat. Greater root density and rooting depth resulted in greater use of stored soil water. (Photograph courtesy Dr. J. B. Fehrenbacher, University of Illinois.)

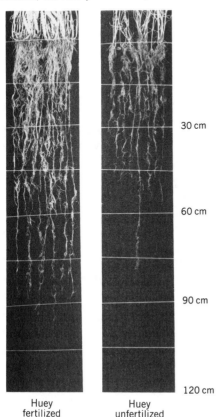

Huey
fertilized

Huey
unfertilized

30 cm

60 cm

90 cm

120 cm

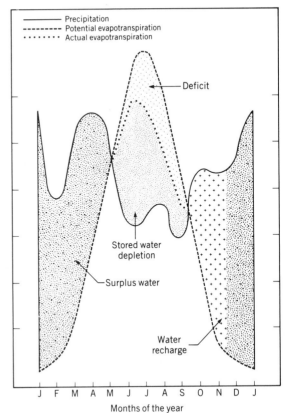

Precipitation
Potential evapotranspiration
Actual evapotranspiration

Deficit

Stored water depletion

Surplus water

Water recharge

J F M A M J J A S O N D J

Months of the year

FIGURE 5.15 Climatic data and average, soil water storage capacity were used to construct the water regime for soils near Memphis, Tennessee. (Data from Thornthwaite and Mather, 1955.)

SOIL WATER REGIME

The *soil water regime* expresses the gradual changes in the water content of the soil. For example, a soil in a humid region may at some time during the year have surplus water for leaching, soil water depletion by plant uptake, and water recharge of dry soil. These conditions are illustrated in Figure 5.15 for a soil with an average water-storage capacity in a humid region, such as near Memphis, Tennessee.

The early months of the year (January to May) are characterized by surplus water (see Figure 5.15). The soil has been recharged with water by rains from the previous fall, when the soil was brought to field capacity. The precipitation in the early part of the year is much greater than the potential for water evaporation (potential evapotranspiration), and this produces *surplus water*. This water (gravitational) cannot be held in a soil at field capacity, and it moves through the soil, producing leaching. During the summer months the *potential evapotranspiration* is much greater than the precipitation. At this time, plants deplete stored soil water. Since there is insufficient stored water to meet the potential demand for water (potential evapotranspiration), a water *deficit* is produced. This means that most of the plants could have used more water than was available. Abundant rainfall in some years prevents a deficit; in very dry years, the water deficit may be quite large and a drought may occur. In the fall the potential evapotranspiration drops below the level of the precipitation, and water recharge of dry soil occurs. Recharge, in this case, is completed in the fall, and additional pecipitation contributes mainly to surplus (gravitational) water and leaching. Leaching occurs only during the period when surplus water is produced.

The water regime in desert climates is characterized by little water storage, a very small soil water depletion, a very large deficit, and no surplus water for leaching.

SUMMARY

The water in soils has an energy level that varies with the soil water content. The energy level of soil water increases as the water content increases and decreases as the water content of the soil decreases. This produces a soil water-energy continuum.

The energy level of soil water is expressed as the soil water potential, commonly as bars or kilopascals.

Water moves within soil and from soil to roots in response to water potential differences. Water movement is directly related to water potential differences and inversely related to the distance of flow.

The rate of water flow is also directly related to soil hydraulic conductivity. The hydraulic conductivity decreases rapidly as soils dry.

Plant uptake of water reduces the water content of the soil and greatly decreases the hydraulic conductivity. Thus, as soil dries, water movement to roots becomes slower. Plants wilt when water movement to roots is too slow to satisfy the plant's demand for water.

During the course of the year the major changes in soil water content include: recharge of dry soil, surplus water and leaching, and depletion of stored water. Water deficits occur when water loss from the soil by evapotranspiration exceeds the precipitation plus stored water for a period of time.

REFERENCES

Classen, M. M. and R. H. Shaw. 1970. "Water Deficit Effects on Corn:II Grain Components." *Agron. J.* **62:**652–655.

Dittmer, H. J. 1937. "A Quantative Study of the Roots and Root Hairs of a Winter Rye Plant." *Am. J. Bot.* **24:**417–420.

Erie, L. J., O. F. French, and K. Harris. 1965. "Consumptive Use of Water by Crops in Arizona." *Arizona Agr. Exp. Sta. Tech. Bull.* 169.

Gardner, W. H. 1968. "How Water Moves in the Soil." *Crops and Soils.* November, pp. 7–12.

Hanks, R. J. 1965. "Estimating Infiltration from Soil Moisture Properties." *J. Soil Water Conserv.* **20:** 49–51.

Hanks, R. J. and G. L. Ashcroft. 1980. *Applied Soil Physics.* Springer-Verlag, New York.

Slayter, R. O. 1967. *Plant-water Relationships.* Academic Press, New York.

Taylor, S. A. 1957. "Use of Moisture by Plants." *Soil, USDA Yearbook.* Washington, D.C.

Thornthwaite, C. W. and J. R. Mather. 1955. "Climatology and Irrigation Scheduling." *Weekly Weather and Crop Bulletin.* National Summary of June 27, 1955.

CHAPTER 6

SOIL WATER MANAGEMENT

The story of water, in a very real sense, is the story of humankind. Civilization and cities emerged along the rivers of the Near East. The world's oldest known dam in Egypt is more than 5,000 years old. It was used to store water for drinking and irrigation, and perhaps, to control flood-waters. As the world's population and the need for food and fiber increased, water management became more important. Now, in some areas of rapidly increasing urban development and limited water supply, competition for the use of water has arisen between agricultural and urban users.

There are three basic approaches to water management on agricultural land: (1) conservation of natural precipitation in subhumid and arid regions, (2) removal of water from wetlands, and (3) addition of water to supplement the amount of natural precipitation (irrigation). Two important needs for water management in urban areas include land disposal of sewage effluent and prescription athletic turf.

WATER CONSERVATION

Water conservation is important in areas in which large water deficits occur in soils within arid and subhumid climates. Techniques for water conservation are designed to increase the amount of water that enters the soil and to make efficient use of this water.

Modifying the Infiltration Rate

Infiltration is the downward entry of water through the soil surface. The size and nature of the soil pores and the water potential gradient near the soil surface are important in determining the amount of precipitation that infiltrates and the amount that runs off. High infiltration rates, therefore, not only increase the amount of water stored for plant use but also reduce flooding and soil erosion.

Surface soils protected by vegetation in a natural forest or grassland tend to have a higher infiltration rate during rains than exposed soils in cultivated fields or at construction sites. In an experiment in which excessive water was applied, the protection of the soil surface by straw or burlap maintained high and constant infiltration rates, as shown in Figure 6.1. When the straw and burlap were removed, the infiltration rates decreased rapidly. The impact of raindrops on ex-

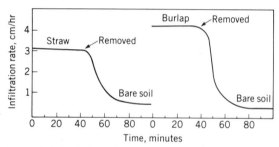

FIGURE 6.1 Effect of straw and burlap and their removal on soil infiltration rate. (Source: Miller and Giffore, 1974.)

posed soil breaks up soil peds and forms a crust. The average pore size in the surface soil decreases, causing a reduction in the infiltration rate. Infiltration is also decreased by overgrazing and deforestation, or by any practices that remove plant cover and/or cause soil compaction. The bare field in Figure 6.2 shows the detrimental effects of raindrop impact on soil structure at the soil surface, decreased water infiltration and, then, increased water runoff and soil erosion.

Since water movement into and through a soil pore increases exponentially as the radius increases, a single earthworm channel that is open at the soil surface may transmit water 100 times faster than the same volume of smaller pores. In natural forest and grassland areas, an abundant food supply at the soil surface encourages earthworms, resulting in large surface soil burrows that are protected from raindrop impact by the vegetative cover.

In fields, it is common for infiltration rates to decrease over time during rainstorms. This is brought about by a combination of factors. First, a reduction in hydraulic conductivity occurs at the soil surface, owing to physical changes. Second, as the soil becomes wet, and the distance of water movement increases to wet the underlying dry soil, the velocity of water flow decreases. Infiltration rates from irrigation furrows, however, a few hours after the start of irrigation have been caused by earthworms that moved toward the moist soil and entered the waterfilled furrows, leaving behind channels that increased infiltration.

Many farmers modify the soil surface conditions with contour tillage and terraces. These practices temporarily pond the water and slow down the rate of runoff, which results in a longer time for the water to infiltrate the soil. In some situations, where there is little precipitation, runoff water is collected for filling reservoirs. Then, surface soils with zero infiltration rates are desir-

FIGURE 6.2 Effects of raindrop impact on soil structure at the soil surface and on water infiltration and water runoff. The soil is unprotected by vegetation or crop residues, resulting in a greatly reduced infiltration rate and greatly increased runoff and soil erosion. Such effects accentuate the differences in the amount of water stored in soils for plant growth on the ridge tops, side slopes, and in the low areas where runoff water accumulates and sedimentation of eroded soil occurs.

able. In these cases, the important product produced by the soil is runoff water.

Summer Fallowing

Most of the world's wheat production occurs in regions having a subhumid or semiarid climate. In the most humid parts of these regions, soils have enough stored water (water from precipitation in excess of evaporation during the winter and early spring), plus that from precipitation during the growing season, to produce a good crop of wheat in almost every year. In the drier part of these regions, the water deficit is larger and in most years insufficient water is available to produce a profitable crop. Here, a practice called *summer fallowing* is used. Summer fallowing means that no crop is grown and all vegetative growth is prevented by shallow tillage or herbicides for a period of time, in order to store water for use by the next crop.

After wheat is harvested, the land is left unplanted. No crop is grown and weeds are controlled by shallow cultivation or herbicides to minimize the loss of water by transpiration. During the year of the fallow period, there is greater water recharge than if plants were allowed to grow and transpire. This additional stored water, together with the next year's precipitation, is used by the wheat. Fallowing produces a landscape of alternating dark-colored fallow strips of soil and yellow-colored strips where wheat was produced, as shown in Figure 6.3. In order to explain how water storage occurs in fallowed land, both evaporation of water from the soil surface and water movement within the soil must be considered.

After a wheat crop has been harvested, the water storage period begins. The surface soil layer is air dry and the underlying soil is at the approximate permanent wilting point (see Figure 6.4). Rain on dry soil immediately saturates and infiltrates the soil surface, and water moves downward as a wetting front, similar to water moving out of an irrigation furrow as shown in Figure 5.8. The depth of water penetration depends on the amount of infiltration and the soil texture. Each centimeter of water that infiltrates will moisten about 6 to 8 centimeters of loamy soil, which is near the permanent wilting point. A heavy rainstorm will produce an upper soil layer at field capacity, whereas the underlying soil remains at the wilting point (see Figure 6.4).

After the rain, some water will evaporate from the soil surface. This creates a water potential gradient between the surface of the drying soil and underlying moist soil, and some water in the underlying moist soil will migrate upward. This phase of soil drying is characterized by rapid water loss. As drying near the soil surface occurs, a rapid decrease in hydraulic conductivity occurs at the soil surface. Water migration upward is greatly reduced, and the evaporative demand at the sur-

FIGURE 6.3 Characteristic pattern of alternate fallow and wheat strips where land is summer fallowed for wheat production.

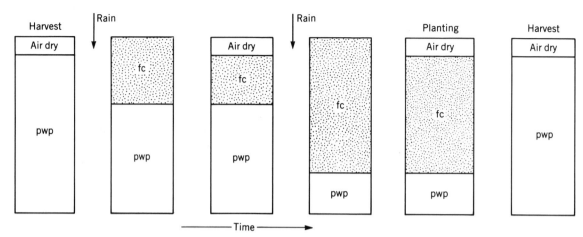

pwp = Permanent wilting point
fc = Field capacity

FIGURE 6.4 Changes in soil water during a summer fallow period for wheat production.

face greatly exceeds the water movement upward from moist soil on a hot summer day. As a result, the immediate surface of the soil becomes air dry within a few hours and the hydraulic conductivity for liquid water approaches zero. This phase is associated with a sharp decline in water loss.

Once a thin layer of surface soil is air dry, and few water films bridge particles, water can move from the underlying moist soil to and through the surface as vapor flow. Vapor movement under these conditions is so slow as to be largely discounted, unless cracks exist. The result is a *capping effect* that protects the stored water from being lost by evaporation. When another rain occurs, the surface soil is remoistened and additional water moves through the previously moistened soil to increase the depth of the recharged soil. Repetition of this sequence of events during a fallow period progressively increases the depth of recharged soil and the amount of stored soil water. The sequence of events during a fallow-cropping cycle is shown in Figure 6.4.

Many studies have shown a good correlation between water stored at planting time and grain yield. The fallowing system, however, is not 100 percent efficient. The capping feature does not

work perfectly, some evaporation occurs each time the soil surface is wetted and dried, and some runoff occurs. A good estimate is that about 25 percent of the rainfall during the fallow period will be stored in the soil for use in crop production. This extra quantity of water, however, has a great effect on yields, as is illustrated in Table 6.1. The frequency of yields over 1,345 or 2,690 kg/hectare (20 or 40 bushels/acre) was significantly increased by summer fallowing. Fallowing was also more effective at Pendleton, Oregon, where the maximum rainfall occurs in winter, unlike Akron, Colorado, and Hays, Kansas, where maximum rainfall occurs in summer. Stored soil moisture has been shown to be as effective as precipitation during the growing season for wheat production.

Saline Seep Due to Fallowing

Most of the world's spring wheat is grown in sub-humid or semiarid regions where winters are severe, such as the northern plains areas of United States, Canada, and the Soviet Union. Low temperature results in low evapotranspiration, which increases the efficiency of water storage during

TABLE 6.1 Percentage Distribution of Wheat Yield Categories at Two Locations in the Great Plains and One Location in the Columbia River Basin

Yield Category	Wheat After Wheat, %			Wheat on Fallowed Land, %		
	Akron, Colo.	Hays, Kans.	Pendleton Ore.	Akron, Colo.	Hays, Kan.	Pendleton, Ore.
Under 336 kg/ha (5 bu/A)[a]	44	33	0	15	17	0
Under 1,345 kg/ha (20 bu/A)	88	62	100	61	36	0
Over 1,345 kg/ha (20 bu/A)	12	38	0	39	64	100
Over 2,690 kg/ha (40 bu/A)	0	5	0	5	12	83

From Mathews, O. R., 1951. [a] A = acre.

the fallow period. In fact, if a fallow period is followed by a wet year, surplus water may occur and percolate downward and out of the root zone. Where surplus water encounters an impermeable layer, soil saturation occurs and water tends to move laterally because of gravity above the impermeable layer. This may create a spring (seepage spot) at a lower elevation. The evaporation of water from the area wetted by the spring results in the accumulation of salts from the water, and an area of salt accumulation called a *saline seep* is formed. The osmotic effect of soluble salt reduces the soil water potential, water uptake, and wheat yields. In some instances the soils become too saline for plant growth, because of very low water potential.

Control of a saline seep depends on reducing the likelihood that surplus water will occur. Management practices to control saline seeps include: (1) reduced fallowing frequency, (2) use of fertilizers to increase plant growth and water use, and (3) the production of deep-rooted perennial crops, such as alfalfa, to remove soil water. When alfalfa is grown and has dried out the soil, the alfalfa grows very slowly, which indicates that it is once again time to begin wheat production.

Effect of Fertilizers on Water Use Efficiency

Plants growing in a soil that contains a relatively small quantity of nutrients grow slower and tran-spire more water per gram of plant tissue produced than those growing where the plant nutrients are in abundance. Since it has been shown that water loss (or use) is mainly dependent on the environment, any practice that increases the rate of plant growth will tend to result in production of more dry matter over a given period of time and per unit of water used. The data in Table 6.2 show that the yield of oats increased from 86 to 145 kg/hectare (2.4 to 4.0 bushels/acre) for each 2.5 centimeters of water used in conjunction with more fertilizer. In humid regions, it has commonly been observed that fertilized crops are more drought resistant. This may be explained on the basis that increased top growth results in increased root growth and root penetration, so that the total amount of water consumed is greater.

Fertilizer use in the subhumid region may occasionally decrease yields. If fertilizer causes a crop to grow faster in the early part of the season, the greater leaf area at an earlier date represents a greater potential for water loss by transpiration. If no rain occurs, or no irrigation water is applied, plants could run out of water before harvest, causing a serious reduction in grain yield.

Because water use in crop production is mainly environmentally determined, fertilizer use and the doubling or tripling of yields has little effect on the sufficiency of the water supply. In underdeveloped countries, large increases in crop yields can be produced without creating a need for significantly more water. Conversely, if water

TABLE 6.2   Oats Production per 2.5 Centimeters of Water Used

| | Low Nitrogen | | High Nitrogen | |
Year	Kilograms per Hectare	Bushels per Acre	Kilograms per Hectare	Bushels per Acre
1949	75	2.1	158	4.4
1950	97	2.7	133	3.7
Average	86	2.4	145	4.0

From Hanks and Tanner, 1952.

supplies become very limited, it may be desirable to use less land and more fertilizer to obtain the same amount of total production with less water used.

SOIL DRAINAGE

About A.D. 1200, farmers in the Netherlands became engaged in an interesting water control problem. Small patches of fertile soil, affected by tides and floodwaters, were enclosed by dikes and drained. Windmills provided the energy to lift gravitational water from drainage ditches into higher canals for return to the sea. Now, 800 years later, about one third of the Netherlands is protected by dikes and kept dry with pumps and canals. Soil dewatering or drainage is used to lower the water table when it occurs at or near the soil surface, or to remove excess water that has accumulated on the soil surface. The use of drainage is commonplace to dewater the soil in the root zone of crops, to dewater the soil near building foundations to maintain dry basements and for the successful operation of septic sewage disposal systems, and to remove excess surface water (see Figure 6.5).

FIGURE 6.5   Dewatering, surface and subsurface drainage, is required to make this land suitable for both building construction and cropping. (Photograph courtesy Soil Conservation Service, USDA.)

Water Table Depth Versus Air and Water Content of Soil

If a well is dug and the lower part fills with water, the surface of the water in the well is the top of the water table. Water does not rise in a well above the water table. In the adjacent soil, however, capillarity causes water to move upward in soil pores. A water-saturated zone is created in the soil above the water table, unless the soil is so gravelly or stony that capillarity does not occur. The water-saturated zone above the water table is the *capillary fringe* and is illustrated in Figure 6.6.

Soils typically have a wide range in pore sizes. Some pores are large enough so that they are air-filled above the capillary fringe, depending on pore size and height above the capillary fringe.

This creates a zone above the capillary fringe that has a decreasing water content and increasing air content toward the soil surface. This capillary-affected zone, plus the capillary fringe, generally ranges in thickness from several centimeters to several meters. For all practical purposes, soil more than 2 meters above the water table is scarcely affected by it. This means that most plants, even those in humid regions, do not use water from the water table.

Generally, the soil above the capillary fringe will have a water content governed by the balance between infiltration and loss by evapotranspiration and percolation. Soils in deserts and semiarid regions typically have a permanently dry zone between the root zone and the water table.

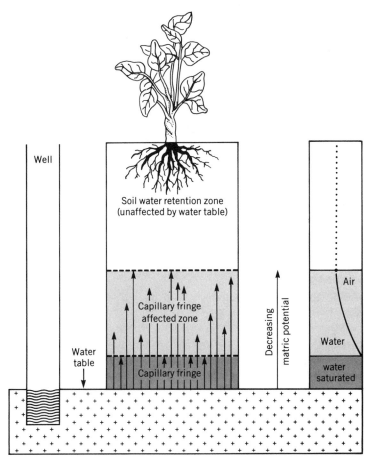

FIGURE 6.6 Generalized relationship of the water table to the soil water zones, and the air and water content of soil above the water table. The capillary fringe is water saturated. Upward from the capillary fringe the water content of the soil decreases and the air content increases within the capillary fringe affected zone.

Benefits of Drainage

Root tips are regions of rapid cell divison and elongation, having a high oxygen demand. Roots of most economic crop plants do not penetrate water-saturated soil or soil that is oxygen deficient. The major purpose of drainage in agriculture and forestry is lowering the water table to increase the plant rooting depth.

Some of the largest increases in plant growth resulting from drainage occur in forests. Drainage on the Atlantic Coastal Plain increased the growth of pine from 80 percent to 1,300 percent (see Table 6.3). Few management practices are as effective as drainage for increasing tree growth. Additional benefits of drainage in forests are easier logging, with less soil disturbance, and easier site preparation for the next crop.

Drainage of wet soil also increases the length of the growing season in regions with low soil temperature and frost hazard. Soils with a lower water content warm more rapidly in the spring. Seed germination is more rapid and roots grow faster. The net result is a greater potential for plant growth.

Surface Drainage

Surface drainage is the collection and removal of water from the surface of the soil via open ditches.

Two conditions favoring the use of surface drainage are: (1) low areas that receive a large amount of water from the surrounding higher land, and (2) impermeable soils that have an insufficient hydraulic conductivity to dispose of the excess water by movement downward and through the soil to drainage tubing. Surface drainage is favored on the Sharkey clay soils along the lower Mississippi River, because the soil is slowly permeable and receives water from surrounding higher land. Sugarcane, which is widely grown and sensitive to poor aeration, is planted on ridges to improve aeration in the root zone.

It is difficult to irrigate without applying excess water in many instances. Surface drainage ditches are used to dispose of excess water and to prevent saturation of the soil on the low end of irrigated fields. Surface drainage is also widely used along highways and in urban areas for surface water control. Rapid removal of surface water from low areas of golf courses and parks is essential to permit their use soon after a rain.

Subsurface Drainage

Ditches for the removal of gravitational water can be made quickly and inexpensively. Drainage ditches, however, require periodic cleaning and are inconvenient for the use of machinery. There

TABLE 6.3 Effect of Drainage on Mean Annual Growth of Pine on Poorly Drained Soils of the Atlantic Coastal Plain

| Species | Ages | Mean Annual Growth[a] | | Percent Increase Over Undrained |
		Drained	Undrained	
Planted loblolly	0–17	17.9	1.28	1,298
Natural pond	0–22	0.74	0.41	80
Natural slash	19–22	9.0	4.9	84
Planted slash	0–5	14.6	5.7	156
Planted slash	0–5	13.3	5.7	133
Planted loblolly	0–5	7.6	1.3	585
Planted loblolly	0–5	4.9	1.3	277
Planted loblolly	0–13	4.3	0.54	696

Adapted from Terry and Hughes, 1975.
[a] Cubic meters per hectare.

are also many situations in which ditches are not satisfactory, for example, water removal around the basement walls of a building.

Drain lines are installed in fields with trenching machines. Perforated plastic tubing is very popular and is automatically laid at the bottom of a trench as the trenching machine moves across the field (see Figure 6.7). Unless some unusual condition prevails, plastic tubing should be laid at least 1 meter deep to have a sufficiently low water table between the drains to permit crops to develop an adequate root system, as illustrated in Figure 6.8.

Drainage in Soil of Container-Grown Plants

Many plants are grown indoors in containers. One of the most common problems of growing plants

FIGURE 6.7 Installation of perforated plastic tubing to remove water from a field during the wet season. The plastic tubing carries water to an outlet where the water is disposed. (Photo courtesy Soil Conservation Service, USDA.)

in containers is poor soil aeration caused by overwatering. How can this be true, since most flowerpots have a large drainage hole in the bottom? Knowledge of changes in air and water content of soil above the water table can help provide the answer.

Consider a flowerpot filled with soil that is water saturated. Water is allowed to drain out of the large hole in the bottom of the pot. Surplus water will exit rapidly from the largest soil pores at the top of the pot and air will be pulled into these pore spaces. Gradually the zone of water saturation will move progressively downward. Shortly thereafter, the loss of water by gravity may essentially stop. At this time, some of the soil at the bottom may still be water saturated, if all the pores in the soil are sufficiently small to attract water more strongly than gravity. This creates a "capillary fringe" in the bottom of the pot. Then, the air and water content of the soil above the saturated soil at the bottom is analogous to soil above the saturated capillary fringe. Compare the drawing in Figure 6.9 with Figure 6.6 and note, in particular, the decrease in water and increase in air with distance upward in the soil above soil that is water saturated.

If all soil pores are very small, all the soil in a flowerpot could remain saturated after watering. A good soil for container-grown plants has many pores too large to retain water against gravity. After wetting and drainage, these large pores will become air filled, even at the bottom of the container. Special attention must be given to soil used for container-grown plants. Ordinary loamy field or garden soils are unsatisfactory. A mix of one-half fine sand and one-half sphagnum peat moss, by volume, is recommended by the University of California to produce a mix with excellent physical properties for plant growth. When sand and soil are mixed together, the small soil particles fill the spaces between the sand particles, producing a low-porosity mix with few large aeration pores.

Placing large gravel in the bottom of a flowerpot or in a hole dug outdoors will create a barrier to the downward flow of water from an overlying finer-textured soil layer or soil mix. The gravel

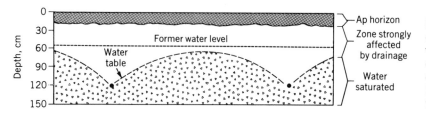

FIGURE 6.8 The effect of drainage lines in lowering the water table. The benefit of drainage is first evident directly over the drainage lines, and it gradually spreads to the soil area between them.

acts in the same manner as a large hole in a container that is filled with a soil mix. A saturated soil zone tends to form above the gravel layer, whatever the depth of the gravel.

IRRIGATION

Irrigation is an ancient agricultural practice that was used 7,000 years ago in Mesopotamia. Other ancient, notable irrigation systems were located in Egypt, China, Mexico, and Peru. Today, approx-

imately 11 percent of the world's cropland is irrigated. Some of the densest populations are supported by producing crops on irrigated land, as in the United Arab Republic (Egypt), where 100 percent of the cropland is irrigated. This land is located along the Nile River. More than 12 percent of the cropland, or about 24,000,000 hectares (61,000,000 acres) are irrigated in the United States. Irrigation is extremely important in increasing soil productivity throughout the world.

Water Sources
Most irrigation water is surface water resulting from rain and melting snow. Many rivers in the world have their headwaters in mountains and flow through arid and semiarid regions where the water is diverted for irrigation. Examples include the Indus River, which starts in the Himalaya Mountains, and the numerous rivers on the western slope of the Andes Mountains, which flow through the deserts of Chile and Peru to the Pacific Ocean. Much of the water in these rivers comes from the melting of snow in the high mountains. In fact, the extent of the snowpack is measured in order to predict the stream flow and amount of water that will be available for irrigation the next season. Many farmers who use well water have their own source of irrigation water. About 20 percent of the water used for irrigation in the United States comes from wells.

FIGURE 6.9 Illustration of conditions in a flowerpot after a thorough watering and drainage. At the bottom of the pot, the soil is saturated. Soil-air content increases and water content decreases with distance upward above the water-saturated layer. The soil pore spaces operate in a manner similar to a series of capillary tubes of varying sizes.

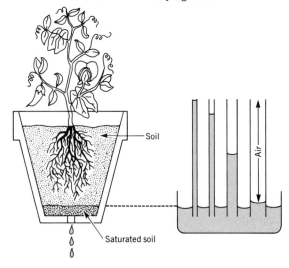

Important Properties of Irrigated Soils
Soils well suited for irrigation have an intermediate texture to a depth of 1 to 2 meters and without compact or water-impermeable layers.

Loamy soils that are underlain by coarse sands and gravel have greater retention of water in the overlying soil than soils that are loamy-textured throughout. The texture, together with the structure, should result in an infiltration rate of about 0.5 to 8 centimeters per hour and a moderate water-storage capacity. The soil should not have an injurious soluble salt content.

Water Application Methods

Choice of various methods of applying irrigation water is influenced by a consideration of: (1) infiltration rate, (2) slope and general nature of the soil surface, (3) supply of water and how it is delivered, (4) crop rotation, and (5) seasonal rainfall. The methods of distributing water can be classified as flood, furrow, subsurface, sprinkler, and drip or trickle.

Flood Irrigation Flood irrigation floods the land by distributing water by gravity. Water is let into the upper end of the field, usually from a large ditch. Sometimes the water is distributed into basins, which is a common practice for orchards, pastures, and some grain crops, including rice (see Figure 6.10). Flood irrigation can easily over-water some parts of a field while leaving some parts underirrigated.

Furrow Irrigation Furrow irrigation is similar to flood irrigation, the difference being that the water distribution is down furrows instead of being allowed to flood over the land. In furrow irrigation, which is used for row crops, water is distributed down between the rows. Furrows that have a slight grade or slope are made across the field. The slope of the furrow bottoms depends on soil infiltration rate, which should be sufficient to distribute water along the entire length of the furrow (see Figure 6.11). Furrow irrigation is unsatisfactory on sand soils, because of the very high infiltration rate and limited distance of water flow down the furrows.

Many methods are used to distribute water into the furrows. Gated pipe is shown in Figure 6.11, and the use of siphon tubes is shown in Figure 6.12. Siphon tubes require extensive manual labor. A major advantage of flood and furrow irrigation is that the water is distributed by gravity, so water distribution costs tend to be low.

Sprinkler Irrigation Sprinklers are widely used to irrigate lawns. Sprinkler systems are versatile and have some special advantages. On sands, where high infiltration rates permit water to flow only a short distance down a furrow, sprinklers are essential to spread water uniformly over the field. On an uneven land surface, expensive land

FIGURE 6.10 Flood irrigation of grapes in the Central Valley in California.

FIGURE 6.11 Planned land with proper grade to give uniform distribution of water down the rows. Gated pipe is being used to distribute water into the furrows.

leveling and grading can be avoided by the use of sprinklers. The portable nature of many sprinkler systems makes them ideally suited for use where irrigation is used to supplement natural rainfall. When connected to soil water-measuring apparati, sprinkler systems can be made automatic.

Large self-propelled sprinkler systems, which pivot in a large circle around a well, have become popular. Such systems can irrigate most of the land in a quarter section (160 acres, or 65 hectares). In the western part of the United States, large green circular irrigated areas produced by pivot sprinklers are clearly visible from the air. Their use on many rolling and sandy soils has greatly increased the production of corn and other grain crops on the Great Plains. In some areas, however, underground water reservoirs are being depleted, and some irrigated areas are again being used as grazing land and for fallowed wheat production.

Sprinkler irrigation modifies the environment by completely wetting plant leaves and the soil surface and by affecting the surrounding air temperature. The high specific heat of water makes

FIGURE 6.12 Use of siphon tubes to transfer water from the head ditch into furrows.

sprinkling an effective means of reducing frost hazard. Sprinkler irrigation has been used for frost protection in such diverse situations as for winter vegetables in Florida, strawberries in Michigan, and citrus in California.

Subsurface Irrigation Subirrigation is irrigating by water movement upward from a water table located some distance below the soil surface. Some areas are naturally subirrigated where water tables are within a meter or two of the soil surface. This occurs in the poorly drained areas of the Sandhills in north-central Nebraska. Artifical subirrigation is practiced in the Netherlands, where subsurface drainage lines are used to drain soils in the wet seasons and then used during drought for subirrigation. The subirrigation depends on the use of drainage ditches and pumps to create and control a water table. Subirrigation works well on the nearly level Florida coastal plains. The sandy soils have high saturated hydraulic conductivity, and the natural water table is 1 meter or less from the soil surface much of the time.

Subirrigation is more efficient in the use of water than other systems, and is adapted only for special situations. In arid regions, the upward movement of water and its evaporation at the soil surface results in salt accumulation. Subirrigation works best where natural rainfall leaches any salts that may have accumulated.

Drip Irrigation Drip irrigation is the frequent or almost continuous application of a very small amount of water to a localized soil area. Plastic hoses about 1 or 2 centimeters in diameter, with emitters, are placed down the rows or around trees. Only a small amount of the potential root zone is wetted. Roots in the localized area can absorb water rapidly because a high water potential, or water availability, is maintained. A major advantage of drip irrigation is the large reduction in the total amount of water used. Other advan-

tages include adaptability of the system to very steep land, where other methods are unsuited, and the ability to maintain a more uniform soil water potential during the growing season. Potatoes develop an irregular shape and become less desirable if tubers alternately grow rapidly and slowly because of large changes in available water during the period of tuber development.

Rate and Timing of Irrigation

An ideal application of water would be a sufficient quantity to bring the soil in the root zone to field capacity. More water may result in waterlogging of a portion of the subsoil or in the loss of water by drainage. If water percolates below the root zone over a period of many years, the water may accumulate above an impermeable layer and produce a perched water table. Conversely, unless enough water is applied to result in some drainage, it is difficult to remove the salt in the irrigation water and prevent the development of soil salinity.

An accepted generalization is that the time to irrigate is when 50 to 60 percent of the plant-available water in the root zone has been used. Potentiometers are used to measure the matric potential in various parts of the root zone, and the timing of irrigation is based on water potential readings. Computer programs are available for irrigation scheduling that are based on maintaining a record of gains and losses of soil water. Farmers commonly irrigate according to the appearance of the crops. Such crops as beans, cotton, and peanuts develop a dark-green color under moisture stress. Some species of lawn grasses turn to an almost black-green color when wilting. In most instances, crops should be irrigated before marked wilting occurs. When irrigation water is in short supply, it may be better to curtail water use early in the season (delay irrigation), so that sufficient water is available during the reproductive stages of plant growth, when water stress is the most injurious to crop yields.

Water Quality

Water in the form of rain and snow is quite pure. By the time the water has reached farm fields, however, the water has picked up soluble materials (salts) from the soils and rocks over and through which the water moved. Four important factors in determining water quality are: (1) total salt concentration, (2) amount of sodium relative to other cations, (3) concentration of boron and other toxic elements, and (4) the bicarbonate concentration.

Total Salt Concentration

An acre-foot of water weighs about 2,720,000 pounds. If it contained 1 ppm of salt, it would contain 2.72 pounds of salt. Water containing 735 ppm of salt would contain 1 ton of salt per acre-foot of water. An annual application of 4 acre-feet of irrigation water containing 735 ppm of salt would involve adding salt equal to 4 tons per acre each year. Since such water would be of good quality, there is great potential for developing sufficient soluble salt content in soils so as to seriously reduce soil productivity. The longtime use of soils for irrigation necessitates some provision for the leaching and removal of salt in drainage water in order to maintain a favorable salt balance in the soil, where the salts are not naturally leached out of the soil.

The total salt concentration, or salinity, is expressed in terms of the electrical conductivity of the water, and it is easily and precisely determined. The electrical conductivity of irrigation water is expressed as decisiemens per meter (dS/m); in earlier literature as millimhos per centimeter. Four water quality classes have been established with conductivity ranging from 0.1 to more than 2.25 dS/m, as shown in Figure 6.13. The salinity hazard ranges from low to very high.

The four salinity classes of water have the following general characteristics. Low salinity water (C1) can be used for irrigation of most crops, with little likelihood that soil salinity will develop. Some leaching is required, but this occurs under normal irrigation practices, except in soils with extremely low hydraulic conductivity.

Medium salinity hazard water (C2) can be used if a moderate amount of leaching occurs. Plants with moderate salt tolerance usually can be grown without special practices for salinity control.

High salinity hazard water (C3) cannot be used on soils with restricted drainage. Even with adequate drainage, special management for salinity control may be required. Plants with good salt tolerance should be selected.

Very high salinity hazard water (C4) is not suitable for irrigation under ordinary conditions, but may be used occasionally under very special circumstances. The soil must be permeable and have adequate drainage. Excess irrigation water must be applied to bring about considerable leaching of salt, and very salt-tolerant crops should be selected.

Sodium Adsorption Ratio

The water quality classification in Figure 6.13 is also based on the sodium hazard, along the left vertical side of the figure. The sodium hazard is based on the sodium adsorption ratio (SAR), which is the amount of sodium relative to other cations.

The negative electrical charge of clay and humus particles is neutralized by cations that are adsorbed and balance the negative charge. This charge, which is discussed in Chapter 10, is different and much stronger than the electronegativity that attracts water molecules. When 15 percent or more of the charge is neutralized by sodium (Na^+), clay and humus particles tend to disperse and decrease the soil's hydraulic conductivity. The SAR expresses the relative amount of sodium in irrigation water relative to other cations and the likelihood that the irrigation water will increase the amount of adsorbed sodium in the soil.

The SAR of irrigation water is calculated as follows:

$$SAR = \frac{(sodium)}{(calcium + magnesium)^{1/2}}$$

The ion concentrations, denoted by parentheses, are expressed in moles per liter. The classifi-

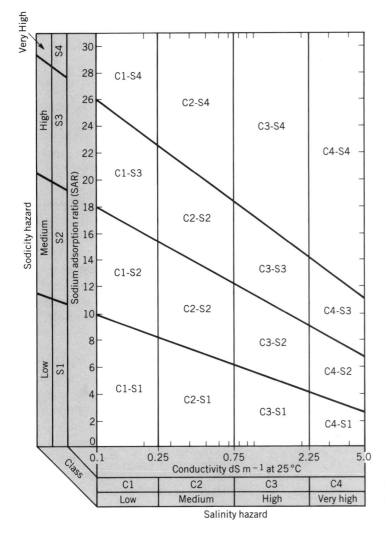

FIGURE 6.13 Diagram for the classification of irrigation waters. (Data from USDA Handbook 60, 1954.)

cation of irrigation waters with respect to SAR is based primarily on the effect of the exchangeable sodium on the physical condition of the soil. Irrigation water with SAR values of 18 to 26 has a high sodium hazard and above 26 the sodium hazard is very high. The four SAR classes combine with the four classes of salinity hazard to give 16 classes of water as shown in Figure 6.13.

Boron Concentration Boron is essential to plant growth, but the amount needed is very small. A small amount of boron in soil may be

toxic. Plants have a considerable range in tolerance. The occurrence of boron in toxic concentration in certain irrigation waters makes it necessary to consider this element in assessing water quality. The permissible limit of boron in several classes of irrigation water is given in Table 6.4, and the relative tolerance of some plants is given in Table 6.5.

Bicarbonate Concentration In waters containing high concentrations of bicarbonate ion (HCO_3^-), calcium and magnesium tend to pre-

TABLE 6.4 Permissible Limits of Boron for Several Classes of Irrigation Waters in Parts per Million

Boron Class	Sensitive Crops	Semitolerant Crops	Tolerant Crops
1	<0.33	<0.67	<1.00
2	0.33–0.67	0.67–1.33	1.00–2.00
3	0.67–1.00	1.33–2.00	2.00–3.00
4	1.00–1.25	2.00–2.50	3.00–3.75
5	>1.25	>2.50	>3.75

From Agriculture Handbook 60, USDA, 1954.

cipitate as carbonates as the soil solution becomes more concentrated. This reaction does not go to completion under ordinary circumstances, but when it does proceed, the concentrations of calcium and magnesium are reduced and the relative proportion of sodium in the soil solution is increased. This results in an increase in the SAR and a decrease in water quality.

In making an estimate of water quality, the effect of salts on both the soil and the plant must be considered. Various factors such as drainage and soil texture influence the effects on soil. The finer the size of clay particles and the greater the tendency for clay particles to expand and contract with wetting and drying, the greater is the effect of a given SAR on dispersion of clay particles and

TABLE 6.5 Tolerance of Plants to Boron (In each Group, the Plants First Named are Considered as Being More Tolerant, and the Last Named More Sensitive)

Tolerant	Semitolerant	Sensitive
Athel (*Tamarix aphylla*)	Sunflower (native)	Pecan
Asparagus	Potato	Black walnut
Palm (*Phoenix canariensis*)	Acala cotton	Persian (English) walnut
Date palm (*P. dactylifera*)	Pima cotton	Jerusalem artichoke
Sugar beet	Tomato	Navy bean
Mangel	Sweetpea	American elm
Garden beet	Radish	Plum
Alfalfa	Field pea	Pear
Gladiolus	Ragged Robin rose	Apple
Broadbean	Olive	Grape (Sultanina and
Onion	Barley	Malaga)
Turnip	Wheat	Kadota fig
Cabbage	Corn	Persimmon
Lettuce	Milo	Cherry
Carrot	Oat	Peach
	Zinnia	Apricot
	Pumpkin	Thornless blackberry
	Bell pepper	Orange
	Sweet potato	Avocado
	Lima beans	Grapefruit
		Lemon

From *Agriculture Handbook 60*, USDA, 1954.

reduction in hydraulic conductivity. The ultimate effect on the plant is the result of these and other factors that operate simultaneously. Consequently, no method of interpretation is absolutely accurate under all conditions, and interpretations are based commonly on experience.

Salt Accumulation and Plant Response

Soil salinity caused by the application of irrigation water usually is not a problem in humid regions where rainfall leaches the soil some time each year. Most irrigation, however, is done in low rainfall areas where leaching is limited and salt tends to accumulate in irrigated soils. Today, crop production is reduced on 50 percent of the world's irrigated land as a result of salinity or drainage problems.

Saline soil contains sufficient salt to impair plant growth. The salts are mainly chlorides, bicarbonates, and sulfates of sodium, potassium, calcium, and magnesium. The best method for assessing soil salinity is to measure the electrical conductivity of the saturated soil extract, the EC_e. This procedure involves preparing a water-saturated soil paste by adding distilled water while stirring the soil until a characteristic end point is reached. A suction filter is used to obtain a sample of the extract for making an electrical conductivity measurement. The saturated-extract electrical conductivity is directly related to the field moisture range from field capacity to the permanent wilting point; this is a good parameter for evaluating the effect of soil salinity on plant growth.

Salinity effects on plants, as measured by the electrical conductivity of the saturated extract, EC_e, in decisiemens per meter, dS/m, are as follows:

1. 0–1: Salinity effects mostly negligible.
2. 2–4: Yields of very salt-sensitive crops may be restricted.
3. 4–8: Yields of many crops restricted.
4. 8–16: Only salt-tolerant crops yield satisfactorily.
5. Over 16: Only a few very salt-tolerant crops yield satisfactorily.

The salt-tolerance figures of crops given in Table 6.6 are arranged according to major crop divisions. Within each group the crops are listed in order of decreasing salt tolerance. The EC_e values given represent the salinity level at which 10, 25, and 50 percent decreases in yields may be expected, as compared with yields on nonsaline soils under comparable growing conditions. Studies of the effects of salt on plants have shown significant differences in varieties of cotton, barley, and smooth brome, whereas variety differences were of no consequence for green beans, lettuce, onions, and carrots.

The effect of salts on plants is mainly indirect, that is, the effect of salt on the osmotic water potential. Decreasing water potential due to salt reduces the rate of water uptake by roots and germinating seeds. Thus, an increase in soil salinity produces the same net effect on water uptake as that produced by soil drying. In fact, the effects of decreases in osmotic and matric water potentials appear to be additive.

Salinity Control and Leaching Requirement

For permanent agriculture, there must be a favorable salt balance. The salt added to soils in irrigation water must be balanced by the removal of salt via leaching. The fraction of the irrigation water that must be leached through the root zone to control soil salinity is termed the *leaching requirement (LR)*. Assuming that the goal is to irrigate a productive soil with no change in salinity, and where soil salinity will not be affected by leaching from rainfall or other factors, the leaching requirement is the ratio of the depth of drainage water, D_{dw}, to the depth of irrigation water, D_{iw}:

$$LR = \frac{D_{dw}}{D_{iw}}$$

TABLE 6.6   Salt Tolerance of Crops Expressed as the EC_e at 25° C for Yield Reductions of 10, 25, and 50 Percent, as Compared to Growth on Normal Soils

Crop	10%	25%	50%
Field crops			
Barley	12	16	18
Sugar Beet	10	13	16
Cotton	10	12	16
Safflower	8	11	14
Wheat	7	10	14
Sorghum	6	9	12
Soybean	5.5	7	9
Sesbania	4	5.5	9
Rice (paddy)	5	6	8
Corn	5	6	7
Broadbean	3.5	4.5	6.5
Flax	3	4.5	6.5
Beans	1.5	2	3.5
Vegetable crops			
Beet	8	10	12
Spinach	5.5	7	8
Tomato	4	6.5	8
Broccoli	4	6	8
Cabbage	2.5	4	7
Potato	2.5	4	6
Corn	2.5	4	6
Sweet potato	2.5	3.5	6
Lettuce	2	3	5
Bell pepper	2	3	5
Onion	2	3.5	4
Carrot	1.5	2.5	4
Beans	1.5	2	3.5
Forage crops			
Bermuda grass	13	16	18
Tall wheat grass	11	15	18
Crested wheat	6	11	18
grass	7	10.5	14.5
Tall fescue	8	11	13.5
Barley hay	8	10	13
Perennial rye grass	8	10	13
Harding grass	6	8	10
Birdsfoot trefoil	4	7	11
Beardless wild rye	3	5	8
Alfalfa	2.5	4.5	8
Orchard grass	2	3.5	6.5
Meadow foxtail	2	2.5	4
Clovers, alsike and red	2	3.6	5.7

TABLE 6.6 (*continued*)

Crop	10%	25%	50%
Fruit crops			
Date palm	8		16
Pomegranate	4.6		9
Fig	4.6		9
Olive	4.6		9
Grape (Thompson)	4		8
Muskmelon	3.5		No data
Orange, grapefruit, lemon	2.5		5
Apple, pear	2.5		5
Plum, prune, peach, apricot, almond	2.5		5
Boysenberry, blackberry, raspberry	1.5–2.5		4
Avocado	2		4
Strawberry	1.5		3

Adapted from *Agr. Information Bull. Nos. 283 and 292* and *Western Fertilizer Handbook*, 1975.

LR, under the conditions specified (no change in soil salinity over time), is also equal to

$$LR = \frac{EC_{iw}}{EC_{dw}}$$

If the electrical conductivity of the drainage water (EC_{dw}) that can be tolerated without adversely affecting the crop is 8 dS/m:

$$LR = \frac{EC_{iw}}{8 \text{ dS/m}}$$

For irrigation water with an electrical conductivity (EC_{iw}) of 1, 2, and 3 dS/m, LR = 0.13, 0.25, and 0.38, respectively. Of the irrigation water applied to the soil, 13, 25, and 38 percent, respectively, represent the amounts of the water that must be leached through the soil. Under these conditions, the salt in the irrigation water is equal to the salt in the drainage water and the salt content of the soil remains unchanged. When the irrigation need is 10 centimeters, the total water application needed is calculated as follows for an *LR* of 25 percent:

$$LR = 10 \text{ cm} + 0.25 \, (10 \text{ cm}) = 12.5 \text{ cm}$$

The need for leaching soils to control salt high-lights a major disadvantage of drip irrigation: the method is not suitable for application of large amounts of water. Leaching of salts without provision for drainage in upland soils, as on the high plains of Texas, and other irrigated areas, results in salt being deposited below the root zone. Most of the irrigated land in arid regions is located in basins or alluvial valleys containing stratified soil materials. Long and continued leaching for salt control commonly results in the accumulation of surplus water above impermeable layers or the formation of perched water tables. When water tables rise to within 1 to 2 meters of the soil surface, water moves upward by capillarity at a rate sufficient to deposit salt on top of the soil from the evaporation of water, as shown in Figure 6.14. It is essential in these cases that subsurface drainage systems be installed to remove drainage water and maintain deep water tables. Consequently, salinity control in arid regions is dependent on drainage.

Archaeological studies in Iraq showed that the Sumerians grew almost equal amounts of wheat and barley in 3500 B.C. One thousand years later, only one-sixth of the grain was the less salt-

FIGURE 6.14 Water tables (water-saturated soil) a short distance below the soil surface result in the upward movement and evaporation of water and the accumulation of salt at the soil surface. The process results in the development of saline soil.

tolerant wheat; by 1700 B.C., the production of wheat was abandoned. This decline in wheat production and increase in salt-tolerant barley coincided with soil salinization. Records show widespread land abandonment, and that salt accumulation in soils was a factor in the demise of the Sumerian civilization.

Even though good irrigation practices are followed, important differences can exist in the salt content of soil that is immediately under the irrigation furrow and the tops of beds. Downward water movement under the furrow leaches the soil and produces low salinity. At the same time, water moves to the top of the beds and salt is deposited as water evaporates. Important differences in salt concentrations are produced (see Figure 6.15).

FIGURE 6.15 Salt accumulation on the ridges resulting from the upward movement and evaporation of water in a cotton field. Salt content under the furrows is less than on the ridges due to leaching under the furrows. (Photo courtesy Soil Conservation Service, USDA.)

Effect of Irrigation on River Water Quality

About 60 percent of the water diverted for irrigation is evaporated or consumed. The remainder appears as irrigation return flow (which includes drainage water) and is returned to rivers for disposal. As a consequence, the salt concentration of river water is increased. A 2- to 7-fold increase in salt concentration is common for many rivers. Along the Rio Grande in New Mexico, water diverted at Percha Dam for Rincon Valley irrigation had an average of 8 milliequivalents per liter of salt from 1954 to 1963. Salt content increased to 9 at Leasburg, 13 at the American Diversion Dam, and 30 milliequivalents per liter at the lower end of El Paso Valley. Rivers serve as sinks for the deposition of salt in drainage water from irrigated land. The Salton Sea, whose surface is below sea level, serves as a salt sink for the Imperial Valley in southern California.

Nature and Management of Saline and Sodic Soils

The development of saline soils is a natural process in arid regions where water tables are near the soil surface. Large areas have become saline from irrigation and from saline seeps as a result of summer fallowing. When sodium is an important component of the salt, a significant amount of adsorbed sodium may occur and may disperse soil colloids and develop undesirable physical properties. Such soils are *sodic*. The quantity, proportion, and nature of salts that are present may vary in saline and sodic soils. This gives rise to three kinds of soils: saline, sodic, and saline-sodic.

Saline Soils Saline soils contain enough salt to impair plant growth. White crusts of salt frequently accumulate at the soil surface. The electrical conductivity of the saturated-soil extract is 4 or more dS/m. The pH is 8.5, or less, and less than 15 percent of the adsorbed cations are sodium.

Soil permeability, or hydraulic conductivity, has not been adversely affected by adsorbed sodium.

Reclamation of saline soils requires the removal of salt by leaching and maintenance of the water table far enough below the soil surface to minimize upward capillary water flow. Important considerations are soil hydraulic conductivity and an adequate drainage system. It is not economical to reclaim some saline soils that have high clay content and a very low hydraulic conductivity.

Several practices are recommended for using saline soils. The soil should not be allowed to dry completely, because the osmotic potential retards water uptake. Sufficient water should be applied with each irrigation to result in leaching of some salt in drainage water. Emphasis must be placed on the selection and growth of salt-tolerant crops. Considerable research effort is being directed toward the development of cultivars with greater salt tolerance.

Sodic Soils *Sodic* soils have been called alkali soils. Sodic soils are nonsaline and have an adsorbed or exchangeable sodium percentage (ESP) of 15 or more. The adsorbed cations (sodium, calcium, magnesium, and potassium) are considered exchangeable, because they compete for adsorption on the negatively charged soil colloids (clay and humus), are in motion, and exchange sites with each other. The sodium adsorption ratio (SAR) of a soil-saturated paste extract is about equal to the ESP. A SAR of 13 or more is the same as ESP of 15 or more. The pH is usually in the range of 8.5 to 10.0. Sodium dissociates from the colloids and small amounts of sodium carbonate form. Deflocculation or dispersion of colloids occurs, together with a breakdown of the soil structure as shown in Figure 6.16. This results in a massive or puddled soil with low water infiltration and very low hydraulic conductivity within the soil. The soil is difficult to till, and soil crusting may inhibit seedling emergence. Organic matter in the soil disperses along with the clay, and the humus coats soil particles

FIGURE 6.16 Distintegrated soil structure in an irrigated field caused by high adsorbed sodium. (Photo courtesy Soil Conservation Service, USDA.)

to give a black color. Sodic soils may develop when the leaching of a saline soil results in high exchangeable sodium and low exchangeable calcium and magnesium.

The basis for treatment of sodic soil is the replacement of exchangeable sodium with calcium and the conversion of any sodium carbonate into sodium sulfate. Finely ground gypsum (Ca-$SO_4 \cdot 2H_2O$) is broadcast and mixed with the soil. The calcium replaces or exchanges for the sodium adsorbed on the negatively charged colloid surfaces. The amount of gypsum needed to replace the exchangeable sodium is the *gypsum requirement.*

The calcium sulfate also reacts with any sodium carbonate that is present to form sodium sulfate and calcium carbonate. Then, leaching of the soil removes both the adsorbed sodium and the sodium that was in the carbonate form, as soluble sodium sulfate. Gradually, the exchangeable sodium percentage decreases to less than 15 percent, and the soil's structure and permeability improve.

Saline-Sodic Soils *Saline-sodic* soils are characterized by salinity and 15 percent or more of exchangeable sodium. As long as a large quantity of soluble salts remains in the soil, the exhangeable sodium does not cause dispersion of colloids. The high concentration of ions in the soil solution keeps the colloids flocculated. If the soluble salts are leached out, however, colloids may disperse with an accompanying breakdown of soil structure and loss of water permeability. The management of these soils is a problem until the excess soluble salt has been removed and the exchangeable sodium level is reduced to below 15 percent. If the soluble salts are removed by leaching without lowering the exchangeable sodium, saline-sodic soils may be converted in sodic soils. Consequently, gypsum must be added to the soil at the time leaching of salts is initiated.

WASTEWATER DISPOSAL

The billions of gallons of wastewater produced daily in the United States create a need for environmentally acceptable water disposal methods. The three major wastewater disposal techniques are: (1) treatment and discharge into surface waters, (2) treatment and reuse, and (3) land disposal via septic tanks and spray irrigation. Major consideration will be given to land disposal of sewage effluent from septic systems and the management of municipal wastewater that results in return of treated water to underlying aquifers for reuse.

Disposal of Septic Tank Effluent

People who reside beyond the limits of municipal sewer lines have used septic sewage disposal systems for many years. The expansion of rural residential areas, and the development of summer homes near lakes and rivers have greatly increased the need for waste disposal in rural areas. The sewage enters a septic tank, usually constructed of concrete, where solid material is digested or decomposed. The liquid portion, or effluent, flows out of the top of the septic tank and into drainage lines of a filter field. The drainage lines are laid in a bed of gravel into which the effluent seeps and drains through the soil, as illustrated in Figure 6.17.

The major factor influencing the suitability of soil for filter field use is the water-saturated hydraulic conductivity. The hydraulic conductivity of sands may be too high, because the sewage effluent may move so rapidly that disease organisms are not destroyed before shallow-water supplies become contaminated. Soils with low hydraulic conductivity are not suited for septic drain fields, because the sewage effluent may saturate soil and contaminate the surface soil.

Most local regulations require that trained personnel make a percolation test before approving the installation of a septic system. The steps for making a soil percolation test (perc test) are as follows:

1. Dig six or more test holes about 15 centimeters in diameter and to the depth of the proposed drainage lines or seepage bed; the holes should be uniformly spaced over the area selected for the filter field. Roughen the sides of the holes to remove smeared soil that might interfere with water entry into the soil. Add 5 centimeters of sand or fine gravel to the bottom of the holes to reduce the soil's tendency to seal.

2. Pour enough water into each hole so that the water is at least 30 centimeters deep. Add water as needed to keep the water depth to 30 centimeters for at least 4 hours—preferably overnight. This wets the soil and produces results representative of the wettest season of the year.

3. The next day, adjust the water level in the test holes so that the water is 30 centimeters deep. Measure the drop in water level over a 30-minute period and calculate the percolation rate.

4. Soils with a percolation rate less than 2.5 centimeters (1 in.) per hour are unsuited for filter fields because of the danger of saturation of the soil above the filter bed with effluent and the leakage of effluent from the soil

FIGURE 6.17   The layout of a septic tank and filter field for disposal of wastewater (effluent) through the soil.

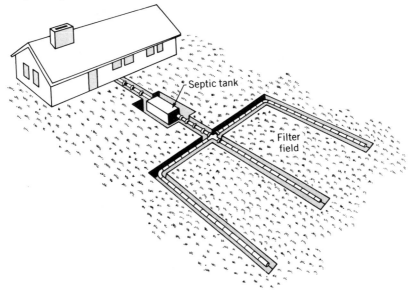

surface. Soils with percolation rates greater than 25 centimeters (10 in.) per hour are unsuited for filter fields because rapid movement of effluent might pollute groundwater.

Within the acceptable range of percolation values, increasing percolation rates result in decreasing seepage-bed area. The more people living in a dwelling, the more effluent for disposal. A graph is used to determine the area of seepage bed needed based on soil percolation rate and number of house occupants.

Several other factors are considered in selection of filter field or seepage-bed sites. They should be located at least 50 feet from any stream, open ditch, or other watercourse into which effluent might escape, and the land slope should be less than 10 percent. Further, septic seepage beds should never be installed in poorly drained soils or where soils become seasonally saturated with water (see Figure 6.5). There is danger that shallow groundwater may be contaminated with disease organisms and that water may seep out of the soil surface and contaminate work or play areas.

Land Disposal of Municipal Wastewater

A relatively small community of 10,000 people may produce about a million gallons of wastewater, or sewage effluent, per day from its sewage treatment plant. The effluent looks much like ordinary tap water; when chlorinated, it is safe for drinking. Discharging the effluent into a nearby stream does not create a health hazard. The effluent, however, is enriched with nutrients, so plant growth may be greatly increased in streams and lakes where the effluent is discharged. Weed growth in lakes and streams may reduce fish populations and interfere with boating and swimming. To combat these problems, researchers at the Pennsylvania State University designed experiments to use the soil as a living filter to remove the nutrients in sewage effluent. The water was applied with sprinklers, as shown in Figure 6.18.

As the effluent water percolates through the soil, nutrients are absorbed by plant roots and micoorganisms. More water is applied than lost by evapotranspiration and the excess water percolates from the bottom of the soil and is added to

FIGURE 6.18 Application of wastewater (effluent) in a forest with a sprinkler system in winter at Pennsylvania State University. (Photo courtesy USDA.)

the underground water reservoir (aquifer). A community of 10,000 people would need 50 hectares (about 125 acres) and a weekly application of 5 centimeters of water to dispose of their effluent. Under these conditions about 80 percent of the effluent water will contribute to aquifer recharge and about 20 percent will be evapotranspired. In the Pennsylvania State University experiment, an almost complete removal of the nitrogen and phosphorus in the effluent occurred—the two nutrients most closely tied to eutrophication. Furthermore, tree and crop growth on the land were greatly stimulated.

Today, more and more cities are utilizing land disposal of sewage effluent. In some areas, the trees produced in the disposal area are chipped and used as fuel to dry and process the solid sewage sludge. Marketing the sludge as a fertilizer results in a considerable cost saving in sewage disposal and conservation of nutrients. Additional benefits include the existence of a protected forested or green zone near the community and recharge of underground water aquifers. Such wastewater management systems can be eco-

nomically and environmentally sound. An example of a wastewater land disposal system that uses nutrients for plant growth and return of water to an underlying aquifer is shown in Figure 6.19.

PRESCRIPTION ATHLETIC TURF

Prescription athletic turf (PAT) is a system of producing vigorous turf for use on athletic fields. It combines several aspects of water management, including irrigation and drainage, and was developed at Purdue University, where football is played on natural grass turf.

The existing turf area is excavated to a depth of about 18 inches and the soil is removed. A plastic liner is laid on the newly exposed surface, 2-inch diameter slitted plastic drain lines are laid on top of the plastic, and sensors for measuring water potential and salinity are installed. The main features of the PAT system are shown in Figure 6.20. The sand layer forms the base for a 2-inch-thick surface layer of sand, peat, and soil. The water potential sensors maintain a thin layer of

FIGURE 6.19 Sewage-treatment system designed to apply wastewater to the land for crop and forest production. The water in excess of evaporation percolates to the water table and contributes to the groundwater (aquifer). (Modified from Woodwell, 1977.)

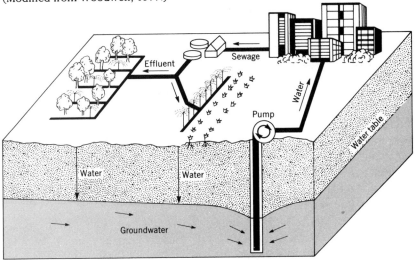

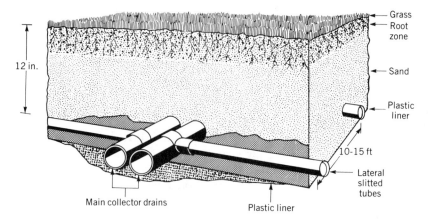

12 in.

Grass
Root
zone

Sand

Plastic liner

10-15 ft

Lateral slitted tubes

Main collector drains

Plastic liner

FIGURE 6.20 Diagram of layout for prescription athletic turf, PAT. The drainage lines remove excess water in wet season and are used for subirrigation in dry seasons. Sensors monitor water and salinity. (Data provided by Prescription Athletic Turf, Inc.)

water above the plastic liner. Grass roots are supplied by water that moves slowly upward by capillarity. During periods of rain, pumps pull excess water out of the soil so that the field is playable with firm traction regardless of the weather. The salt sensors monitor the buildup of salts from irrigation and fertilization. The sandy nature of the soil resists compaction by traffic.

SUMMARY

The three approaches to water management on agricultural fields include the conservation of natural precipitation in arid and subhumid regions, the removal of water (drainage) from wetlands, and irrigation to supplement natural precipitation.

Protection of the land surface by vegetation and plant residues shields the soil surface from raindrop impact and encourages water infiltration. Summer fallowing results in storage of water for crop production at a later time. Fertilizer use may increase the efficiency of water use by plants by increasing the rate of plant growth.

Removal of water by drainage provides an aerated root zone for crops and dry soil for building foundations. Upward from the surface of a water-saturated soil layer, there is an increase in the air

content of soil, a decrease in the water content, and a decrease in the matric water potential.

Irrigation water contains soluble salts and some provision (e.g., drainage) must be made to maintain a favorable salt balance to keep irrigated soils permanently productive. Salt accumulation results in formation of saline soils. The accumulation of exchangeable sodium, greater than about 15 percent of the adsorbed cations, results in the formation of sodic soils. High sodium saturation results in a deterioration of soil structure and a loss of soil water permeability. Sodic soils can be improved by applying gypsum and then leaching the soil.

Wastewater (sewage effluent) disposal on the land results in removal of nutrient ions and return of water to the underlying aquifers.

Prescription athletic turf is an artifical system for the production of vigorous grass on athletic fields by precise control of water, aeration, drainage, salinity, and nutrient supply.

REFERENCES

Baker, K. F., ed. 1957. "The U. C. System for Producing Healthy Container-Grown Plants." *Ca. Agr. Exp. Sta. Manual* 23. Berkeley.

Bender, W. H. 1961. "Soils Suitable for Septic Tank Filter Fields." *Agr. Information Bull*. 243. USDA, Washington, D.C.

Bernstein, L. 1964. "Salt Tolerance of Plants." *Agr. Information Bull*. 283. USDA, Washington, D.C.

Bernstein, L. 1965. "Salt Tolerance of Fruit Trees." *Agr. Information Bull*. 292. USDA, Washington, D.C.

Bower, C. A., 1974. "Salinity of Drainage Waters." *Drainage in Agriculture*. Am. Soc. Agron. Madison. pp. 471–487.

Edminister, T. W. and R. C. Reeve. 1957. "Drainage Problems and Methods." *Soil*. USDA Yearbook of Agriculture. Washington, D.C., pp. 379–385.

Evans, C. E. and E. R. Lemon. 1957. "Conserving Soil Moisture." *Soil*. USDA Yearbook of Agriculture. Washington, D.C.

Hanks, R. J. and C. B. Tanner. 1952. "Water Consumption by Plants as Influenced by Soil Fertility." *Agron. J*. **44**:99.

Jacobsen, T. and R. M. Adams. 1958. "Salt and Silt in Ancient Mesopotamian Agriculture." *Science*. **128**:1251–1258.

Kardos, L. T. 1967. "Waste Water Renovation by the Land-A Living Filter." in *Agriculture and the Quality of Our Environment*. AAAS Pub. 85, Washington, D.C.

Kemper, W. D., T. J. Trout, A. Segeren, and M. Bullock. 1987. "Worms and Water." *J. Soil and Water Conservation*. **42**:401–404.

Mathews, O. R. 1951. "Place of Summer Fallow in the Agriculture of the Western States." *USDA Cir*. 886. Washington, D.C.

Miller, D. E. and Richard O. Gifford. 1974. "Modification of Soil Crusts for Plant Growth," in *Arizona Agr. Exp. Sta. Tech. Bull*. 214.

Soil Improvement Comm. Calif. Fert. Assoc. 1985. *Western Fertilizer Handbook*. 7th ed. Interstate, Danville, Ill.

Terry, T. A. and J. H. Hughes. 1975. "The Effects of Intensive Management on Planted Loblolly Pine Growth on the Poorly Drained Soils of the Atlantic Coastal Plain." *Proc. Fourth North Am. Forest Soils Conf*. Univ. Laval, Quebec.

United States Salinity Laboratory Staff, 1969. "Diagnosis and Improvement of Saline and Alkali Soils." *USDA Agr. Handbook* 60. Washington, D.C.

Woodwell, G. M. 1977. "Recycling Sewage through Plant Communities." *Am. Scientist* **65**:556–562.

CHAPTER 7

SOIL EROSION

Soil formation and soil erosion are two natural and opposing processes. Many natural, undisturbed soils have a rate of formation that is balanced by a rate of erosion. Under these conditions, the soil appears to remain in a constant state as the landscape evolves. Generally, the rates of soil erosion are low unless the soil surface is exposed directly to the wind and rainwater. The erosion problem arises when the natural vegetative cover is removed and rates of soil erosion are greatly accelerated. Then, the rate of soil erosion greatly exceeds the rate of soil formation and there is a need for erosion control practices that will reduce the erosion rate and maintain soil productivity.

Erosion is a three-step process: *detachment* followed by *transport* and *deposition*. The energy for erosion is derived from falling rain and the subsequent movement of runoff water or the wind.

WATER EROSION

Several types of water erosion have been identified: raindrop (splash), sheet, rill, and gully or channel. In addition, undercutting of stream banks by water causes large masses of soil to fall into the stream and be carried away. Very wet soil masses on steep slopes are likely to slide slowly downhill (solifluction).

Raindrops falling on bare soil detach particles and splash them up into the air. When there is little water at the soil surface, the water tends to run over the soil surface as a thin sheet, causing *sheet* erosion. *Rill* and *gully* erosion are due to the energy of water that has concentrated and is moving downslope. As water concentrates, it first forms small channels—called rills—followed by greater concentrations of water. Large concentrations of water lead to gully formation. Some of the highest soil erosion rates occur at construction sites where the vegetation has been removed and the soil is exposed to the erosive effects of rain, as shown in Figure 7.1. The tendency of water to collect into small rills that coalesce to form large channels and produce gully erosion is also shown in Figure 7.1.

Predicting Water Erosion Rates on Agricultural Land

The rate of soil erosion on agricultural land is affected by rainfall characteristics, soil erodibility,

FIGURE 7.1 Removal of the vegetative cover increases water runoff and leaves soil unprotected from the erosive effects of the rain, resulting in accelerated soil erosion. As water concentrates in channels, its erosive power increases. (Photograph courtesy Soil Conservation Service, USDA.)

slope characteristics, and vegetative cover and/or management practices. The first plots in the United States for measuring runoff and erosion, as influenced by different crops, were established at the University of Missouri in 1917 (see Figure 7.2). Since 1930, many controlled studies on field plots and small watersheds have been conducted to study factors affecting erosion. Experiments were conducted on various soils at many different locations in the United States, using the same standard conditions. For example, many erosion plots were 72.6 feet long on 9 percent slopes and were subjected to the same soil management practices.

These quantitative data form the basis for predicting erosion rates by using the Universal Soil-Loss Equation (USLE) developed by Wischmeier and Smith. The USLE equation is

$$A = RKLSCP$$

where A is the computed soil loss per unit area as tons per acre, R is the rainfall factor, K is the soil-erodibility factor, L is the slope-length factor, S is the slope-gradient factor, C is the cropping-management factor, and P is the erosion-control practice factor. The equation is designed to predict water erosion rates on agricultural land surfaces, exclusive of erosion resulting from the formation of large gullies.

R = The Rainfall Factor

The rainfall factor (R) is a measure of the erosive force of specific rainfall. The erosive force, or available energy, is related to both the quantity and intensity of rainfall. A 5-centimeter rain falling at 32 kilometers per hour (20 mph) would have 6 million foot-pounds of kinetic energy. The tremendous erosive power of such a rain is sufficient to raise an 18-centimeter-thick furrow slice to a height of 1 meter. The four most intense storms in a 10-year period at Clarinda, Iowa, accounted for 40 percent of the erosion and only 3 percent of the water runoff on plots planted with corn that were tilled up and down the slope. Anything that protects the soil from raindrop impact, such as plant cover or stones, protects the soil from erosion, as shown in Figure 7.3.

FIGURE 7.2 Site of the first plots in the United States for measuring runoff and erosion as influenced by different crops. The plots were established in 1917 at the University of Missouri. (Photograph courtesy Dr. C. M. Woodruff, University of Missouri.)

FIGURE 7.3 Columns of soil capped by stones that absorbed the energy of raindrop impact and protected the underlying soil from erosion.

The rainfall factor (R) is the product of the total kinetic energy of the storms times the maximum 30-minute intensity of fall and modified by any influence of snowmelt. Rainfall factors have been computed for many locations, with values ranging from less than 20 in western United States to 550 along the Gulf Coast of the southeastern United States. An iso-erodent map of average annual values of R is shown in Figure 7.4.

K = The Soil Erodibility Factor

Soil factors that influence erodibility by water are: (1) those factors that affect infiltration rate, permeability, and total water retention capacity, and (2) those factors that allow soil to resist disper-

FIGURE 7.4 Rainfall factors for the United States. (After Wischmeier and Smith, 1978.)

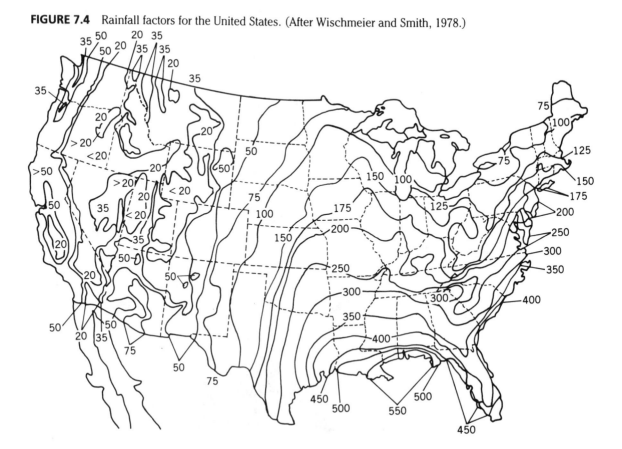

sion, splashing, abrasion, and the transporting forces of rainfall and runoff. The erodibility factor (*K*) has been determined experimentally for 23 major soils on which soil erosion studies were conducted since 1930. The soil loss from a plot 72.6 feet long on a 9 percent slope maintained in fallow (bare soil surface and with no crop planted) and with all tillage up and down the slope, is determined and divided by the rainfall factor for the storms producing the erosion. This is the erodibility factor. Values of *K* for 23 soils are listed in Table 7.1.

LS = The Slope Length and Slope Gradient Factors

Slope length is defined as the distance from the point of origin of overland water flow either to a point where the slope decreases to the extent that deposition occurs, or a point where water runoff enters a well-defined channel. Runoff from the upper part of a slope contributes to the total amount of runoff that occurs on the lower part of the slope. This increases the quantity of water running over the lower part of the slope, thus increasing erosion more on the lower part of the slope than on the upper part. Studies have shown that erosion via water increases as the 0.5 power of the slope length ($L^{0.5}$). This is used to calculate the slope length factor, *L*. This results in about 1.3 times greater average soil loss per acre for every doubling of slope length.

As the slope gradient (percent of slope) increases (*S*), the velocity of runoff water increases, which in turn increases the water's erosive power. A doubling of runoff water velocity increases ero-

TABLE 7.1 Computed K Values for Soils

Soil	Source of Data	Computed K
Dunkirk silt loam	Geneva, N.Y.	0.69[a]
Keene silt loam	Zanesville, Ohio	0.48
Shelby loam	Bethany, Mo.	0.41
Lodi loam	Blacksburg, Va.	0.39
Fayette silt loam	LaCrosse, Wis.	0.38[a]
Cecil sandy clay loam	Watkinsville, Ga.	0.36
Marshall silt loam	Clarinda, Iowa	0.33
Ida silt loam	Castana, Iowa	0.33
Mansic clay loam	Hays, Kans.	0.32
Hagerstown silty clay loam	State College, Pa.	0.31[a]
Austin clay	Temple, Tex.	0.29
Mexico silt loam	McCredie, Mo.	0.28
Honeoye silt loam	Marcellus, N.Y.	0.28[a]
Cecil sandy loam	Clemson, S.C.	0.28[a]
Ontario loam	Geneva, N.Y.	0.27[a]
Cecil clay loam	Watkinsville, Ga.	0.26
Boswell fine sandy loam	Tyler, Tex.	0.25
Cecil sandy loam	Watkinsville,Ga.	0.23
Zaneis fine sandy loam	Guthrie, Okla.	0.22
Tifton loamy sand	Tifton, Ga.	0.10
Bath flaggy silt loam with surface stones 2 inches removed	Arnot, N.Y.	0.08[a]
Freehold loamy sand	Marlboro, N.J.	0.08
Albia gravelly loam	Beemerville, N.J.	0.03

From Wischmeier and Smith, 1965.
[a] Evaluated from continuous fallow. All others were computed from row crop data.

sive power four times, and causes a 32-time increase in the amount of material of a given particle size that can be carried in the runoff water. Splash erosion by raindrop impact causes a net downslope movement of soil. Downward slope movement also increases with an increase in the slope gradient. Combined slope length and gradient factors (LS) for use in the soil-loss prediction equation are given in Figure 7.5.

C = The Cropping-Management Factor

A vegetative cover can absorb the kinetic energy of falling raindrops and diffuse the rain's erosive potential. Furthermore, vegetation by itself retains a significant amount of the rain and slows the flow of runoff water. As a result, the presence or absence of a vegetative cover determines whether erosion will be a serious problem. This is shown in Figure 7.6 where erosion on plots 72.6 feet long with an 8 percent slope had about 1,000 times more erosion when in continuous corn as compared with continuous bluegrass.

The C factor measures the combined effect of all interrelated crop cover and management variables including type of tillage, residue management, and time of soil protection by vegetation. Frequently, different crops are grown in sequence or in rotation; for example, corn is grown one year followed by wheat, which is then followed by a hay or meadow crop. This would be a three-year rotation of corn-wheat-meadow (C-W-M). The benefit of a rotation is that a year of high erosion, during the corn production year, is averaged with years of lower erosion to give a lower average erosion rate, than if corn was planted every year. However, when corn is grown no-till, there may be little soil erosion because of the protection of the soil by previous crop residues, as shown in Figure 7.7.

Complicated tables have been devised for calculating the C factor. As an illustration, the C factor for a 4-year rotation of wheat-alfalfa and bromegrass-corn-corn (W-AB-C-C) in central Indiana with conventional tillage, average residue management, and average yields is 0.119. Some comparison values are as low as 0.03 for corn mulched in no-till with 90 percent of the soil

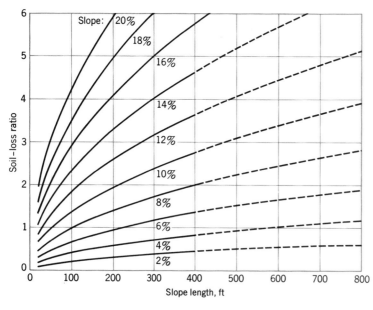

FIGURE 7.5 Slope length and gradient graph for determining the topographic factor, LS.

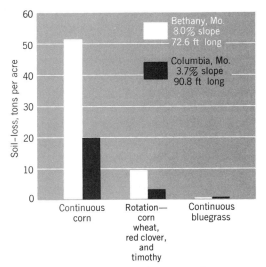

FIGURE 7.6 A continuous plant cover (e.g., bluegrass) is many times more effective for erosion control than cropping systems that leave the land bare most of the time. Continuous corn production resulted in the longest amount of time with the soil unprotected by vegetation and the most erosion.

covered with crop residues and 0.48 for continuous corn with conventional tillage.

P = The Erosion Control Practice Factor

In general, whenever sloping land is cultivated and exposed to erosive rains, the protection of-

fered by sod or closely growing crops in the system needs to be supported by practices that will slow runoff water, thus reducing the amount of soil carried. The most important of these practices for croplands are contour tillage, strip cropping on the contour, and terrace systems. Contour tillage is tillage operations and planting on the contour as contrasted to tillage up and down the slope or straight across the slope. No-till planting of corn on the contour between terraces in an Iowa field is shown in Figure 7.8.

Limited field studies have shown that contour farming alone is effective in controlling erosion during rainstorms of low or moderate intensity, but it provides little protection against the occasional severe storm that causes breakovers by runoff waters of the contoured ridges or rows. *P* values for contouring are given in Table 7.2, from which it is apparent that contouring is the most effective on slopes of 2.1 to 7 percent. Strip cropping is the practice of growing two or more crops in alternating strips along contours, or perpendicular to surface water flow. Strip cropping, together with contour tillage, provides more protection. In cases where both strip cropping and contour tillage are used, the P values listed in Table 7.2 are divided by 2.

Terraces are ditches that run perpendicular to

FIGURE 7.7 No-till corn with the residues of a previous soybean crop providing protection against soil erosion. (Photograph courtesy Soil Conservation Service, USDA.)

FIGURE 7.8 Factors contributing to soil erosion control in this field include no-till tillage on the contour with terraces that greatly reduce the effective length of slope. (Photograph courtesy Soil Conservation Service, USDA.)

the direction of surface water flow and collect runoff water. The grade of the channel depends on the amount of runoff water. In areas of limited rainfall and permeable soil, channels may have a zero grade. In humid regions, the terrace channel usually is graded, and the runoff water that is collected by the terrace is carried to an outlet for disposal. Terraces are an effective way to reduce slope length. To account for terracing, the slope length used to determine the LS factor should represent the distance between terraces. Short slope length between terraces is shown in Figure 7.8.

Application of the Universal Soil-Loss Equation

The USLE was designed to predict the long-term erosion rates on agricultural land so that management systems could be devised that result in ac-

TABLE 7.2 Practice Factor Values for Contouring

Land Slope, %	P value
1.1 to 2	0.60
2.1 to 7	0.50
7.1 to 12	0.60
12.1 to 18	0.80
18.1 to 24	0.90

From Wischmeier and Smith, 1965.

ceptable levels of erosion. Having been developed under the conditions found in the United States, one must be cautious in applying the equation to drastically different conditions, such as exist in the tropics. The procedure for computing the expected, average annual soil loss from a particular field is illustrated by the following example.

Assume that there is a field in Fountain County, Indiana, on Russell silt loam. The slope has a gradient of 8 percent and is about 200 feet long. The cropping system is a 4-year rotation of wheat-alfalfa and bromegrass-corn-corn (W-AB-C-C) with tillage and planting on the contour, and with corn residues disked into the soil for wheat seeding and turned under in the spring for second-year corn. Fertility and residue management on this farm are such that crop yields per acre are about 100 bushels of corn, 40 bushels of wheat, and 4 tons of alfalfa-bromegrass hay. The probability of an alfalfa-bromegrass failure is slight, and there is little risk that serious erosion will occur during the alfalfa-bromegrass year due to a lack of a good vegetative cover.

The first step is to refer to the charts and tables discussed in the preceding sections and to select the values of R, K, LS, C, and P that apply to the specific conditions on this particular field.

The value of the rainfall factor (R) is taken from Figure 7.4. Fountain County, in west-central Indiana, lies between isoerodents of 175 and 200.

By linear interpolation, R equals 185. Soil scientists consider that the soil erodibility factor (K) for Russell silt loam is similar to that for Fayette silt loam, being 0.38, according to Table 7.1.

The slope-effect chart, Figure 7.5, shows that, for an 8 percent slope, 200 feet long, the LS is 1.41. For the productivity level and management practices, assumed in this example, the factor C for a W-AB-C-C rotation was shown earlier to be 0.119. Table 7.2 shows a practice factor of 0.6 for contouring on an 8 percent slope.

The next step is to substitute the selected numerical values for the symbols in the soil erosion equation, and solve for A. In this example:

$$A = 185 \times 0.38 \times 1.41 \times 0.119 \times 0.6$$
$$= 7.1 \text{ tons of soil loss per acre}$$

If planting had been up and downslope instead of on the contour, the P factor would have been 1.0, and the predicted soil loss would have been $185 \times 0.38 \times 1.41 \times 0.119 \times 1.0 = 11.8$ tons per acre (26.4 metric tons per hectare).

If farming had been on the contour and combined with minimum tillage for all corn in the rotation, the factor C would have been 0.075. The predicted soil loss would then have been 4.5 tons per acre (10.1 metric tons per hectare).

The Soil Loss Tolerance Value

The *soil loss tolerance value* (T) has been defined in two ways. This is an indication of the lack of concensus among soil scientists as to what approach should be taken with T values or how much erosion should be tolerated. First, the T value is the maximum soil erosion loss that is offset by the theoretical maximum rate of soil development, which will maintain an equilibrium between soil losses and gains. Shallow soils over hard bedrock have small T values. Erosion on soils with claypans will result in shallower rooting depths and in the need for smaller T values than for thick permeable soils, when both kinds of soils have developed from thick and water-permeable unconsolidated parent materials. The emphasis is on the maintenance of a certain thickness of soil overlying any impermeable layer that might exist.

Second, the T value is the maximum average annual soil loss that will allow continuous cropping and maintain soil productivity without requiring additional management inputs. Fertilizers are used to overcome the decline in soil fertility caused by erosion.

On thick, water-permeable soils, such as Marshall silt loam in western Iowa, artificial removal of the surface soil reduced yields to less than one half of the crop yield obtained from uneroded soil when no fertilizer was applied. The application of fertilizer containing 200 pounds of nitrogen per acre, however, resulted in similar yields of corn on both the normal soil and the soil with the surface soil removed. The Marshall soil is a thick permeable soil that developed in a thick layer of loess (parent material). Loess is a material transported and deposited by wind and consisting predominantly of silt-sized particles. Many soils that have developed from thick sediments of loess are agriculturally productive.

There was some evidence that in years of moisture stress, the normal fertilized Marshall silt loam outproduced the soil where the surface soil had been removed and fertilized. As a result of this and similar experiments, another approach was developed to establish T values. This method considers the maximum soil loss consistent with the maintenance of productivity in terms of the maintenance of a rooting depth with desirable physical properties. Where subsoils have physical properties unsuitable for rooting, erosion results in reductions in soil productivity that cannot be overcome with only fertilizer application. The data in Figure 7.9 show that removing the surface layer from a fine-textured Austin soil in the Blacklands of Texas resulted in longtime or permanent reductions in soil productivity. Soil loss tolerance value guidelines from the Soil Conservation Service of the USDA are given in Table 7.3.

The data in Figure 7.6 show that average annual soil loss with continuous bluegrass was 0.03 tons

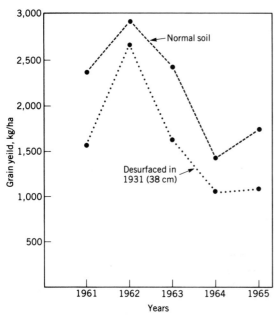

FIGURE 7.9 Effect of removal of 38 centimeters of topsoil on grain sorghum yields on Austin clay at Temple, Texas, 1961–1965. (Data from Smith et al., 1967.)

per acre and resulted in the removal of about 2.5 centimeters (1 in.) of soil every 5,000 years. Obviously, this soil loss is well below the *T* value for the soil. On the other hand, continuous corn re-

sulted in a soil loss exceeding 50 tons per acre or 2.5 centimeters (1 inch) every 2 to 3 years, a loss far exceeding the *T* value. In the Indiana case, the calculated soil losses ranged from about 4 to 12 tons per acre. The average annual rate of soil erosion on cropland land in the United States is 5 tons per acre. It has been estimated that soil erosion is the most important conservation problem on 51 percent of the nation's cropland. In general, *T* values in the United States range from 1 to 5 tons per acre (2 to 11 metric tons per hectare). The USLE can be used to determine the management required to maintain soil losses at or below a certain *T* value.

Water Erosion on Urban Lands

Historic soil erosion rates have changed as land use has changed. Much of the land in the eastern United States, which contributed much sediment to streams and rivers a hundred or more years ago, has been removed from cropping. Lower erosion rates resulted from abandonment of land for cropping and returning it to forest. In the past few decades, much of this land has experienced a high erosion rate because of urbanization, construction of shopping centers, roads, and housing areas, as shown in Figure 7.10.

TABLE 7.3 Guide for Assigning Soil Loss Tolerance Values to Soils with Different Rooting Depths

Rooting Depth	Soil-Loss Tolerance Values			
	Renewable Soil †[a]		Nonrenewable soil ‡[b]	
cm	ton/acre	ton/ha	ton/acre	ton/ha
<25	1.0	2.2	1.0	2.2
25–51	2.0	4.5	1.0	2.2
51–102	3.0	6.7	2.0	4.5
102–152	4.0	9.0	3.0	6.7
>152	5.0	11.2	5.0	11.2

Data from USDA-SCS, 1973.
[a] † Soils that have a favorable substratum and can be renewed by tillage, fertilizer, organic matter, and other management practices.
[b] ‡ Soils that have an unfavorable substratum, such as rock, and cannot be renewed economically.

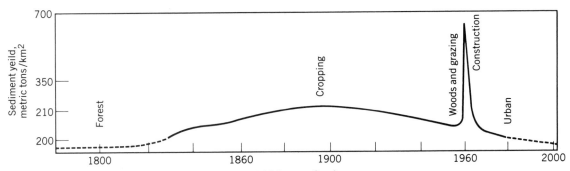

FIGURE 7.10 Sequential land use and sediment yield from a fixed Midatlantic area landscape. (Data of Wolman, 1967.)

Now, about 50 percent of the sediment in streams and waters throughout the country originates on agricultural land; the other 50 percent originates on urban lands. It has been noted that exposed land on steep slopes is very susceptible to erosion. In fact, exposure of land during highway and building construction commonly results in erosion rates many times those that typically occur on agricultural land. Urban erosion rates as large as 350 tons per acre in a year have been reported. It is important to plan construction so that land will be exposed a minmum period of time (see Figure 7.11).

Water Erosion Costs

In northern China, there is an extensive loess deposit that is locally over 80 meters thick. This loess area is the major source of sediment for the Huang Ho (Yellow) River. This sediment is the major parent material for the alluvial soils of the vast north-China plain, which supports millions of people. The loessial landscape is very distinctive, with deeply entrenched streams and gullies. It has been estimated that the loess will be removed by erosion in 40,000 years, which would be an annual soil loss of 15 tons per acre (33 metric tons per hectare). This appears to be an excessive soil erosion rate.

The costs and benefits of the erosion and sedimentation are difficult to assess. In the short term, food production costs are increased in areas where erosion occurs. Excessive water runoff and soil erosion contribute to flooding and addition of fresh sediments to the soils of the north-China plain.

Soil erosion results in the preferential loss of the finer soil components that contribute the most to soil fertility. Loss of organic matter, and the subsequent reduction in nitrogen mineralization, are particularly detrimental to the production of nonleguminous grain crops. As mentioned, this loss can in some cases, be overcome by the addition of nitrogen fertilizer on responsive soils. However, there is an increase in the cost of crop production.

Exposure of clay-enriched Bt horizons (because of surface soil erosion) reduces water infiltration and increases water runoff. This results in more soil erosion and retention of less water for plant growth. Seedbed preparation is more difficult and the emergence of small plant seedlings may be reduced, resulting in a less than desirable density of plants. Summarization of data from many studies showed that 2.5 centimeters (1 in.) of soil loss reduced wheat yield 5.3 percent, corn 6.3 percent, and grain sorghum 5.7 percent. Some effects of soil erosion on crop growth and production costs are given in Table 7.4. Increased production costs justify a long-term investment in erosion control measures to maintain erosion at or below acceptable T values.

Off-site costs result from sediment in water settling and filling reservoirs and navigation chan-

FIGURE 7.11 Slurry truck spraying fertilizer and seed on freshly prepared seedbed along road in Missouri. A thin layer of straw mulch will be secured with a thin spray of tar to protect the soil until vegetation becomes established. (Photograph courtesy Soil Conservation Service, USDA.)

nels. The sediments adversely affect the ecology of streams and lakes, as well as municipal water supplies. In 1985 it was estimated that the cost of soil erosion resulting from all phases of pollution, both agricultural and urban, was $6 billion annually in the United States. This is a loss greater than that estimated for the loss in soil productivity from the eroded land.

WIND EROSION

Wind erosion is indirectly related to water conservation in that a lack of water leaves land barren and exposed to the wind. Wind erosion is greatest in semiarid and arid regions. In North Dakota, wind-erosion damage on agricultural land was reported as early as 1888. In Oklahoma, soil erosion was reported 4 years after breaking of the sod. Blowing dust was a serious hazard of early settlers on the Plains, and was one of the first problems studied by the Agricultural Experiment Stations of the Great Plains.

Types of Wind Erosion

Soil particles move in three ways during wind erosion: *saltation, suspension, and surface creep*. During saltation, fine soil particles (0.1 to 0.5 mm

in diameter) are rolled over the surface by direct wind pressure, then suddenly jump up almost vertically over a short distance to a height of 20 to 30 centimeters. Once in the air, particles gain in velocity and then descend in an almost straight line, not varying more than 5 to 12 degrees from the horizontal. The horizontal distance traveled by a particle is four to five times the height of its jump. On striking the surface, the particles may rebound into the air or knock other particles into the air before coming to rest. The major part of soil carried by wind moves by this process. About 93 percent of the total soil movement via wind takes place below a height of 30 centimeters. Probably 50 percent or more of movement occurs between 0 and 5 centimeters above the soil surface.

Very fine dust particles are protected from wind action, because they are too small to protrude above a minute viscous layer of air that clings to the soil surface. As a result, a soil composed of extremely fine particles is resistant to wind erosion. These dust-sized particles are thrown into the air chiefly by the impact of particles moving in saltation. Once in the air, however, their movement is governed by wind action. They may be carried very high and over long distances via suspension.

Relatively large sand particles (between 0.5 and

TABLE 7.4 Effects of Erosion on Corn Growth and Production Costs

	Degree of Erosion		
	Slight	Moderate	Severe
Depth of topsoil inches	12	7	5
Decrease in height of plant during early stages of growth (from slight), percent	—	13	22
Stand at harvest, percent of planting rate	87	83	76
Corn yield, bushels per acre	112	96	87
Soybean yield, bushels per acre	43	29	16
Reduction in yield for all crops (from slight), percent	—	17	26
Increase in production cost (from slight), percent	—	20	56

From Beasley, 1974.

1.0 mm in diameter) are too heavy to be lifted by wind action, but they are rolled or pushed along the surface by the impact of particles during saltation. Their movement is by surface creep.

Between 50 and 75 percent of the soil is carried by saltation, 3 to 40 percent by suspension, and 5 to 25 percent by surface creep. From these facts, it is evident that wind erosion is due principally to the effect of wind on particles of a size suitable to move in saltation. Accordingly, wind erosion can be controlled if: (1) soil particles can be built up into peds or granules too large in size to move by saltation; (2) the wind velocity near the soil surface is reduced by ridging the land, by vegetative cover, or even by developing a cloddy surface; and (3) strips of stubble or other vegetative cover are sufficient to catch and hold particles moving in saltation.

Wind Erosion Equation

The factors that affect wind erosion are contained in the equation to estimate soil loss by wind erosion:

$$E = f (I, K, C, L, V)$$

where: E equals the soil loss in tons per acre, I equals the soil erodibility, K equals the soil roughness, C equals the climatic factor, L equals the

field length, and V equals the quantity of vegetative cover. Wind erosion tolerance levels (T values) have been established, just as for water erosion. The use of the wind erosion equation is more complex than that used for calculating water erosion losses. Details for use of the wind erosion equation are given in the National Agronomy Manual of the Soil Conservation Service, 1988.

Factors Affecting Wind Erosion

Soil erodibility by wind is related primarily to soil texture and structure. As the clay content of soils increases, increased aggregation creates clods or peds too large to be transported. A study in western Texas showed that soils with 10 percent clay eroded 30 to 40 times faster than soil with 25 percent clay, as shown in Figure 7.12.

Rough soil surfaces reduce wind erosion by reducing the surface wind velocity and by trapping soil particles. Surface roughness can be increased with tillage operations, as shown in Figure 7.13. Tillage and rows are positioned at right angles to the dominant wind direction for maximum effectiveness.

Climate is a factor in wind erosion via its effect on wind frequency and the velocity and wetness of soil during high-wind periods. Wind erosion

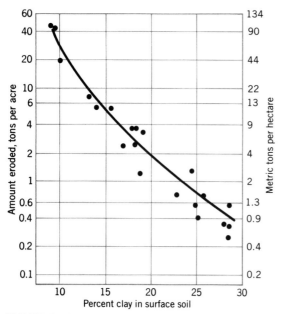

FIGURE 7.12 Amount of wind erosion in relation to the amount of clay in the soil. (Data from Chepil et al., 1955.)

and blowing dust are persistent features of deserts in which plants are widely spaced and dry soil is exposed most of time. When surface soils contain a wide range in particle sizes, wind erosion preferentially removes the finer particles, leaving the larger particles behind. In this way a *desert pavement* is created, consisting of gravel particles that are too large and heavy to be blown away. Consequently, deserts are an important source of loess, which is composed mainly of silt-sized particles. Furthermore, climate affects the amount of vegetative cover. During the Dust Bowl of the 1930s, both wheat yields and wind erosion were related to rainfall. The lowest yields and highest soil erosion rates occurred during years of least rainfall.

Wind erosion increases from zero, at the edge of an open field, to a maximum, with increasing distance of soil exposed to the wind. As particles bounce and skip, they create an avalanche effect on soil movement. Windbreaks of trees and alternate strips of crops can be used to reduce the

FIGURE 7.13 Contour ridges reduce wind velocity and collect much of the blowing soil. (Photograph courtesy soil Conservation Service, USDA.)

effective field length exposed to the wind. Finally, wind erosion is inversely related to the degree of vegetative cover. Growing crops and the residues of previous crops are effective in controlling wind erosion. Stubble mulch tillage—tillage that leaves plant residues on the soil surface—is popular in the wheat producing region of the Great Plains.

Deep Plowing for Wind Erosion Control

Deep plowing has been used to control wind erosion where sandy surface soils are underlain by Bt horizons that contain 20 to 40 percent clay. In Kansas and Texas, plowing to control wind erosion must be deep enough to include at least 1 centimeter of Bt horizon for every 2 centimeters of the surface soil (A horizon) thickness. In many places, the soil must be plowed to a depth of 50 to 60 centimeters. This kind of plowing increases the clay content of the surface soil by 5 to 12 percent, in some cases. Since 27 percent clay in the surface was required to halt soil blowing, deep plowing by itself resulted in only partial and temporary control of wind erosion. When deep plowing was not accompanied by other soil erosion control practices, wind erosion removed clay from the surface soil, and the effects of deep plowing were temporary.

Wind Erosion Control on Organic Soils

Organic soil areas of appreciable size are frequently protected from wind erosion by planting trees in rows—windbreaks. The use of trees as windbreaks must be limited, however, because of the large amount of soil that they take out of crop production. A number of shrubs are also used; for example, spirea is popular as a windbreak. Trees and shrubs also provide cover for wildlife.

The lightness of dry organic soils contributes to the wind erosion hazard. Moist organic soil is heavier than dry organic soils, and particles tend to adhere more, reducing the wind erosion hazard. Accordingly, use of an overhead sprinkler irrigation system is helpful during windy periods, before the crop cover is sufficient to provide protection from the wind. Rolling the muck soil with a heavy roller induces moisture to rise more rapidly by capillarity, thus dampening the soil surface. The water table must be quite close to the soil surface for this practice to be effective.

SUMMARY

The factors that affect soil erosion by water are rainfall, soil erodibility, slope length and gradient, crop management practices, and erosion control practices. The universal soil loss equation is used to estimate soil erosion loss on agricultural land.

The permissible soil loss is the T value. The universal soil loss equation can be used to determine the soil management systems that will result in acceptable erosion levels.

Soil erosion is greatly accelerated when the vegetative cover is removed. Thus, soil erosion is a serious problem in both rural and urban areas.

Cost of soil erosion includes reduced soil productivity and higher crop production costs. Offsite costs are due to silting of watercourses and reservoirs and reduced water quality.

Wind erosion is a major problem when sandy soils are used for crop production in regions with dry seasons, where the soil surface tends to be exposed to the wind because of limited vegetative cover. Wind erosion control is based on one or more of the following:

1. Trapping soil particles with rough surface and/or the use of crop residues and strip cropping.
2. Deep plowing to increase clay content of surface soil and simultaneous use of other erosion control practices.
3. Protecting the soil surface with a vegetative cover.

REFERENCES

Anonymous. 1982. *Soil Erosion: Its Agricultural, Environmental, and Socioeconomic Implications*. Council for Agr. Sci. and Technology Report No. 92, Ames, Iowa.

Beasley, R. P. 1974. "How Much Does Erosion Cost?" *Soil Survey Horizons.* **15:**8–9.

Browning, G. M., R. A. Norton. A. G. McCall, and F. G. Bell. 1948. "Investigation in Erosion Control and the Reclamation of Eroded Land at the Missouri Valley Loess Conservation Experiment Station." *USDA Tech Bull.* 959, Washington, D.C.

Burnett, E., B. A. Stewart, and A. L. Black. 1985. "Regional Effects of Soil Erosion on Crop Productivity-Great Plains." *Soil Erosion and Crop Productivity.* American Society of Agronomy, Madison, Wis.

Chepil, W. S., N. P. Woodruff, and A. W. Zingg. 1955. "Field Study of Wind Erosion in Western Texas." *Kansas and Texas Agr. Exp. Sta. and USDA* Washington, D.C.

Colacicco, D., T. Osborn, and K. Alt. 1989. "Economic Damage from Soil Erosion." *J. Soil and Water Con.* **44:**(1), 35–39.

Massey, H. F. and M. L. Jackson. 1952. "Selective Erosion of Soil Fertility Constitutents." *Soil Sci. Soc. Am. Proc.* **16:**353–356.

Smith, R. M., R. C. Henderson, E. D. Cook, J. E. Adams, and D. O. Thompson. 1967. "Renewals of Desurfaced Austin Clay." *Soil Sci.* **103:**126–130.

Soil Conservation Service, 1988. *National Agronomy Manual.* USDA, Washington, D.C.

Tuan, Yi-Fu. 1969. *China.* Aldine, Chicago.

USDA-SCS. 1973. *Advisory Notice, 6.* Washington, D.C.

Wischmeier, W. H. and D. D. Smith. "Predicting Rainfall-Erosion Losses from Cropping East of the Rocky Mountains." *USDA Agr. Handbook* 282. Washington, D.C.

Wischmeier, W. H. and D. D. Smith. 1978. "Predicting Rainfall Erosion Loss–A Guide to Conservation Planning." *USDA Agr. Handbook* 537. Washington, D.C.

Woodruff, N. P. and F. H. Siddoway. 1965. "A Wind Erosion Equation." *Soil Sci. Soc. Am. Proc.* **29:**602–608.

Wolman, M. G. 1967. "A Cycle of Sedimentation and Erosion in the Urban River Channels." *Geog. Ann.* **49A:**385–395.

Zobeck, T. M., D. W. Fryrear, and R. D. Petit. 1989. "Management Effects on Wind-eroded Sediment and Plant Nutrients." *J. Soil Water Con.* **44:**160–163.

CHAPTER 8

SOIL ECOLOGY

The soil is the home of innumerable forms of plant, animal, and microbial life. Some of the fascination and mystery of this underworld has been described by Peter Farb:

We live on the rooftops of a hidden world. Beneath the soil surface lies a land of fascination, and also of mysteries, for much of man's wonder about life itself has been connected with the soil. It is populated by strange creatures who have found ways to survive in a world without sunlight, an empire whose boundaries are fixed by earthen walls.*

Life in the soil is amazingly diverse, ranging from microscopic single-celled organisms to large burrowing animals. As is true with organisms above the ground, there are well-defined food chains and competition for survival. The study of the relationships of these organisms in the soil environment is called *soil ecology*.

* Peter Farb, *Living Earth,* Harper and Brothers Publishers, New York, 1959.

THE ECOSYSTEM

The sum total of life on earth, together with the global environments, constitutes the *ecosphere*. The ecosphere, in turn, is composed of numerous self-sustaining communities of organisms and their inorganic environments and resources, called *ecosystems*. Each ecosystem has its own unique combination of living organisms and abiotic resources that function to maintain a continuous flow of energy and nutrients.

All ecosystems have two types of organisms based on carbon source: (1) producers, and (2) the consumers and decomposers. The producers use (fix) inorganic carbon from carbon dioxide, and are *autotrophs*. The consumers and the decomposers use the carbon fixed by the producers, such as glucose, and are *heterotrophs*.

Producers

The major primary producers are vascular plants that use solar energy to fix carbon from carbon dioxide during photosynthesis. The tops of plants provide food for animals above the soil-atmosphere interface. Plants produce roots, tubers, and other underground organs within the soil that serve as food for soil-dwelling organisms. A very small amount of carbon is fixed from carbon dioxide by algae during photosynthesis that occurs at or near the soil surface. Some bacteria obtain their energy from chemical reactions, *chemoautotrophs*, and fix a tiny amount of carbon from carbon dioxide. The material produced by the producers serves as food for the consumers and decomposers.

Consumers and Decomposers

Consumers are, typically, animals that feed on plant material or on other animals. For example, very small worms invade and eat living roots. The worms might be eaten or consumed by mites which, in turn might be consumed by centipedes. Eventually, all organisms die and are added to the soil, together with the waste material of animals.

All forms of dead organic materials are attacked by the decomposers, mainly by bacteria and fungi. Through enzymatic digestion (decomposition), the carbon is returned to the atmosphere as CO_2 and energy is released as heat. The nutrients in organic matter are mineralized and appear as the original ions that were previously absorbed by plant roots. The result is that the major function of the consumers and decomposers is the cycling of nutrients and energy. In this context it is appropriate to consider soil as *the stomach of the earth*. Without decomposers to release fixed carbon, the atmosphere would become depleted of CO_2, life would cease, and the cycle would stop.

MICROBIAL DECOMPOSERS

All living heterotrophs, animals and microorganisms, are both consumers and decomposers. They are consumers in the sense that they consume materials for growth and, at the same time, they are decomposers in the sense that carbon is released as CO_2. *Mineralization* is the conversion

of an element from an organic form to an inorganic state as a result of microbial activity. An example is the conversion of nitrogen in protein into nitrogen in ammonia (NH_3). The microorganisms are considered to be the major or ultimate decomposers. They are the most closely associated with the ultimate release of the energy and nutrients in decomposing materials and in making the nutrients available for another cycle of plant growth. Only about 5 percent of the primary production of green plants is consumed by animals. By contrast, about 95 percent of the primary production is ultimately decomposed by microorganisms.

General Features of Decomposers

The decomposers secrete enzymes that digest organic matter outside the cell, and they absorb the soluble end products of digestion. Different enzymes are excreted, depending on the decomposing material. For example, the enzyme cellulase is excreted to decompose cellulose and the enzyme protease is excreted to decompose protein. These enzymes and processes are similar to those that occur in the digestive system of animals.

The microorganisms near plant roots share the same environment with roots, and they compete with each other for the available growth factors. When temperature, nutrients and water supply, pH, and other factors are favorable for plant growth, the environment is generally favorable for the growth of microorganisms. Both absorb nutrient ions and water from the same soil solution. Both are affected by the same water potentials, osmotic effects of salts, pH, and gases in the soil atmosphere. Both may be attacked by the same predators. On the other hand, plants and microorganisms also play roles that benefit each other. Roots slough off cells and leak organic compounds that serve as food for microorganisms, whereas the microbes decompose the organic materials, which results in the mineralization of nutrient ions for root absorption.

In the case of symbiotic nitrogen fixation, the

nitrogen-fixing organisms obtain energy and food from the plant while the plant benefits from the nitrogen fixed. *Nitrogen fixation* is the conversion of molecular nitrogen (N_2) to ammonia and subsequently to organic combinations or to forms utilizable in biological processes. The net effect of the fixation is the transfer of nitrogen in the atmosphere, which is unavailable to plants, to plants and soils where the nitrogen is available.

The major distinguishing feature of microorganisms is their relatively simple biological organization. Many are unicellular, and even multicellular organisms lack differentiation into cell types and tissues characteristic of plants and animals. They form spores, or a resting stage, under unfavorable conditions and germinate and grow again when conditions become favorable.

Bacteria

Bacteria are single celled, among the smallest living organisms, and exceed all other soil organisms in kinds and numbers. A gram of fertile soil commonly contains 10^7 to 10^{10} bacteria. The most common soil bacteria are rod-shaped, a micron (1/25,000 of an inch) or less in diameter, and up to a few microns long (see Figure 8.1). Researchers have estimated that the live weight of bacteria in soils may exceed 2,000 kilograms per hectare (2,000 pounds per acre). The estimated weights and numbers of some soil organisms are given in Table 8.1.

Most of the soil bacteria are heterotrophs and require preformed carbon (such as carbon in organic compounds). Most of the bacteria are aerobes and require a supply of oxygen (O_2) in the soil atmosphere. Some bacteria are anaerobic and can thrive only with an absence of oxygen in the soil atmosphere. Some bacteria, however, are facultative aerobes. They thrive well in an aerobic environment but they can adapt to an anaerobic environment.

Bacteria normally reproduce by binary fission (the splitting of a body into two parts). Some divide as often as every 20 minutes and may multi-

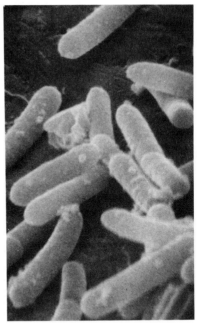

FIGURE 8.1 Rod-shaped bacteria magnified 20,000 times. (Courtesy Dr. S. Flegler of Michigan State University).

ply very rapidly under favorable conditions. If a single bacterium divided every hour, and every subsequent bacterium did the same, 17 million cells would be produced in a day. Such a rapid growth rate cannot be maintained for a long period of time because nutrients and other growth factors are exhausted and waste products accumulate and become toxic. In general, the growth of the decomposers is chronically limited by the small amount of readily decomposable substrate that is available as a food source. Most of the organic matter in soils is humus; organic material that is very resistant to further enzymatic decomposition.

Fungi

Fungi are heterotrophs that vary greatly in size and structure. Fungi typically grow or germinate from spores and form a threadlike structure, called the *mycelium*. Whereas the activity of bacteria is limited to surface erosion in place, fungi readily extend their tissue and penetrate into the surrounding environment. The mycelium is the working structure that absorbs nutrients, continues to grow, and eventually produces reproductive structures that contain spores, as shown in Figure 8.2.

It is difficult to make an accurate determination of the number of fungi per gram of soil, because mycelia are easily fragmented. Mycelial threads or fragments have an average diameter of about 5 microns, which is about 5 to 10 times that of bacteria. A gram of soil commonly contains 10 to 100 meters of mycelial fragments per gram. On this basis, the live weight of fungal tissue in soils is about equal to that of bacteria (see Table 8.1).

The most common fungi are molds and mushrooms. Mold mycelia are commonly seen growing on bread, clothing, or leather goods. *Rhizopus* is a common mold that grows on bread and in soil. It has rootlike structures that penetrate the bread and absorb nutrients, much like the roots of plants. The structure on the surface of the bread forms the working structure, or mycelium, from which fruiting bodies that contain spores develop.

FIGURE 8.2 A soil fungus showing mycelium and reproductive structures that contain spores. (Photograph courtesy of Michigan State University Pesticide Research Electron Microscope Laboratory.)

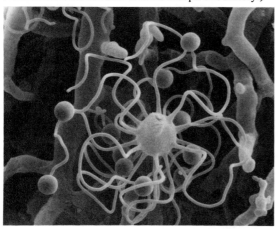

Some molds grow on plant leaves and produce white cotton-colored colonies that produce a disease called downy mildew. Mushroom fungi have an underground mycelium that absorbs nutrients and water, and an above-ground mushroom that contains reproductive spores. Many mushrooms are collected for food, such as the shaggy-mane mushroom shown in Figure 8.3.

Fungi are important in all soils. Their tolerance for acidity makes them particularly important in acidic forest soils. The woody tissues of the forest floor provide an abundance of food for certain fungi that are effective decomposers of lignin. In the Sierra Nevada Mountains, there is a large mushroom fungus, called the train wrecker fungus, because it grows abundantly on railroad ties unless the ties are treated with creosote.

Actinomycetes

Actinomycetes refers to a group of bacteria with a superficial resemblance to fungi. The actinomycetes resemble bacteria in that they have a very simple cell structure and are about the same size in cross section. They resemble filamentous fungi in that they produce a branched filamentous network. The network compared to fungi, however, is usually less extensive. Many of these organisms reproduce by spores, which appear to be very much like bacterial cells.

Actinomycetes are in great abundance in soils, as shown in Table 8.1. They make up as much as 50 percent of the colonies that develop on plates containing artificial media and inoculated with a soil extract. The numbers of actinomycetes may vary from 1 to 36 million per gram of soil. Although there is evidence that actinomycetes are abundant in soils, it is generally concluded they that are not as important as bacteria and fungi as decomposers. It appears that actinomycetes are much less competitive than the bacteria and fungi when fresh additions of organic matter are added to soils. Only when very resistant materials remain do actinomycetes have good competitive ability.

TABLE 8.1 Estimates of Amount of Organic Matter and Proportions, Dry Weight, and Number of Living Organisms in a Hectare of Soil to a Depth of 15 Centimeters in a Humid Temperate Region

Item	Dry Weight		Estimated Number of Individuals
	%	kg/ha	
Organic matter, live and dead	6	120,000	—
Dead organic matter	5.28	105,400	—
Roots of higher plants	0.5	10,000	—
Microorganisms (protists)			
Bacteria	0.10	2,600	2×10^{18}
Fungi	0.10	2,000	8×10^{16}
Actinomycetes	0.01	220	6×10^{17}
Algae	0.0005	10	3×10^{14}
Protozoa	0.005	100	7×10^{16}
Nonarthropod animals			
Nematodes	0.001	20	2.5×10^{9}
Earthworms (and potworms)	0.005	100	7×10^{3}
Arthropod animals			
Springtails (Collembola)	0.0001	2	4×10^{5}
Mites (Acarine)	0.0001	2	4×10^{5}
Millipedes and centipedes (Myriapoda)	0.001	20	1×10^{3}
Harvestman (Oliliones)	0.00005	1	2.5×10^{4}
Ants (Hymenoptera)	0.0002	5	5×10^{6}
Diplopoda, Chilopods, Symphyla	0.0011	25	3.8×10^{7}
Diptera, Coleoptera, Lepidoptera	0.0015	35	5×10^{7}
Crustacea (Isopods, crayfish)	0.0005	10	4×10^{7}
Vertebrate animals			
Mice, voles, moles	0.0005	10	4×10^{5}
Rabbits, squirrels, gophers	0.0006	12	10
Foxes, badgers, bear, deer	0.0005	10	<1
Birds	0.0005	10	100

Adapted and reprinted by permission from *Soil Genesis and Classification* by S. W. Buol, F. D. Hole, and R. J. McCracken, copyright© 1972 by Iowa State University Press, Ames, Iowa 50010.

Vertical Distribution of Decomposers in the Soil

The surface of the soil is the interface between the lithosphere and the atmosphere. At or near this interface, the quantity of living matter is greater than at any region above or below. As a consequence, the A horizon contains more organic debris or food sources than do the B and C horizons. Although other factors besides food supply influence activity and numbers of microorganisms, the greatest abundance of decomposers typically occurs in the A horizon (see Figure 8.4).

SOIL ANIMALS

Soil animals are numerous in soils (see Table 7.1). Soil animals can be considered both consumers and decomposers because they feed on or consume organic matter and some decomposition occurs in the digestive tract. Animals, however, play a minor consumer-decomposer role in organic matter decomposition. Some animals are parasitic vegetarians that feed on roots, whereas others are carnivores that prey on each other.

FIGURE 8.3   Mushroom fungal caps that contain spores—an edible type.

Worms

There are two important kinds of worms in soils. Microscopic roundworms, *nematodes*, are very abundant soil animals. They are of economic importance because they are parasites that invade living roots. The other important worm is the ordinary earthworm.

Earthworms   Earthworms are perhaps the best known of the larger soil animals. The common earthworm, *Lumbricus terrestris*, was imported

into United States from Europe. This worm makes a shallow burrow and forages on plant material at night. Some of the plant material is dragged into the burrow. In the soil, the plant materials are moistened and more readily eaten. Other kinds of earthworms exist by ingesting organic matter found in the soil. Earthworms eat their way through the soil by ingesting the soil en masse. Excreted materials (earthworm castings) are deposited both on and within the soil. Although castings left on the soil surface of lawns are objectionable, earthworms feed on thatch, thus helping to prevent thatch buildup in turf grass. The intimate mixing of soil materials in the digestive track of earthworms, the creation of channels, and production of castings, alter soil structure and leave the soil more porous. Channels left open at the soil surface greatly increase water infiltration.

Earthworms prefer a moist environment with an abundance of organic matter and a plentiful supply of available calcium. As a result, they tend to be least abundant in dry and acidic sandy soils. They normally avoid water-saturated soil. If they emerge during the day, when it is raining, they are killed by ultraviolet radiation unless they quickly find protection. Obviously, the number and activity of earthworms vary greatly from one location to another. Suggestive numbers of earthworms in an Ap horizon vary from a few hundred to more than a million per acre. Estimated weights of earthworms are 100 to 1,000 pounds per acre (kilograms per hectare). Attempts to increase earth-

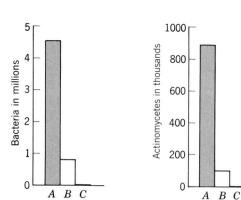

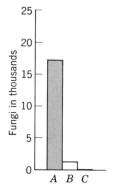

FIGURE 8.4   Distribution of microorganisms in the A, B, and C horizons of a cultivated grassland soil. All values refer to the number of organisms per gram of air-dry soil. (Data from Iowa Res. Bul. 132.)

worm activities in soils have been disappointing, because the numbers present probably represent an equilibrium population for the existing conditions.

Nematodes   *Nematodes* (roundworms) are microscopic worms and are the most abundant animals in soils. They are round shaped with a pointed posterior. Under a 10-power hand lens they appear as tiny transparent threadlike worms, as illustrated in Figure 8.5. Most nematodes are free-living and inhabit the water films that surround soil particles. They lose water readily through their skin, and when soils dry or other unfavorable conditions occur, they encyst, or form a resting stage. Later, they reactivate when conditions become favorable. Based on food preferences, nematodes can be grouped into those that feed on: (1) dead and decaying organic matter, (2) living roots, and (3) other living organisms as predators.

Parasitic nematodes deserve attention owing to their economic importance in agriculture. Many plants are attacked, including tomatoes, peas, carrots, alfalfa, turf grass, ornamentals, corn, soybeans, and fruit trees. Some parasitic nematodes have a needlelike anterior end (stylet) used to pierce plant cells and suck out the contents. Host plants respond in numerous ways: plants, for example, develop galls or deformed roots. In the case of a root vegetable such as carrots, the market quality is seriously affected.

Investigations indicate that nematode damage is much more extensive than previously thought. Nematodes are a very serious problem for pineapple production in Hawaii, where the soil for a new planting is routinely fumigated to reduce nematode populations. Parasititic nematodes attack a wide variety of plants throughout the world, reducing the efficiency of plant growth, analogous to worms living in the digestive track of animals. Agriculturally, the most important role of nematodes in soils appears to be economic, as parasitic consumers.

Arthropods

A high proportion of soil animals is *arthropods*; they have an exoskeleton and jointed legs. Most have a kind of heart and blood system, and usually a developed nervous system. The most abundant arthropods are mites and springtails. Other important soil arthropods include spiders, insects (including larvae), millipedes, centipedes, wood lice, snails, and slugs.

Springtails   *Springtails* are primitive insects less than 1 millimeter ($\frac{1}{25}$ in.) long. They are distributed worldwide and are very abundant (see Table 8.1). Springtails have a springlike ap-

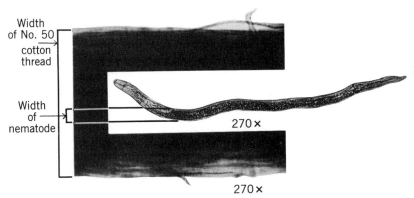

Width of No. 50 cotton thread

Width of nematode

270×

270×

FIGURE 8.5   A nematode and a piece of ordinary cotton thread photographed at the same magnification (270). The nematode is $\frac{1}{15}$ as thick as the thread and is not visible to the naked eye. (Courtesy H. H. Lyon, Plant Pathology Department, Cornell University.)

pendage under their posterior end that permits them to flit or spring in every direction. If the soil in which they are abundant is disturbed and broken apart, they appear as small white dots darting to and fro. They live in the macropores of the litter layers and feed largely on dead plant and animal tissue, feces (dung), humus, and fungal mycelia. Water is lost rapidly through the skin, so they are restricted to moist layers. Springtails in the lower litter layers are the most primitive and are without eyes and pigment.

Springtails are in the order *Collembola,* and they are commonly called Collembola. Their enemies include mites, small beetles, centipedes, and small spiders. Collembola that were grown in the laboratory, and are in various stages of development, are shown in Figure 8.6.

Mites *Mites* are the most abundant air-breathing soil animals. They commonly have a saclike body with protruding appendages and are related to spiders. They are small, like Collembola; a mite photographed on the top of a pinhead is shown in Figure 8.7.

Some mites are vegetarians and some are carnivores. Most feed on dead organic debris of all kinds. Some are predaceous and feed on nema-

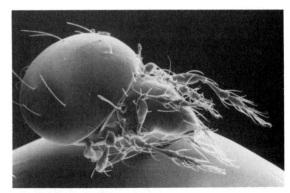

FIGURE 8.7 Electron microscope photograph of a mite on the top of a pinhead; magnified about 50 to 100 times. (Photograph courtesy Michigan State University Pesticide Research Laboratory.)

todes, insect eggs, and other small animals, including springtails. Activities of mites include the breakup and decomposition of organic material, movement of organic matter to deeper soil layers, and maintenance of pore spaces (runways).

Millipedes and Centipedes *Millipedes* and *centipedes* are elongate, fairly large soil animals, with many pairs of legs. They are common in forests, and overturning almost any log or stone will send them running for cover. Millipedes have many pairs of legs and are mainly vegetarians. They feed mostly on dead organic matter, but some browse on fungal mycelia. Centipedes typically have fewer pairs of legs than millipedes, and are mainly carnivorous consumers. Centipedes will attack and consume almost any-sized animal that they can master (see Figure 8.8). Worms are a favorite food of some centipedes. The data in Table 8.1 show that the numbers of millipedes and centipedes are small compared with springtails and mites, but that their biomass may be larger.

FIGURE 8.6 Springtails and eggs photographed through a light microscope.

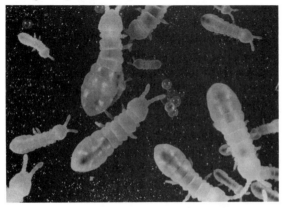

White Grubs Some kinds of insect larvae play a more important role in soils than do the adults as

FIGURE 8.8 A centipede, a common carnivorous soil animal.

does, for example, the *white grub*. White grubs are larvae of the familar May beetle or june bug. The grubs are round, white, about 2 to 3 centimeters long, and curl into a C shape when disturbed. The head is black and there are three pairs of legs just behind the head. White grubs feed mainly on grass roots and cause dead spots in lawns. A wide variety of other plants are also attacked, making white grubs an important agricultural pest. Moles feed on insect larvae and earthworms, resulting in a greater likelihood of mole damage in lawns when white grubs are present.

Ants and termites are important soil insects and will be considered later in relation to the earth-moving activities of soil animals.

Interdependence of Microbes and Animals in Decomposition

Microorganisms and the animals work together as a decomposing team. When a leaf falls onto the forest floor, both micoorganisms and animals attack the leaf. Holes made in the leaf by mites and springtails facilitate the entrance of microorganisms inside the leaf. Soil animals also break up organic debris and increase the surface area for subsequent attack by bacteria and fungi. Soil animals ingest bacteria and fungi, which continue to function within the digestive tracks of the

animals. The excrement of animals is attacked by microbes and ingested by animals. Thus, material is subjected to decomposition in several stages and in widely different environments. Organic matter, and the soil en masse, may be ingested by earthworms, thereby producing an intimate mixing of organic and mineral matter. Again, digestion results in some decomposition and, ultimately, the material is excreted as castings. The net result of these activities is the mineralization of elements in organic matter, conversion of carbon to CO_2, and release of energy as heat. The major credit for nutrient mineralization and recyling of nutrients and energy, however, goes to the microbes (bacteria and fungi).

NUTRIENT CYCLING

Nutrient cycling is the exchange of nutrient elements between the living and nonliving parts of the ecosystem. Plants and microbes absorb nutrients and incorporate them into organic matter, and the microbes (with aid of animals) digest the organic matter and release the nutrients in mineral form. Nutrient cycling conserves the nutrient supply and results in repeated use of the nutrients in an ecosystem.

Nutrient Cycling Processes

Two simultaneous processes, *mineralization* and *immobilization*, are involved in nutrient cycling. Immobilization is the uptake of inorganic elements (nutrients) from the soil by organisms and conversion of the elements into microbial and plant tissues. These nutrients are used for growth and are incorporated into organic matter. Mineralization is the conversion of the elements in organic matter into mineral or ionic forms such as NH_3, Ca^{2+}, $H_2PO_4^-$, SO_4^{2-}, and K^+. These ions then exist in the soil solution and are available for another cycle of immobilization and mineralization.

Mineralization is a relatively inefficient process

in that much of the carbon is lost as CO_2 and much of the energy escapes as heat. This typically produces a supply of nutrients that exceeds the needs of the decomposers; the excess of nutrients released can be absorbed by plant roots. Only in unusual circumstances is there serious competition between the plants and microbes for nutrients.

A Case Study of Nutrient Cycling

One approach to the study of nutrient cycling is to select a small watershed and to measure changes in the amounts and locations of the nutrients over time. An example is a 38-acre (15-hectare) watershed in the White Mountains of New Hampshire. In this mature forest, the amount of nutrients taken up from the soil approximated the amount of nutrients returned to the soil. For calcium, this was 49 kilograms per hectare (44 lb per acre) taken up from the soil and returned to the soil, as illustrated in Figure 8.9. About 9 kilograms of calcium per hectare were added to the system by mineral weathering and 3 kilograms per hectare were added in the precipitation. These additions were balanced by losses due to leaching and runoff. This means, in effect, that 80 percent of the calcium in the cycle was recycled or reused by the forest each year. Plants were able to take up 49 kilograms of calcium per hectare, whereas only 12 kilograms were added to the system annually. Researchers concluded that northern hardwood forests have a remarkable ability to hold and to recirculate nutrients.

The data in Table 8.2 show that 70 to 86 percent of four major nutrients absorbed from the soil were recycled by return of the nutrients in leaves and wood to the litter layer each year. Without this extensive recycling, the productivity of the forests would be much lower. Nutrient cycling accounts for the existence of enormous forests on some very infertile soils. In some very infertile soils, most of the nutrients exist in the organic matter and are efficiently recycled.

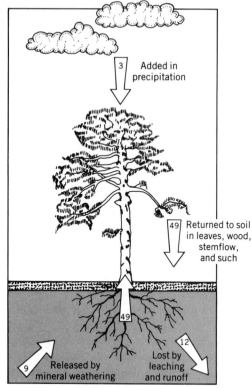

FIGURE 8.9 Calcium cycling in a hardwood forest in New Hampshire in kilograms per hectare per year. (Data from Bormann and Likens, 1970.)

Effect of Crop Harvesting on Nutrient Cycling

The grain and stalks (stover) of a corn (*Zea maize*) crop may contain nitrogen equal to more than 200 kilograms per hectare (200 lb per acre). It is not uncommon for farmers in the Corn Belt region of the United States to add this amount of nitrogen to the soil annually to maintain high yields. According to the data in Table 8.2, by contrast, only 10 kilograms of nitrogen per hectare were stored annually in wood or removed from the system by tree growth in a beech forest. Naturally occurring nitrogen fixation and the addition via precipitation can readily supply this amount of nitrogen. Basically, food production represents

TABLE 8.2 Annual Uptake, Retention, and Return of Nutrients to the Soil in a Beech Forest

| | Nutrients | | | | | | | |
| | Kilograms per Hectare | | | | Pounds per Acre | | | |
	N	P	K	Ca	N	P	K	Ca
Uptake from soil	50	12	14	96	45	11	13	86
Stored in wood or lost from soil	10	2	4	13	9	2	4	12
Returned to soil in litter	40	10	10	83	36	9	9	74
Percent recycled	80	82	70	86	80	82	70	86

Stoeckler, J. H., and H. F. Arneman, "Fertilizers in Forestry," Adv. in *Agron.* 12:127–195, 1960. By permission of author and Academic Press.

nutrient harvesting. The natural nutrient cycling processes cannot provide enough nutrients to grow high-yielding crops. Yields of grain in many of the world's agricultural systems stabilize at only 500 to 600 kilograms per hectare (8 to 10 bushels per acre) where there is little input of nutrients as manure or fertilizer.

When the agricultural system is based on feeding most of the crop production to animals, as in dairying, most of the nutrients in the feed are excreted by the animals as manure. On the average, 75 to 80 percent of the nitrogen, 80 percent of the phosphorus, and 85 to 90 percent of the potassium in the feed that is eaten by farm animals is excreted as manure. Crops can be harvested with only a modest drain on the nutrients in the cycle if the nutrients in the manure are returned to the land. Careful management of manure and the use of legumes to fix nitrogen, together with lime to reduce soil acidity, were the most important practices used to develop a prosperous agriculture in dairy states like New York and Wisconsin.

Specialization in agriculture has resulted in very large crop and animal farms. Today, many large livestock enterprises have more manure, and nutrients, than they can dispose of by applica-

tion of manure to the land. Waste disposal is an important problem. Conversely, large crop farms without animals have no manure to apply to the land, so they balance the nutrient cycle deficit with the application of fertilizer.

SOIL MICROBE AND ORGANISM INTERACTIONS

The role of soil organisms in nutrient cycling has been stressed, and it has been shown that some organisms are parasites that feed on plant roots. In this section the specialized roles of microorganisms living near and adjacent to plant roots will be considered.

The Rhizosphere

Plant roots leak or exude a large number of organic substances into the soil. Sloughing of root caps, and other root cells, provide much new organic matter. More than 25 percent of the photosynthate produced by plants may be lost from the roots to the soil. These substances are food for microorganisms, and they create a zone of intense biological activity near the roots in an area called the *rhizosphere*. The rhizosphere is the zone of soil immediately adjacent to plant roots in which the kinds, numbers, or activities of microorganisms differ greatly from that of the bulk soil. Although many kinds of organisms inhabit the rhizosphere, it appears that bacteria are benefited the most. The bacteria appear mainly in colonies and may cover 4 to 10 percent of the root surface. The intimate association of bacteria and roots may result in mutually beneficial or detrimental effects. Nutrient availability may be affected and plant-growth-stimulating or toxic substances may be produced. Since roots tend to grow better in sterile soil inoculated with organisms than in sterile soil that is not inoculated, it appears that the net effect of bacteria growing in the rhizosphere is beneficial.

Disease

Soils may contain organisms that cause both plant and animal disease. Bacteria are responsible for wilt of tomatoes and potatoes, soft rots of a number of vegetables, leaf spots, and galls. Some fungi cause damping-off of seedlings, cabbage yellows, mildews, blight, and certain rusts. The catastrophic potato famine in Ireland from 1845 to 1846 was caused by a fungus that produced potato blight. Certain species of actinomycetes cause scab in potatoes and sugar beets. The soil used for bedding and greenhouse plants is routinely treated with heat or chemicals to kill plant pathogens.

The organisms that produce disease in humans must be able to tolerate the soil environment, at least for a period long enough to cause infection. Many disease organisms occur in fecal matter and human sewage, such as the virus that produces infectious hepatitis and the protozoan (amoeba) that produces amebic dysentery. When materials containing animal disease organisms are added to soil, the organisms encounter a very different environment than the one they occupied in an animal. Consequently, many animal disease organisms quickly die when added to soils. Some human disease organisms, however, are very persistent. The disposal of human feces by backpackers in wilderness areas poses a health threat, because some organisms are not readily killed by shallow burial. There is also danger of causing disease in humans when human feces are used for fertilizing gardens or fields, because of resistant spore forms of some disease organisms. In China, much of the human excrement is used for fertilizer, but this material is composted for a sufficiently long period of time to kill the disease organisms before the compost is applied to the fields.

Historically, human excrement (night soil) has played an important role in the maintenance of soil fertility. In 1974 the value of the nitrogen, phosphorus, and potassium in the human excreta in India was estimated to be worth more than $700 million as compared with the cost of the nutrients as fertilizer. In countries where most of the food is consumed by humans, careful use of human excrement can be very helpful in maintaining soil fertility. In highly industrialized societies, the nutrients are disposed of through sewage systems and lost from the food production enterprise. The low cost of energy means that the cost of nutrients in the form of fertilizer is less expensive than processing waste to recover and reuse the nutrients.

Mycorrhiza

Some fungi infect the roots of most plants. Fortunately, most of the fungi form a symbiotic relationship—one that benefits both fungi and higher plants. After the appropriate fungal spores germinate, hyphae, or small segments of the mycelium, invade young roots and grow a mycelium both on the exterior and interior of roots. Fungal hyphae on the exterior of roots serve as an extension of roots for water and nutrient absorption. These external hyphae function as root hairs and are called *mycorrhiza*. Mycorrhiza literally means "fungus roots."

There are two types of hypha or mycorrhiza: ectotrophic and endotrophic. The ectotrophic hyphae exist between the epidermal cells, using pectin and other carbohydrates for food. Continued growth of the hyphae outside the root results in the formation of a sheath (mantle) that completely surrounds the root, as shown in Figure 8.10. Literally thousands of species belong in the group of fungi that produce ectotrophic mycorrhiza and produce mushroom or puffball fruiting bodies. They are associated typically with trees and, thus are beneficial to trees, as illustrated in Figure 8.11.

Endomycorrhiza are more abundant than ectomycorrhiza. They benefit most of the field and vegetable crop plants. Hyphae invade roots and ramify both between and within cells, usually avoiding the very central core of the roots. The host plants provide the fungi with food. The benefits to the host plant include: (1) increased effective surface area of roots for the absorption of water and nutrients, particularly phosphorus; (2) increased drought and heat resistance; and (3) reduced infection by disease organisms.

Nitrogen Fixation

We live in a "sea" of nitrogen, because the atmosphere is 79 percent nitrogen. In spite of this, nitrogen is generally considered the most limiting nutrient for plant growth. Nitrogen exists in the air as N_2 and, as such, is unavailable to higher plants and most soil microbes. There are some species of bacteria (nitrogen fixers) that absorb N_2 gas from the air and convert the nitrogen into ammonia that they and the host plant can use. This process of nitrogen fixation is symbiotic. The bacteria obtain food from the host plant and the host plant benefits from the nitrogen fixed. The bacteria responds to and invades the roots of the host

FIGURE 8.10 Ectomycorrhiza. (*a*) Uninfected pine root. (*b*) Infected root with fungal mantle. (*c*) Diagram of the anatomy. The Hartig net is the hyphae between root cells. (Courtesy D. H. Marx, University of Georgia.)

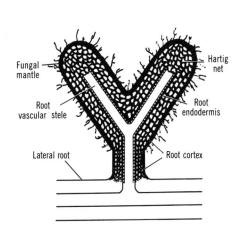

Anatomy of an ectotrophic mycorrhiza

Fungal mantle
Hartig net
Root vascular stele
Root endodermis
Lateral root
Root cortex

(*a*)   (*b*)   (*c*)

FIGURE 8.11 Effect of mycorrhizal fungi on the growth of 6-month-old Monterey pine seedlings. (*a*) Fertile Prairie soil. (*b*) Fertile Prairie soil plus 0.2 percent by weight of Plainfield sand from a forest. (*c*) Plainfield sand alone. Only plants of *b* and *c* were infected with mycorrhiza. (Photograph courtesy of the late Dr. S. A. Wilde, University of Wisconsin.)

plant, and the host responds by forming a nodule that surrounds the bacteria, and in which nitrogen is fixed, as shown in Figure 8.12.

In aquatic ecosystems, blue-green algae fix nitrogen. They are the land counterpart by being the major means by which nitrogen from the atmosphere is added to aquatic ecosystems. A detailed consideration of nitrogen fixation occurs in Chapter 12, where the nitrogen cycle is discussed.

SOIL ORGANISMS AND ENVIRONMENTAL QUALITY

During the billions of years of organic evolution, organisms evolved that could decompose the compounds naturally synthesized in the environment. Today, humans have become important contributors of both natural and synthetic compounds to the environment. The problems of pesticide degradation and contamination of soils from oil spills will be considered as they relate to environmental quality. The application of sewage sludge to the land is covered in Chapter 15.

Pesticide Degradation

Pesticides include those substances used to control or eradicate insects, disease organisms, and weeds. One of the first successful synthetic pesticides, DDT, was used to kill mosquitoes for malaria control. After decades of use it was found in the cells of many animals throughout the world. This dramatized the resistance of DDT to biodegradation and its persistence in the environment. The structure of DDT appears to be different from naturally occurring compounds and, therefore, few if any organisms have enzymes that can degrade DDT. The difference in structure and degradability of 2,4-D and 2,4,5-T is illustrated in Figure 8.13.

Generally, the pesticides used have almost no significant effect on soil microbes. The soil microbes attack the organic pesticides similar to their attack of other organic substrates. In some instances, the microbes detoxify the pesticide, for example, by the shifting of chlorine atoms. In other cases, the pesticide is metabolized or degraded. In general, those pesticides that resist detoxification and/or degradation and persist in the environment long after they are applied, are called *hard* pesticides. The others that degrade easily are *soft* pesticides. The challenge for scientists is to develop pesticides that can perform their useful function and disappear from the ecosystem with minimal undesirable side effects.

Oil and Natural Gas Decontamination

Many species of microbes can decompose petroleum compounds. Crude oil and natural gas are a good energy source, or substrate, for many microbes, and their growth is greatly stimulated.

The first result of an oil spill or natural gas leak is the displacement of soil air and creation of an anaerobic soil. Any vegetation is likely to be killed by a lack of soil oxygen. The reducing conditions result in large increases in available iron and manganese. It is suspected that toxic levels of manganese for higher plants are produced in some cases. In time, the oil or natural gas is

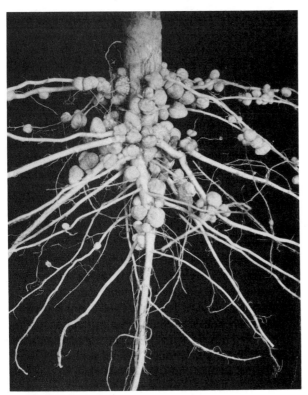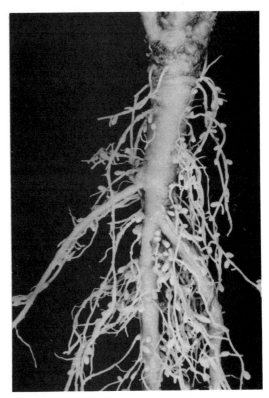

FIGURE 8.12 Symbiotic nitrogen-fixing nodules on the roots of soybean roots (*left*), and on sweet clover roots (*right*). (Photographs courtesy Nitragin Company.)

FIGURE 8.13 Structure representing 2,4-D on the left and 2,4,5-T on the right. The structures are very similar except for the additional Cl at the meta position of 2,4,5-T, which is metabolized with great difficulty (if at all) by soil microorganisms.

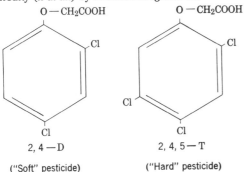

O — CH₂COOH   O — CH₂COOH

2, 4 — D   2, 4, 5 — T

("Soft" pesticide)   ("Hard" pesticide)

decomposed and the soil returns to near normal conditions. A study of 12 soils exposed to contamination near Oklahoma City showed that organic matter and nitrogen contents had been increased about 2.5 times. The increased nitrogen supply in soils that had been contaminated probably explains the increased crop yields obtained on old oil spills.

Natural gas leaks are common in cities. The destruction of grass and trees is due to a lack of normal soil air, which is displaced by the natural gas (methane). Sometimes the problem can be diagnosed by detecting the odor of a pungent compound in the natural gas that escapes from a crack in a nearby sidewalk or driveway. Because methane (CH_4) is odorless, an odor has been added to public gas supplies.

EARTH MOVING BY SOIL ANIMALS

All soil animals participate as consumers and play a minor role in the cycling of nutrients and energy. Many of the larger animals move soil to such an extent that they affect soil formation.

Earthworm Activity

Earthworms are probably the best known earth movers. Darwin made extensive studies and found that earthworms may deposit 4 to 6 metric tons of castings per hectare (10 to 15 tons per acre) annually on the soil surface. This could result in the buildup of a 2.5-centimeter-thick layer of soil every 12 years. This activity produces thicker than normal, dark-colored surface layers in soils. Stones and artifacts are buried over time, which is a subject of great interest to archaeologists.

As a result of their earth-moving activities, earthworms leave channels. Where these channels are open at the soil surface, they can tranport water very rapidly into and through the soil. Earlier it was reported that infiltration of water from irrigation furrows increased after earthworms had burrowed to and entered the irrigation furrows. Renewed interest in the role of earthworms has been created in regard to no-till systems, which leave plant residues on the soil surface. The residues provide a source of food for the earthworms, which forage on the soil surface, and the residues and growing plants protect the soil surface from raindrop impact so that the channels remain open and functional for water infiltration.

Ants and Termites

The activities of ants and termites are, perhaps, more important than the activities of earthworms. Ants transport large quantities of material from within the soil, depositing it on the surface. Some of the largest ant mounds are about 1 meter tall and more than 1 meter in diameter. The effect of this transport is comparable to that of earthworms in creating thick A horizons and in burying objects lying on the soil surface.

A study of ant activity on a prairie in southwestern Wisconsin showed that ants brought material to the surface from depths of about 2 meters and built mounds about 15 centimeters high and more than 30 centimeters in diameter (see Fig. 8.14). Furthermore, it was estimated that 1.7 percent of the land was covered with ant mounds. Assuming that the average life of a mound is 12 years, the entire land surface would be reoccupied every 600 years. The researchers believe that the earth movement activity of ants resulted in a greater than normal content of clay in the A horizon, because of the movement to the surface of material from a Bt horizon, a clay-enriched B horizon.

Ants also harvest large amounts of plant material and in some ecosystems are important consumers. Their gathering of seeds retards natural reseeding on range lands. Leaf harvester ants march long distances to cut fragments of leaves and stems and bring them to their nests to feed fungi. The fungi are used as food. Organic matter is incorporated into the soil depths and nutrients concentrate in the nest sites.

Termites inhabit tropical and subtropical soils. They exhibit great diversity in food and nesting habits. Some feed on wood, some feed on organic refuse, and others actively cultivate fungi. Protozoa (single-celled microscopic animals) in the digestive tract of many termites aid in the digestion of woody materials. Some species build huge nests up to 3 meters high and several meters in diameter. Most mounds are on a smaller scale. Some termites have nests in the soil and tunnnels on the surface to permit foraging for food. Soil material is brought to the surface from depths as great as 3 meters. Termites have had an enormous effect on many tropical soils, where their earth-moving activity has been occurring for hundreds of thousands of years.

(a)

(b)

FIGURE 8.14 Ant (*Formica cinera*) in a Prairie soil in southwestern Wisconsin. Upper photo shows ant mounds more than 15 centimeters high and more than 30 centimeters in diameter. The lower sketch shows soil horizons and location of ant channels; numbers refer to the number of channels observed at the depth indicated. (Photographs courtesy Dr. F. D. Hole, Wisconsin Geological and Natural History Survey, University of Wisconsin.)

In summary, ants and termites create channels in soils and transport soil materials to such an extent that soil horizons are greatly affected and sometimes obliterated. A concentration of nutrients occurs at the nest sites, which is why some farmers in Southeast Asia make use of the higher soil fertility in areas occupied by mounds.

Rodents

Many rodents, including mice, ground squirrels, marmots, gophers, and prairie dogs inhabit the soil. A characteristic microrelief called *mima mounds*, which consists of small mounds of earth, is the work of gophers in Washington and California. Mima mounds occur on shallow soils. They appear to be the result of gophers building nests in the dry soil near the tops of the mounds. Successive generations of gophers at the same site create mounds that range from 0.5 to 1 meter high and more than 5 to 30 meters in diameter.

Extensive earth moving by paririe dogs has been documented. An average of 42 mounds per hectare (17 per acre) were observed near Akron, Colorado. Each mound consisted of an average of 39 tons of soil material. The upper 2 to 3 meters of the soil formed from loess, a silty wind-deposited material, and was underlain by sand and gravel. All of the mounds observed contained sand and gravel that had been brought up from depths of more than 2 meters. Prairie-dog activity had changed the surface soil texture from silt loam to loam on one third of the area. Abandoned burrows filled with dark-colored surface soil, called *crotovinas,* are common in grasslands.

SUMMARY

Higher plants are the major producers contributing to the supply of soil organic matter. The microorganisms (bacteria and fungi) are the major decomposers and are mainly responsible for the cycling of nutrients and energy in soil ecosystems.

Soil animals play a minor role in the cycling of nutrients and energy, but play an important role in earth-moving activities.

Nutrient cyling results in reuse of the nutrients in an ecosystem. Nutrients are efficiently recycled in natural ecosystems. Interference of the cycle, such as cropping and removal of nutrients in food, results in reduced soil fertility. Manures and fertilizers are used to maintain soil fertility in agriculture.

Soil organisms and higher plants engage in many interactions related to disease, mycorrhiza, and nitrogen fixation, and soil organisms and higher plants compete for the same growth factors.

A zone adjacent to plant roots with a high population of microorganisms is the rhizosphere.

Microorganisms play important environmental quality roles, such as detoxification of chemicals and decomposition of oil from spills.

Earthworms, ants, termites, and rodents move large quantities of soil and may greatly alter the nature of soil horizons.

REFERENCES

Alexander, M. 1977. *Soil Microbiology*, 2nd ed. John Wiley, New York.

Arkley, R. J. and H. C. Brown. 1954. "The Origin of Mima Mound Microrelief in the Far Western States." *Soil Sci. Soc. Am. Proc.* **18**:195–199.

Barley, K. P. 1961. "The Abundance of Earthworms in Agricultural Land and Their Possible Significance in Agriculture." *Adv. in Agronomy.* **13**:249–268. Academic Press, New York.

Baxter, F. P. and F. D. Hole. 1967. "Ant Pedoturbation in a Prairie Soil." *Soil Sci. Soc. Am. Proc.* **31**:425–428.

Bormann, F. F. and G. E. Likens. 1970. "The Nutrient Cycles of an Ecosystem." *Sci. Am.* **223**:92–101.

Bowen, E. 1972. "The High Sierra." *Time,* Inc., New York

Buol, S. W., F. D. Hole, and R. J. McCracken. 1973. *Soil Genesis and Morphology.* Iowa State University Press, Ames.

Ellis, R., Jr. and R. S. Adams, Jr. 1961. "Contamination of Soils by Petroleum Hydrocarbons." *Adv. in Agronomy.* **13**:197–216. Academic Press, New York.

Flaig, W., B. Nagar, H. Sochtig, and C. Tietjen. 1977. Organic Materials and Soil Productivity. *FAO Soils Bull.* 35. United Nations, Rome.

Richards, B. N. 1974. *Introduction to the Soil Ecosystem.* Longmans, New York.

Thorp, J. 1949. "Effect of Certain Animals that Live in Soils." *Sci. Monthly.* **42**:180–191.

Schaller, F. 1968. *Soil Animals.* University of Michigan Press, Ann Arbor.

Smucker, A. J. M. 1984. "Carbon Utilization and Losses by Plant Root Systems." *Roots, Nutrient and Water Flux and Plant Root Systems.* ASA Spec. Pub. 49, Soil Sci. Soc. Am., Madison.

Stoeckeler, J. H. and H. F. Arneman. 1960. "Fertilizers in Forestry." *Adv. in Agronomy.* **12**:127–195.

Temple. K. L. A. K. Camper, and R. C. Lucas. 1982. "Potential Health Hazard from Human Wastes in Wilderness." *J. Soil and Water Con.* **37**:357–359.

Walters, E. M. P. 1960. *Animal Life in the Tropics.* George Allen and Unwin, London.

CHAPTER 9

SOIL ORGANIC MATTER

Almost all life in the soil is dependent on organic matter for nutrients and energy. People have long recognized the importance of organic matter for plant growth. Although organic matter in soils is very beneficial, Justus von Liebig, a famous German chemist, pointed out about 150 years ago that soils composed entirely of organic matter are naturally infertile. This chapter considers the origin, nature, and importance of the organic matter in soils.

THE ORGANIC MATTER IN ECOSYSTEMS

The organic matter in an ecosystem consists of the organic matter above and below the soil surface. The distribution of organic matter in a ponderosa pine forest ecosystem is shown in Table 9.1. The distribution of organic matter, expressed as organic carbon, was 38 percent in the trees and ground cover and 9 percent in the forest floor. The remaining 53 percent of the organic matter was in the soil and included the roots plus the organic matter associated with soil particles. In the forest

there is a continual growth of plants and additions to these three pools of organic matter: standing crop, forest floor, and soil. These three pools of organic matter represent a supply of nutrients for future growth.

In grassland ecosystems, much more of the organic matter is in the soil and much less occurs in the standing plants and grassland floor. Although approximately 50 percent of the total organic matter in the forest ecosystem may be in the soil, over 95 percent may be in the soil, where grasses are the dominant vegetation.

DECOMPOSITION AND ACCUMULATION

The primary or original source of soil organic matter (SOM) is the production of the primary producers—the higher plants. This organic material is subsequently consumed and decomposed by soil organisms. The result is the decomposition and the accumulation of organic matter in soils that has great diversity and a highly variable composition.

TABLE 9.1   Quantity and Distribution of Organic Carbon (organic matter) in a Ponderosa Ecosystem in Arizona

Component	Organic Carbon		
	kg/ha	pds/A	Percent of Total
Standing crop			
Tree tops	74,680	66,689	37.7
Herbage	43	38	trace
Forest floor			
Needles, branches	6,686	5,971	3.4
Wood, humus	10,393	9,281	5.2
Soil			
Roots	17,000	15,181	8.6
Humus, etc.	89,300	79,745	45.1
Totals	198,102	176,905	100.0

Adapted from Klemmedson, 1975.

Decomposition of Plant Residues

The plant residues, including tops and roots, contain a wide variety of compounds: sugar, cellulose, hemicellulose, protein, lignin, fat, wax, and others. Observe from Table 9.2 that the composition of SOM is distinctly different than that of the original plant material added to the soil.

In an experiment, wheat straw was added to soil and the changes in the major components in the straw were followed over time. It was found that proteins, the soluble fraction, and the cellulose and hemicellulose disappeared or decomposed very rapidly, whereas lignin decomposed very slowly, as shown in Figure 9.1a. As a result of the rapid decomposition of proteins and the soluble fraction, cellulose and hemicellulose, and other readily degradable compounds, there was a corresponding and rapid increase in microbial products (see Figure 9.1b).

Microbial products include living and dead microbial cells and their waste or excretion products. Some of the organic compounds that are synthesized in the soil during decomposition react with each other and with mineral soil components. Consequently, the decomposition of plant residues results in: (1) the production of a considerable mass of microbial products, and (2) the production of a wide variety of materials of varying resistance to decomposition. As a result of the addition and decomposition of plant and animal residues in soils, *labile* and *stable* fractions of organic matter are produced. Labile organic matter is organic matter that is liable to change or is unstable. The liable and stable organic matter fractions correspond, in general, to the organic residues and humus fractions, respectively.

TABLE 9.2   Partial Composition of Mature Plant Tissue and Soil Organic Matter

Component	Percent	
	Plant Tissue	Soil Organic Matter
Cellulose	20–50	2–10
Hemicellulose	10–30	0–2
Lignin	10–30	35–50
Protein	1–15	28–35
Fats, waxes, etc.	1–8	1–8

Labile Soil Organic Matter

The labile fraction of SOM consists of any readily degradable materials from the plant and animal residues, and readily degradable microbial prod-

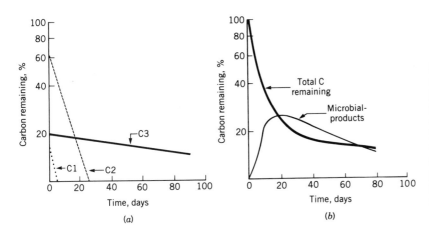

FIGURE 9.1 Decomposition of wheat straw in soil in the laboratory. (*a*) Decreases over time: proteins and solubles (C1), cellulose and hemicellulose (C2), and lignin (C3). (*b*) Changes over time in amount of carbon in original plant material and in microbial products. (Data from Van Veen and Paul, 1981.)

ucts. The labile organic matter composes about 10 to 20 percent of the total SOM. Labile organic matter is an important reservoir of nutrients because the nutrients, especially nitrogen, are rapidly recycled in the soil ecosystem. The addition to the soil of a relatively small amount of easily degraded organic matter, such as manure, organic waste materials, and plant residues, may contribute more to the available nutrient supply than the slow release of nutrients from the much larger amount of stable organic matter. The labile organic matter is rapidly degraded when conditions are favorable. As a result, any labile organic matter incorporated within soils tends to disappear quickly if conditions are favorable for decomposition. Therefore, the decomposers are frequently quite inactive, owing to the rapid disappearance of readily decomposable substrates or labile organic matter.

Stable Soil Organic Matter

The stable SOM fraction consists of resistant compounds: (1) in the decomposing residues, (2) in microbial products, and (3) that formed as a result of interaction of organic compounds with each other and with the mineral components of the soil, especially the clay. The stable organic matter is equivalent to humus. The nutrients within it are recycled very slowly. Radioactive studies have shown that certain stable SOM fractions are commonly more than a few hundred or thousands of years old. The stable organic matter has a long residence time in the soil and plays important roles in structure formation and stability, water adsorption, and adsorption of nutrient cations.

Stable soil organic matter, or humus, originates from several processes. Lignin, a 6-carbon ring structure, is a plant compound and is very resistant per se. Thus, lignin accumulates in soils because of its resistant or stable nature. Organic compounds of varying resistance interact with lignin and form new compounds that are resistant. This accounts for the significant quantity of protein or proteinlike material in humus (see Table 9.2).

Much of the stable organic matter is intimately associated with the mineral particles. As such, stable organic matter plays an important role in the formation of soil structure and the ability of soil peds to resist crushing and breakdown from tillage and raindrop impact. Studies have shown that some readily decomposable materials (sugars and proteins) are decomposed by microbes in less than a day under laboratory conditions. When the same materials are added to soils, the decomposition time is greatly increased. The adsorption of organic compounds onto soil particles makes these compounds less decomposable. Scientists believe that the organic matter is protected from decomposition; this protection helps to increase the stability of both readily decomposable and resistant organic matter. Much of the humus is associated intimately with clay and silt particles, and the humus plays an important role in formation of soil structure, as shown in Figure 9.2.

Polysaccharides include microbial gum and two other abundant compounds found in microbial cell walls: hemicellulose and cellulose. These polysaccharides are organic polymers (giant organic molecules formed by the combination of smaller molecules) having high molecular weight and are present as ropes and nets. These polymers are widely and intimately associated with the soil fabric. On drying, soil particle surfaces are brought into very close contact with the netlike arrangement of the organic compounds. The hydroxyls of the polymers and the exposed oxygen atoms of the mineral soil particles bind strongly to each other. The result is an intimate

FIGURE 9.2 Some arrangements of organic matter and mineral soil particles. Organic matter intimately associated with mineral particles is protected from decomposition and contributes to the formation and stability of soil structure. (a) Quartz-organic matter-quartz. (b) Quartz-organic matter-clay. (c) Organic matter-clay-organic matter.

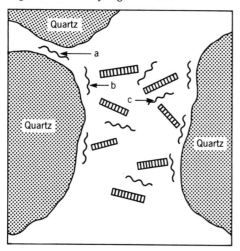

association of mineral and organic matter that protects the organic matter from decomposition and forms stable soil peds. One half or more of the SOM is likely to be protected.

Decomposition Rates

Although SOM has been characterized as having two forms—labile and stable—there is great variety in the nature and decomposition rates of various organic fractions of the SOM. In a native grassland soil, about 9 percent of the SOM were shown to consist of roots and plant residues and 4 percent consists of unprotected and decomposable liable organic matter with a decomposition rate of 8 percent per day. This rate, which assumes optimal conditions, is shown in Table 9.3. Soil organisms were 1.3 percent of the total organic matter with a decomposition rate of 3 percent per day.

The other fractions of the SOM, 84 percent, had much slower decomposition rates. The bulk of the SOM (54 percent) consisted of decomposable organic matter that was protected by attachment to clays and other mineral components, and had a decomposition rate of 0.08 percent per day. Two fractions, composed mainly of ligneous material, decomposed at 0.0008 and 0.000,008 percent per day, for unprotected and protected ligneous materials, respectively. Thus, the decomposition rate of the most readily decomposable material is about 1 million times that of the most stable (see Table 9.3). The most resistant organic matter fractions have been called *recalcitrant*, meaning stubborn.

Assuming that plant residues, roots, decomposable unprotected organic matter, and soil organisms are liable organic matter, the data in Table 9.3 show that the soil contained about 15 percent liable organic matter and 85 percent stable organic matter.

Properties of Stable Soil Organic Matter

Stable soil organic matter (humus) is a heterogeneous mixture of amorphous compounds that are resistant to microbial decomposition and possess

TABLE 9.3 Soil Organic Matter and Decomposition Rates under Optimum Conditions for a Native Grassland Soil

| | Organic carbon, upper 16 inches | | | |
Organic matter	kg/ha	pds/A	% of total	Decay rate, % per day
Plant residues, roots	8,378	7,480	9.2	8.0
Decomposable, unprotected	3,293	2,940	4.0	8.0
Soil organisms	1,098	980	1.3	3.0
Decomposable, protected	44,128	39,400	54	0.08
Ligneous unprotected	12,432	11,100	15	0.0008
Ligneous, protected	12,432	11,100	15	0.000,008

Adapted from Van Veen and Paul, 1981.

a large surface area per gram. The large surface area enables humus to absorb water equal to many times its weight.

During decomposition of plant and animals residues, there is a loss of carbon as carbon dioxide and the conservation and reincorporation of nitrogen into microbial products, which eventually are incorporated into humus. As a consequence, humus contains about 50 to 60 percent carbon and about 5 percent nitrogen, producing a C : N ratio of 10 or 12 to 1. The more decomposed the humus is the lower or narrower will be the C : N ratio. Compared with plant materials, humus is relatively rich in nitrogen and, when mineralized, humus is a good source of biologically available nitrogen. Humus is also a significant source of sulfur and phosphorus. The ratio of C:N:P:S is about 100:10:1:1.

Humus contains phenolic groups, 6-C rings with OH, that dissociate H^+ at pH above 7.0. During humus formation there is an increase in carboxyl groups, R-COOH, and these groups tend to dissociate H^+ at pH less than 7:

$$R—COOH = R—COO^- + H^+$$

As a result, humus has a considerable amount of negative charge. Cations in the soil solution tend to be weakly adsorbed onto the negatively charged phenolic and carboxyl groups. However, adsorbed cations can be readily exchanged by other cations remaining in solution. The total negative charge of the SOM, or ability to adsorb exchangeable cations, is called the *cation exchange capacity* (CEC). The CEC of organic matter acts similarly to that of clay particles, in that the charge adsorbs and holds nutrient cations in position to be available to plants, while effectively holding these cations against loss by leaching. The most abundant nutrient ions adsorbed are Ca^{2+}, Mg^{2+}, and K^+. Adsorbed H^+ affects the pH of the soil solution, depending on the amount adsorbed and the degree of dissociation into the soil solution.

Clay and SOM are the source of most of the negative charge in soils. The sum of the CEC of the SOM plus the CEC of the clay is generally considered equivalent to the CEC of the total soil. The source and amount of CEC of soil clays are discussed in Chapters 10 and 11.

Protection of Organic Matter by Clay

The amount of organic matter present in a soil at any given time is the net effect of the amount that has been added minus the amount that has been decomposed. The major factors affecting the SOM content are soil texture, vegetation type, water content or aeration of the soil, and soil temperature.

Locally, within a given landscape, a correlation has been found between soil texture and the organic matter content of soils as a result of the role of the soil matrix, especially clay, in protecting organic matter from decomposition. Organic matter is positively correlated with the clay content and negatively correlated with the sand content in some grassland soils in Wyoming (see Figure 9.3). These soils had a mean average soil temperature of 8 to 15° C (47 to 59° F).

In the same study, soils with lower average temperature showed a weaker relationship between organic matter content and texture. This suggests that low soil temperature, by decreasing the rate of decomposition, appeared to have had an important effect on the organic matter content of the soils. The data in Figure 9.4 are generally representative of the plains of the United States between Canada and Mexico, and show that high precipitation at low temperature results in much more organic matter in soils as compared to high precipitation and high temperature. In most local (farm) situations, where temperature is a constant, clay content is positively associated with organic matter content. Thus, soils with apprecia-

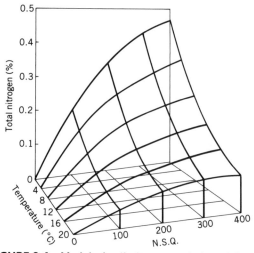

FIGURE 9.4   Model of soil nitrogen variation of the Great Plains from Canada to Mexico (back to front curves) and from the desert to the humid regions (left to right curves). The NSQ is the precipitation divided by the absolute saturation deficit of air; it is a measure of the effectiveness of the precipitation. Decreases in temperature and increases in humidity are associated with increases in total soil-nitrogen content. (From Jenny and used by permission of Springer-Verlag, New York, copyright © 1980.)

ble clay content tend to have a higher content of organic matter and, therefore tend to have a good combination of soil fertility and physical properties.

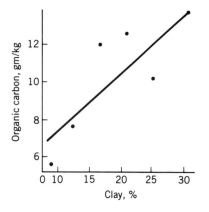

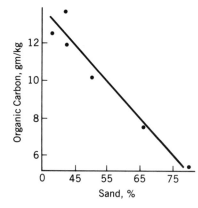

FIGURE 9.3   Relationships between the content of organic matter to a depth of 40 centimeters and the clay and sand content of some Wyoming grassland soils. (Adapted from McDaniel and Munn, 1985.)

ORGANIC SOILS

The emphasis thus far has been on mineral soils, soils with a volume composition consisting of about 50 percent solids, and these solids composed of 45 percent mineral matter and 5 percent organic matter. In such soils the mineral fraction has a dominating influence on soil properties. Organic soils, by contrast, have properties that are controlled mainly by the organic matter.

Organic Soil Materials Defined

Organic soil materials are saturated with water for long periods or are artificially drained and, excluding roots, (1) contain 18 percent (dry-weight basis) organic carbon if the mineral fraction is 60 percent or more clay; (2) contain 12 percent or more organic carbon if the mineral fraction has no clay; and (3) contain a proportional content of organic carbon between 12 and 18 percent, depending on clay content, as shown in Figure 9.5. If the soils are never water saturated for more than a few days each year, the organic carbon content must be 20 percent or more.

Organic soils are those soils composed of organic soil materials that are of a significant thickness, depending on the depth to rock and other lithic materials.

Formation of Organic Soils

High organic matter contents in soils are the result of slow decomposition rates rather than high rates of organic matter addition. Organic soils are very common in tundra regions where the rate of plant growth is very slow. In the shallow water of lakes and ponds, plant residues accumulate instead of decomposing because of the anaerobic conditions. Consequently, organic soils may consist almost entirely of organic matter. In some cases, an infusion of mineral sediment from soil erosion may result in the accumulation of some mineral matter, together with the organic matter. Some organic soils, peat, for example, are mined and used for fuel.

Formation of many organic soils consists of the accumulation of organic matter in a body of water. After glaciation, many depressions received clay and other materials that sealed the bottoms and created ponds and lakes. As the plant material accumulated on the bottom of the bodies of

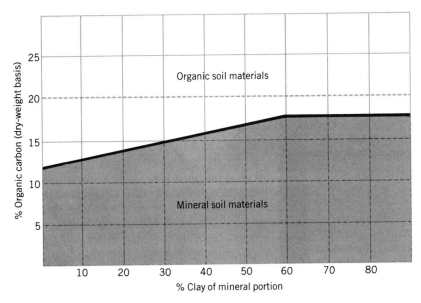

FIGURE 9.5 Organic carbon content (dry weight basis) of organic and mineral soil materials. Mineral soil materials below the heavy solid line and organic soil materials above the heavy solid line. Organic soil materials above the dashed line are water saturated for no more than a few days per year.

water, the water depth decreased and the kinds of plants contributing organic matter changed. Another important organic component is the remains and fecal products of small aquatic animals.

Minor climatic changes have occurred since the last glaciation, and this has had an effect on the water levels and types of vegetation and the animal populations. Thus, a series of layers or soil horizons develop from the bottom upward that reflect the conditions at the time each layer was formed. Pollen grains are preserved in the peat, which can be readily identified. The vegetation type can be verified by pollen analysis. A total of 5 meters of peat soil accumulated in Sweden in 9,000 years or at a rate equal to 1 centimeter every 18 years. This is illustrated in Figure 9.6.

Properties and Use

Organic soils are composed mainly of plant and animal residues and some microbial decomposition products. The soils are very light and have low bulk density, high CEC, and a high nitrogen, phosphorus, and sulfur content. The absence or low content of sand and silt particles, composed of minerals containing essential nutrients, may result in low amounts of available potassium.

Organic soils must be drained before they can be used for crop production. The upper soil layers then become aerobic and the organic matter begins to decompose. This converts undecomposed peat into well-decomposed organic soil called muck. In time, the entire soil may disappear. This is a serious problem in a warm climate with a rapid decomposition rate, as in the Florida Everglades, where sugarcane is a major crop, together with a wide variety of vegetable crops (see Figure 9.7). Organic soils typically occupy the lowest elevation in the landscape, and crops grown in such soils are the most likely to be killed by frost. Their low bulk density makes them susceptible to wind erosion and desirable for sod production (see Figure 9.8).

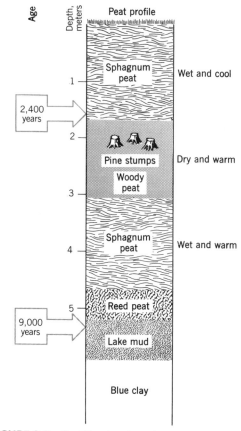

FIGURE 9.6 Stratification found in a peat soil in Sweden, where changes in climate produced changes in the kind of vegetation that formed the peat. (Data from Lucas, 1982.)

Archaeological Interest

The excellent preservation ability in a peat bog is illustrated by examination of the famous Tollund man, who was found in 1950 in a peat bog in Denmark. The facial expression at death and bristles of the beard were well perserved. Excellent fingerprints were made, and an autopsy revealed the contents of the last meal, which consisted mainly of seeds, many of them weed seeds. When found, the Tollund man was buried under 2 meters of peat that was estimated to have formed in the 2,000 years since his burial.

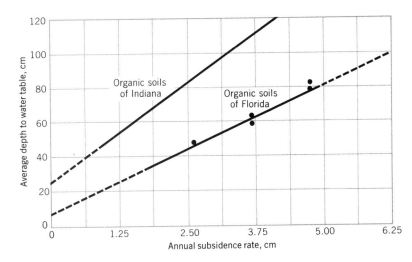

FIGURE 9.7 The subsidence rates for organic soils depends on the depth to the water table; the lower the water table the greater is the soil aeration and decomposition of organic matter. Subsidence is greater in Florida than in Indiana because of higher temperatures in Florida.

THE EQUILIBRIUUM CONCEPT

During the early phase of soil formation, soil organic matter content increases. After a few centuries, the soil attains a dynamic equilibrium content of organic matter; additions balance losses. Even though the organic matter content remains constant from year to year, there is a significant turnover of organic matter and recycling of nutrients.

FIGURE 9.8 Organic soil landscape. Crop on left is grass sod used for landscaping and in the rear is a windbreak of willow trees. The lightness of the soil makes the soil desirable for sod production; however, it also contributes to susceptibilty to wind erosion.

A Case Example

The pool, or amount of organic matter in a soil, can be compared to a lake. Changes in the water level in a lake depend on the difference between the amount of water entering and leaving it. This idea, applied to soil organic matter, is illustrated at the top of Figure 9.9, showing that the change in soil organic matter is the difference between the amount of organic matter added and the amount that is decomposed.

Soil organic matter, labile and stable considered together, decomposes in mineral soils at a rate equal to about 1 to 3 percent per year. Assuming a 2 percent decomposition rate, and 40,000 kilograms in the plow layer of a hectare, 800 kilograms of soil organic matter would be decomposed or lost each year. Conversely, if 800 kilograms of humus or stable organic matter were formed from decomposition of residues added to the soil, the organic matter content of the soil would remain the same from one year to the next. The soil would be at the equilibrium level illustrated in Figure 9.9.

Decomposition of 800 kilograms of organic matter containing 5 percent nitrogen would result in mineralization of 40 kilograms per hectare of nitrogen. On an acre basis, this would amount to about 40 pounds of available nitrogen per acre.

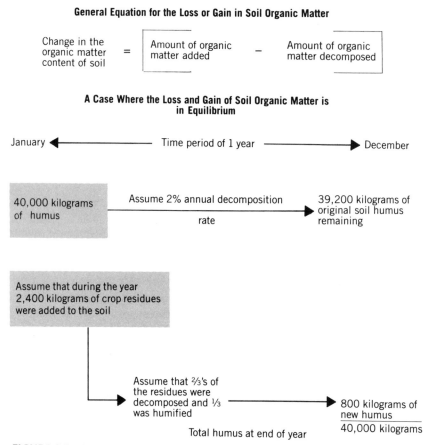

FIGURE 9.9 Schematic illustration of the equilibrium concept of soil organic matter as applied to a hectare plow layer weighing 2 million kilograms and containing 2 percent organic matter. (Values are similar on a pounds per acre basis for a 2-million pound acre-furrow-slice.)

Assuming that the ratio of N:P:S in the soil organic matter is 10:1:1, there would be 4 kilograms per hectare (4 lb/acre) of phosphorus and sulfur mineralized. Other nutrients are also mineralized, but their availability is more closely related to the soil mineral fraction and mineral weathering.

Effects of Cultivation

Even on nonerosive land that is converted from grassland or forest to cultivated land, a rapid loss of organic matter resulting from cropping occurs.

At the Missouri Agricultural Experiment Station, as a result of 60 years of cultivation, the soil lost one third of its organic matter (see Figure 9.10). The organic matter content decreased 25 percent during the first 20 years, 7 percent during the next 20 years, and only 3 percent during the third 20-year period. The organic matter content of the soil was moving to a new equilibrium level. In general, cultivated soils today contain about one half as much organic matter as they did before cultivation. Theoretically, if a soil is fallowed, so that no plants grow and decomposition continues, the

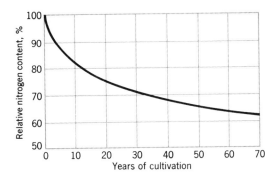

FIGURE 9.10 Decline in soil nitrogen (or organic matter) with length of time of cultivation under average farming practices in Missouri. (Data from Jenny, Missouri Agr. Exp. Sta. Bull. 324, 1933.)

organic matter content would eventually approach zero. On the other hand, if cultivated fields are allowed to revert to their original native vegetation, the organic matter content will gradually increase toward the equilibrium level that existed prior to cultivation.

Soils in arid regions naturally have a low organic matter content. Irrigating such soils and producing crops that return a large amount of plant residues to the soil results in an increase in the SOM content. In this case, converting land to agricultural use results in an equilibrium level that is higher than that of the virgin land.

Maintenance of Organic Matter in Cultivated Fields

The organic matter content of a soil is more difficult to determine than the carbon content. As a result, changes in the soil carbon content are used as a measure of changes in organic matter content. Multiplying the percent of soil carbon by 1.72 is generally considered to equal the percent of organic matter.

Soil organic matter is constantly subject to decomposition and loss. Therefore, each soil and cropping situation requires the addition of a certain amount of organic matter each year to maintain the status quo. In a 16-year experiment in Minnesota, 5 tons of crop residues per acre annually were required to maintain the original carbon content of the soil at 1.8 percent (see Figure 9.11). Addition of more than 5 tons per acre increased SOM content and lesser amounts resulted in a decline in the organic matter content. While land is being farmed (excluding the desert land), it is

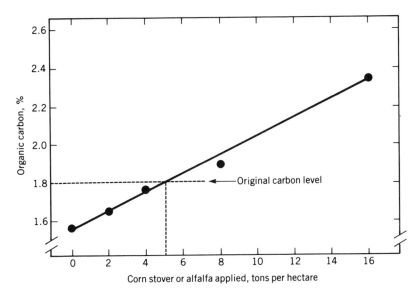

FIGURE 9.11 Relationship between annual application rates of crop residues and content of organic carbon in soils after 11 years. (Data of Frye, Bennett, and Buntley, 1985).

virtually impossible, to maintain the organic matter content of the virgin soil. It is, therefore, prudent to accept the organic matter content that results from economical or profitable farming.

In Figure 9.12, the effect of different yields of corn grain on soil carbon is given for a typical well-drained agricultural soil in southern Michigan. The grain was harvested and all of the remaining crop residues were returned to the soil, including the roots that equal about 15 to 20 percent of the total crop residues. Note that a corn grain yield of 3,150 kilogram per hectare (kg/ha) (50 bushels per acre) is predicted to maintain an equilibrium level equal to 1.0 percent organic carbon. By contrast, yields of 6,300 and 9,450 kilograms per hectare (100 and 150 bushels per acre) are predicted to maintain equilibrium carbon levels of 1.7 and 2.3 percent, respectively. Higher yields return more residues and maintain a higher organic matter content, which in turn makes still higher yields more likely.

Another observation can be made from Figure 9.12. A yield of 6,300 kilograms per hectare would result in about a 400 kilograms per hectare loss of carbon if the soil contained 2.5 percent organic carbon. Conversely, if the soil contained only 0.5 percent organic carbon, the same yield of 6,300

FIGURE 9.12 Estimated annual soil-humus carbon changes as related to corn (maize) yields and organic carbon content of loamy soils in southern Michigan. (Adapted from Lucas and Vitosh, 1978.)

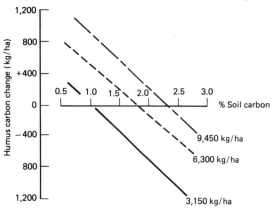

kilograms per hectare would increase the soil carbon content by about 800 kilograms per hectare. Since World War II there have been large increases in crop yields in many countries in spite of claims of declining soil organic matter and soil depletion. It appears that soil management systems that result in high crop yields maintain SOM at acceptable contents. Many soil management programs focus on the production of high crop yields and proper crop residue management rather than the organic matter content of soils, per se. One of the most serious causes of organic matter loss is soil erosion that tends to preferentially remove the organic matter fraction. High crop yields also mean more vegetative cover and reduced soil erosion.

Effects of Green Manure

Green manuring is the production of a crop for the purpose of soil improvement. These crops are easily and inexpensively established and, frequently, are grown during the fall, winter, and/or spring when the land would normally be unprotected by vegetation. Benefits include erosion control and the accumulation of nutrients (into organic matter) that might otherwise be lost by leaching. There is also less likelihood of pollution of water tables with nitrate nitrogen. In addition, organic matter from the green manure crop is eventually incorporated into the soil, and the following crop benefits from the rapid release of nutrients from a large amount of readily decomposable (labile) material. The humus that is formed contributes to the maintenance of the content of soil organic matter. However, this is usually a secondary reason why green manure crops are grown.

HORTICULTURAL USES OF ORGANIC MATTER

Organic materials used for horticultural purposes consist mainly of naturally occurring peat and moss and the products of composting. High-content organic matter materials are used as a component of soil mixes and as a mulch.

Horticultural Peats

Peats that are used for soil amendments are generally classified as moss peat, reed-sedge peat, and peat humus. Moss peat forms from moss vegetation and reed-sedge peat from reeds, sedges, cattails, and other associated plants. Some properties of common horticultural peats are given in Table 9.4. The low nitrogen content of sphagnum peat means that it may not have enough nitrogen to satisfy the needs of decomposers and, thus, its use creates competition between decomposers and plants for any available nitrogen that might be present in the soil mix. The low pH of sphagnum peat, however, makes it desirable as a mulch for plants, such as azaleas and rhododendrons, that grow well in acid soil. Peat moss makes a neat-looking surface that sets off plants and protects the soil from the disruptive impact of raindrops. The soil remains more porous and receptive to rain or irrigation water (see Figure 9.13).

Composts

Many gardners have organic residues with wide C : N ratios such as tree leaves, weeds, grass clip-

FIGURE 9.13 Mulching a rose garden with peat moss. The mulch protects the soil surface from raindrop impact and maintains a high water-infiltration rate.

pings, or other plant and garbage wastes. These materials can be *composted* with a resulting decrease in the C:N ratio and production of an organic material used for mulching and as a component of soil mixes. Composting consists of storing or maintaining the organic materials in a pile while providing favorable water content, aeration, and temperature. As the organic matter decomposes, much of the carbon, hydrogen, and oxygen is released as carbon dioxide and water. Nutrients, like nitrogen, are continuously reused and recycled by the microbes and conserved

TABLE 9.4 Characteristics of Common Horticultural Peats

Type	Range in Nitrogen[a], %	Range of Water-Absorbing Capacity[a], %	Range in Ash Content[a], %	Range in Dry Densities[a], lb/ft^3	Range in pH
Sphagnum moss peat	0.6–1.4	1,500–3,000	1.0–5.0	4.5–7.0	3.0–4.0
Hypnum moss peat	2.0–3.5	1,200–1,800	4.0–10.0	5.0–10.0	5.0–7.0
Reed-sedge peat (low lime)	1.5–3.0	500–1,200	5.0–15.0	10.0–15.0	4.0–5.0
Reed-sedge peat (high lime)	2.0–3.5	400–1,200	5.0–18.0	10.0–18.0	5.1–7.5
Decomposed peat	2.0–3.5	150–500	10.0–50.0	10.0–40.0	5.0–7.5

From Lucus, et al., Ext. Bull., 516, Michigan State University, East Lansing, Michigan.
[a] Oven dry basis.

TABLE 9.5 Materials Recommended for
Making Compost

	Cups per Tightly Packed Bushel
For general purposes, including acid-loving plants:	
Ammonium sulfate	1
Superphosphate (20 percent)	$\frac{1}{2}$
Epsom salt	$\frac{1}{16}$
or:	
10-6-4 fertilizer	$1\frac{1}{2}$
For plants not needing acid soil:	
Ammonium sulfate	1
Superphosphate (20 percent)	$\frac{1}{2}$
Dolomitic limestone or wood ashes	$\frac{2}{3}$
or:	
10-6-4 fertilizer	$1\frac{1}{2}$
Dolomitic limestone or wood ashes	$\frac{2}{3}$

From Kellogg, 1957.

within the composting pile. Thus, while there is a loss of carbon, oxygen, and hydrogen, other nutrient elements are being concentrated. The occasional turning of the compost pile increases aeration and generally hastens rotting. The rotted material has good physical properties and is a good source of nutrients. Small amounts of materials, from the weeding of a garden, for example, can be composted in a black plastic bag.

The low nitrogen content of many composting materials may retard their decomposition. For this reason, most composters add some nitrogen fertilizer to the compost pile. The nutritive value of the compost can also be increased by the addition of other materials. Some recommendations of the U.S. Department of Agriculture are given in Table 9.5.

SUMMARY

The total soil organic matter is composed of a labile fraction and a stable fraction. The labile fraction consists of readily decomposable plant,

microbial, and animal products. Even though it is a minor amount of the total soil organic matter, labile organic matter is very important in nutrient cycling. The stable organic matter is humus, which contributes to cation exchange capacity, water adsorption, and soil structure stability. Stable organic matter decomposes very slowly, and as a result, releases (mineralizes) nutrients very slowly.

Organic soils consist mainly of organic matter. They have light weight and are susceptible to wind erosion. Most organic soils require drainage for cropping, which aerates the soil and greatly increases the rate of decomposition. The absence or very low content of minerals is associated with very low potassium supplying power.

The organic matter content of a soil tends to approach an equilibrium. When the losses of organic matter exceed additions, the soil organic matter content declines. Cultivated soils contain about half as much organic matter as they did before the soils were converted to cropping. The exception is desert soils where cropping may result in an increase in organic matter content.

Horticultural peats and composts are organic materials used for mulching and for soil mixes.

REFERENCES

Alexander, M. 1977. *Soil Microbiology.* 2nd ed. John Wiley, New York.

Allison, F. E. 1973. *Soil Organic Matter and Its Role in Crop Production.* Elsevier, New York.

Frye, W. W., O. L. Bennett, and G. J. Buntley. 1985. "Restoration of Crop Productivity on Eroded or Degraded Soils." *Soil Erosion and Crop Productivity.* pp. 335–356. Am. Soc. Agronomy. Madison, Wis.

Glob, P. V. 1954. "Lifelike Man Preserved 2,000 Years in Peat." *Nat. Geog. Mag.* **105:**419–430.

Jenny, Hans. 1933. "Soil Fertility Losses Under Missouri Conditions." *Missouri Agr. Exp. Sta. Bull.* 324. Columbia, Mo.

Jenny, Hans. 1980. *The Soil Resource.* Springer-Verlag, New York.

Kellogg, C. E. 1957. "Home Gardens and Lawns". *USDA Agricultural Yearbook.* pp. 665–688. Washington, D.C.

Klemmedson, J. O. 1975. "Nitrogen and Carbon Regimes in an Ecosystem of Young Dense Ponderosa Pine in Arizona." *Forest Science.* **21:**163–168.

Lucas, R. E. 1982. "Organic Soil (Histosols)." *Mi. Agr. Exp. Sta. Res. Report* 435. East Lansing, Mi.

Lucas, R. E. and P. E. Rieke, and R. S. Farnham. 1971. "Peats for Soil Improvement and Soil Mixes." *Mi. Agr. Ext. Bull.* 516. East Lansing, Mi.

Lucas, R. E. and M. L. Vitosh. 1978. "Soil Organic Matter Dynamics." *Mi. Agr. Exp. Sta. Res. Report* 358. East Lansing, Mi.

McDaniel, P. A. and L. C. Munn. 1985. "Effect of Temperature on Organic Carbon-texture Relationships in Mollisols and Aridisols." *Soil Sci. Soc. Am. J.* **49:** 1486–1489.

Paul, E. A. 1984. "Dynamics of Organic Matter in Soils." *Plant and Soil.* **76:**275–285.

Schreiner, O. and B. E. Brown. 1938. "Soil Nitrogen." *Soils and Men.* USDA Yearbook of Agriculture, Washington, D.C.

Van Veen, J. A. and E. A. Paul. "1981. Organic Carbon Dynamics in Grassland Soils." *Can. J. Soil Sci.* **61:**185–210.

CHAPTER 10

SOIL MINERALOGY

Minerals are natural inorganic compounds, with definite physical and chemical properties, that are so conspicuous in many granitic rocks. They are broadly grouped into primary and secondary minerals. *Primary minerals* have not been altered chemically since their crystallization from molten lava. Disintegration of rocks composed of primary minerals (by physical and chemical weathering) releases the individual mineral particles. Many of these primary mineral particles become sand and silt particles in parent materials and soils. Primary minerals weather chemically (decompose) and release their elements to the soil solution. Some of the elements released in weathering react to form *secondary minerals*. A secondary mineral results from the decomposition of a primary mineral or from the precipitation of the decomposition products of minerals. Secondary minerals originate when a few atoms react and precipitate from solution to form a very small crystal that increases in size over time. Because of the generally small particle size of secondary minerals, they dominate the clay fraction of soils.

Consideration of the characteristics of minerals found in soils, and their transformation from one form to another, is essential to understanding both the nature of the soil's chemical properties and the origin of its fertility. Since soils develop from parent material composed of rocks and minerals from the earth's crust, attention is directed first to the chemical and mineralogical composition of the earth's crust.

CHEMICAL AND MINERALOGICAL COMPOSITION OF THE EARTH'S CRUST

About 100 chemical elements are known to exist in the earth's crust. Considering the possible combinations of such a large number of elements, it is not surprising that some 2,000 minerals have been recognized. Relatively few elements and minerals, however, are of great importance in soils.

Chemical Composition of the Earth's Crust

Approximately 98 percent of the mass of the earth's crust is composed of eight chemical elements (see Figure 10.1). In fact, two elements,

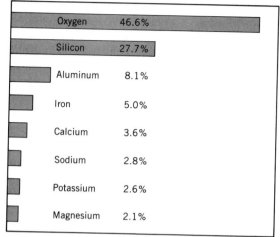

Oxygen	46.6%
Silicon	27.7%
Aluminum	8.1%
Iron	5.0%
Calcium	3.6%
Sodium	2.8%
Potassium	2.6%
Magnesium	2.1%

FIGURE 10.1 The eight elements in the earth's crust comprising over 1 percent by weight. The remainder of elements make up 1.5 percent. (Data from Clark, 1924.)

oxygen and silicon, compose 75 percent of it. Many of the elements important in the growth of plants and animals occur in very small quantities. Obviously, these elements and their compounds are not evenly distributed throughout the earth's surface. In some places, for example, phosphorus minerals (apatite) are so concentrated that they are mined; in other areas, there is a deficiency of phosphorus for plant growth.

Mineralogical Composition of Rocks

Most elements of the earth's crust have combined with one or more other elements to form the minerals. The minerals generally exist in mixtures to form rocks, such as the igneous rocks, granite and basalt. The mineralogical composition of igneous rocks, and the sedimentary rocks, shale and sandstone, are given in Table 10.1. Limestone is also an important sedimentary rock, which is composed largely of calcium and magnesium carbonates, with varying amounts of other minerals as impurities. The dominant minerals in these rocks are feldspars, amphiboles, pyroxenes, quartz, mica, apatite, clay, iron oxides (goethite), and carbonate minerals.

WEATHERING AND SOIL MINERALOGICAL COMPOSITION

The soil inherits from the parent material a mineral suite (a set of minerals) that typically contains both primary and secondary minerals.

TABLE 10.1 Average Mineralogical Composition of Igneous and Sedimentary Rocks

Mineral Constituent	Origin	Igneous Rock, %	Shale, %	Sandstone, %
Feldspars	Primary	59.5	30.0	11.5
Amphiboles and Pyroxenes	Primary	16.8	—	a
Quartz	Primary	12.0	22.3	66.8
Micas	Primary	3.8	—	a
Titanium minerals	Primary	1.5	—	a
Apatite	Primary or secondary	0.6	—	a
Clays	Secondary	—	25.0	6.6
Iron oxides	Secondary	—	5.6	1.8
Carbonates	Secondary	—	5.7	11.1
Other minerals	—	5.8	11.4	2.2

Data from Clarke, 1924.
a Present in small amounts.

Weathering in soils results in the destruction of existing minerals and the synthesis of new minerals. The minerals with the least resistance to weathering disappear first. The minerals most resistant to weathering such as quartz, make up a greater proportion of the remaining soil and appear to accumulate. As a consequence, the mineralogy of a soil changes over time, when viewed on a time scale of hundreds of thousands of years.

Weathering Processes

Numerous examples of weathering abound and can be observed every day. These include the rusting of metal and the wearing down and disintegration of brick walls. Mineral weathering is stimulated by acid conditions. Respiration of roots and microorganisms produces carbon dioxide that reacts with water to form carbonic acid (H_2CO_3). An acid environment stimulates the reaction of water with minerals and is one of the most important weathering reactions. For example, the reaction of a feldspar (albite) with water (hydrolysis) and H^+ is as follows:

$$2NaAlSi_3O_8 + 9H_2O + 2H^+ =$$
(albite)

$$H_4Al_2Si_2O_9 + 4H_4SiO_4 + 2Na^+$$
(kaolinite)

In the reaction a primary mineral, albite, was converted to *kaolinite*. The first-formed kaolinite particles are very small. They increase in size as additional ions attach or join the crystal, resulting in an gradual increase in particle size. The clay fraction of soils tends to be dominated by particles formed in this manner and particles like kaolinite are called *clay minerals*. Clay minerals tend to be resistant to further weathering. Therefore, as primary minerals, like albite, weather and disappear, clay minerals, like kaolinite, increase in abundance.

In the reaction, sodium was released as an ion, Na^+. In addition, some of the silicon in the albite was incorporated into kaolinite and some into H_4SiO_4 (silicic acid). The silicic acid tends to be leached from the soil, particularly in humid regions.

The kaolinite, even though very resistant to further weathering, may be decomposed and disappear from the soil. The reaction:

$$H_4Al_2Si_2O_9 + 5H_2O = 2Al(OH)_3 + 2H_4SiO_4$$
(kaolinite)   (gibbsite)

The decomposition of kaolinite resulted in the formation of gibbsite, a mineral more resistant to weathering than the kaolinite. Both kaolinite and gibbsite are clay minerals that have distinctly different properties. Thus, as weathering proceeds and the mineralogical composition of the soil changes, the chemical and physical properties also change.

The loss of silicic acid by leaching results in the progressive loss of silicon. There is a progressive increase in the accumulation of aluminum, because the aluminum tends to be incorporated into resistant secondary minerals that accumulate in the soil. For this reason, an indicator of weathering intensity is the silica:alumina ratio (SiO_2/Al_2O_3). As silicon is lost from the soil by leaching and aluminum accumulates, the silica:alumina ratio decreases. The SiO_2/Al_2O_3 ratio decreases from 3:1 for albite to 1:1 for kaolinite.

Other weathering reactions include oxidation, dissolution, hydration, reduction, and carbonation. An example of mineral weathering and the formation of clay minerals on the surface of a piece of basalt is shown in Figure 10.2.

Summary Statement

1. Mineral weathering is stimulated by an acid soil environment.
2. Plant-essential nutrient ions are released to the soil solution and clay minerals (secondary) are formed.
3. Primary minerals disappear and secondary minerals, especially the clay minerals, accumulate.
4. Much of the silicon released in weathering is removed from the soil by leaching, unless

FIGURE 10.2 Minerals in this 10-centimeter-long piece of basalt have weathered into clay. The thin light-colored clay layer has cracked as a result of drying.

there is limited precipitation and leaching is ineffective. The silica:alumina ratio decreases as soils become more weathered.

Weathering Rate and Crystal Structure

Particle size, through its effect on specific surface, is an important factor affecting the weathering rate of minerals. That is, a given mineral in the silt fraction weathers faster than if the mineral is in the sand fraction. Chemical bonding within the mineral crystal, however, is the major factor affecting weathering rate.

Oxygen, silicon, and aluminum are the three most abundant elements in the earth's crust, com-prising about 47 percent, 28 percent, and 8 per-cent by weight, respectively (see Figure 10.1). Many soil minerals are silicate or aluminosilicate minerals. Oxygen is abundant on a weight basis, and because of the large size of the oxygen atom, oxygen occupies more than 90 percent of the vol-ume of the earth's crust (and soil), as shown in Table 10.2.

Note the generally small atomic size of the abundant cations listed in Table 10.2 as com-pared with the oxygen. In the minerals, the nega-tive charge of oxygen is balanced by the positive charge of cations. It is helpful to think of soil minerals as being composed of Ping-Pong balls with much smaller balls (like marbles), in the interstices. The Ping-Pong balls represent large negatively charged oxygen atoms, and the small balls represent the positively charged cations in the interstices.

Silicon is a very abundant cation, which per-forms a role in the mineral world similar to the role carbon plays in the organic world. The silicon ion fits in an interstice formed by four oxygen ions. The covalent bonding (electron sharing) be-tween oxygen and silicon forms a tetrahedron as shown in Figure 10.3. The silicon-oxygen tetrahe-dron is the basic unit in silicate minerals. The valence of the silicon is $+4$ and that of the oxygen is -2. Therefore, each tetrahedron (SiO_4^{4-}) has a net charge of -4.

TABLE 10.2 Size, Percent Volume in Earth's Crust, Coordination Number and Valence of the Most Abundant Elements in Soil Minerals

Element	Atomic Radius, Nanometers	Volume, Percentage in Earth's Crust	Coordination Number with Oxygen	Valence
O	0.132	93.8	—	-2
Si	0.042	0.9	4	$+4$
Al	0.051	0.5	6 (and 4)	$+3$
Fe	0.083(Fe^{+3} 0.067)	0.4	6	$+2$
Mg	0.066	0.3	6	$+2$
Na	0.098	1.3	8	$+1$
Ca	0.099	1.0	8	$+2$
K	0.133	1.8	8 (and 12)	$+1$

The *coordination number* is the number of ions that can be packed around a central ion. Four oxygen ions can be packed around a silicon ion, resulting in a coordination number of 4 (as shown in Figure 10.3). Note that larger cations have a larger coordination number. Aluminum, iron, and magnesium ions are larger than silicon ions, and aluminum, iron, and magnesium typically have a coordination number of 6. Potassium has a coordination number of 8 and 12 (see Table 10.2).

Different silicate minerals are formed, depending on the way the silicon-oxygen tetrahedra are linked together and how the net charge of the tetrahedra is neutralized. In olivine, the silicon-oxygen (Si-O) tetrahedra are bonded together by magnesium and/or iron. The crystal is neutral and has the formula of $MgFeSiO_4$. The ratio of magnesium and iron, however, is highly variable, resulting in several olivine species. The model of olivine in Figure 10.4 shows that each magnesium ion, or iron ion, is surrounded by six oxygen ions; three oxygen ions form each of the faces of two adjacent Si-O tetrahedrons. Magnesium and iron ions are larger than silicon and more oxygen ions are required to form an interstice large enough for them to occupy. The coordination number of magnesium and iron is 6.

The oxygen-silicon bonds are much stronger than the magnesium and iron bonds with oxygen. Consequently, reaction of olivine with water results in the H^+ of the water replacing magnesium and iron ions from the crystal face, rather than the silicon. The weathering of olivine can be viewed as the separation of the Si-O tetrahedra with the release of the Mg^{2+} and Fe^{2+}, as illustrated in Figure 10.4. The Si-O tetrahedra and the magnesium and iron ions encounter other ions in the soil solution, where the ions can react and form new minerals. The magnesium and iron ions can be absorbed by roots. Olivine weathers rapidly and soils with a high content of olivine may have so much Mg^{2+} in solution as to be detrimental to some plants. Conversely, olivine is absent in many soils because of its low resistance to weathering.

The net charge of individual Si-O tetrahedra may be partly or entirely neutralized by the common sharing of oxygen between adjacent tetrahedra. If adjacent Si-O tetrahedra share one oxygen, single chains are formed as in pyroxene and augite (see Figure 10.5). As shown in Figure 10.5, as more sharing of oxygen by silicon occurs, minerals with double chains, sheets, and three-dimensional structures are formed. Mica, and kaolinite, have a sheet structure, and quartz has a three-dimensional structure. In the case of quartz (SiO_2) all of the charge is neutralized by the sharing of all of the oxygen with silicon. The result is a mineral with strong chemical bonding within the crystal and great weathering resistance. As more

FIGURE 10.3 Models showing the tetrahedral (four-sided) arrangement of silicon and oxygen ions in silicate minerals. The silicon-oxygen tetrahedron on the left has the apical oxygen set off to one side to show the position of the silicon ion.

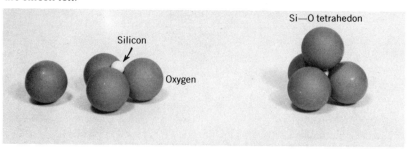

FIGURE 10.4 Models representing the weathering of olivine. Olivine is composed of silicon-oxygen tetrahedra held together by iron and/or magnesium. Every other tetrahedron is "inverted," as shown by the light-colored tetrahedra in the olivine model on the left. During weathering, the silicon-oxygen tetrahedra separate with the release of iron and magnesium.

oxygen sharing occurs, the oxygen-silicon ratio (ratio of oxygen ions to silicon ions) of the mineral decreases. This is associated with increased weathering resistance.

Mineralogical Composition Versus Soil Age

The difference in the weathering resistance of minerals has been used to develop a series of weathering stages that parallel soil age (amount of weathering and soil formation). The weathering stages are based on the representative minerals in the fine silt and clay fractions. Table 10.3 contains 13 weathering stages grouped into minimal, moderate, and intensive weathering stages.

Weathering stage 1 represents soils that have been subjected to little, if any, weathering and leaching, because the soils contain weatherable minerals such as gypsum, $(CaSO_4 \cdot 2H_2O)$, and halite $(NaCl)$. Each higher weathering stage is dominated by minerals with greater weathering resis-

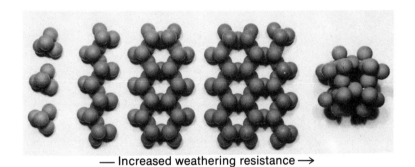

FIGURE 10.5 Models showing the common arrangements of silicon-oxygen tetrahedra in silicate minerals and their relation to weathering resistance.

— Increased weathering resistance →

Arrangement of Si-O tetrahedra and representative minerals

Individual	Single chain	Double chain	Sheet	3-dimensional
Olivine	Pyroxene augite	Amphibole hornblende	Biotite (mica)	Quartz
		Oxygen-silicon ratio		
4	3	2.7	2.5	2

TABLE 10.3 Representative Minerals and Soils Associated with Weathering Stages

Weathering Stage	Representative Minerals	Typical Soil Groups
	Minimal weathering stages	
1	Gypsum (also halite, sodium nitrate)	Soils dominated by these minerals in the fine silt and clay fractions are the minimally weathered soils all over the world, but are mainly soils of the desert regions where limited water keeps chemical weathering to a minimum.
2	Calcite (also dolomite, apatite)	
3	Olivine-hornblende (also pyroxenes)	
4	Biotite (also glauconite, nontronite)	
5	Albite (also anorthite, microcline, orthoclase)	
	Moderate weathering stages	
6	Quartz	Soils dominated by these minerals in the fine silt and clay fractions are mainly those of temperate regions developed under grass or trees. Includes the major soils of the wheat and corn belts of the world.
7	Muscovite	
8	2:1 layer silicates (including vermiculite)	
9	Smectite (montmorillonite)	
	Intensive weathering stages	
10	Kaolinite	Many intensely weathered soils of the warm and humid equatorial regions have clay fractions dominated by these minerals. The soils are frequently characterized by acidity and infertility.
11	Gibbsite	
12	Hematite (also goethite)	
13	Anatase (also rutile, zircon)	

Based on Jackson and Sherman, 1953.

tance, and in effect, more developed or older soils. Soils in which the fine silt and clay fractions are dominated by minerals of the first five weathering stages are considered to be minimally weathered. These soils have sand and silt fractions that are rich in primary minerals. The soils tend to occur where there has been a lack of time or water for weathering or where temperatures are too low for effective weathering. Minimally weathered soils are generally fertile.

Soils dominated by minerals represented by moderate weathering are in stages 6 thru 9. These soils tend to have a significant amount of both primary minerals (quartz and muscovite mica) and secondary minerals (vermiculite and smectite). Many temperate region soils are mod-

erately weathered and they are quite fertile soils. Most of the world's corn and wheat are produced on these soils.

Soils of weathering stages 10 through 13 are considered to be intensively weathered. Minerals representative of the intensive weathering stages are secondary minerals with great weathering resistance (see Table 10.3). These soils tend to occur in warm and humid tropical regions where there has been sufficient time to weather almost all of the primary minerals. Many intensively weathered soils have clay fractions dominated by kaolinite, gibbsite $(Al(OH)_3)$, and iron oxides, including hematite. The world's oldest and most weathered soils are high in content of oxides of aluminum (gibbsite), iron (hematite), and/or tita-

FIGURE 10.6 Nipe soil in Puerto Rico developed from parent material rich in iron minerals. The soil is 60 percent iron oxide and chunks of ironstone occur in the soil (by the hammer).

nium (anatase). Some of these soils are mined as ore. The soils are characterized by great infertility, especially in the A and B horizons. Even so, old and infertile soils of the humid tropics support most of the rain forests in the world, in part because of efficient nutrient cycling.

Theoretically, given sufficient time and an intense weathering and leaching environment, the soil becomes an accumulation of the most insoluble constituent that was formed during weathering. An intensively weathered soil from Puerto Rico that is about 60 percent iron oxide (hematite) is shown in Figure 10.6. Fates that await such old soils, as well as all other soils, include geologic uplift followed by stripping as a result of

erosion and the exposure of fresher underlying parent materials or the burial of soils by volcanic deposits or other sediments. In all cases, another cycle of soil genesis and weathering is initiated. Thus, in the overall geomorphoric cyle, soil genesis and weathering never end.

Summary Statement

1. Soils inherit from the parent material a mineral suite that typically includes both primary and secondary minerals.
2. Gradually, the primary minerals weather and disappear and secondary minerals are formed and accumulate. Over time, the mineralogical composition of the soil changes.
3. Minimally and moderately weathered soils are generally considered to be fertile.
4. The ultimate effect of weathering and leaching is the formation of intensively weathered soils dominated by highly resistant secondary minerals. Very few nutrient ions are released to the soil solution, so the soils are very infertile.

SOIL CLAY MINERALS

Weathering of primary minerals releases many elements for plant growth. The accompanying secondary minerals that form and accumulate in

FIGURE 10.7 On the left, aluminum in 6-coordination with bydroxyl (OH). The unit has eight sides; is an octahedron. On the right, adjacent octahedra share common hydroxyls and form a sheet; the A1-OH octahedral sheet.

FIGURE 10.8   Models illustrating a 2 : 1 structure consisting of two tetrahedral sheets and one octahedral sheet. On the left are two silicon-oxygen tetrahedral "sheets" facing each other. The center model shows the addition of aluminum, Al^{3+}, that is shared by the four surrounding or apical oxygens. The right model shows the addition of two hydroxyls (one in rear is invisible) to place the aluminum in 6-coordination in an octahedral arrangement. The aluminum shares one half valence bond from each of the six surrounding anions (four oxygen and two hydroxyl).

soils play an entirely different role. This section considers in greater detail the formation, nature, and importance of the soil's secondary or clay minerals.

Mica and Vermiculite

The basic mica structure can be visualized as the polymerization of sheets of silicon-oxygen tetrahedra and sheets of aluminum-hydroxyl (Al-OH) octahedra. A silicon-oxygen tetrahedral sheet representative of mica is shown in Figure 10.5.

The aluminum in Al-OH octahedra is in 6-coordination with hydroxyl and shares half a bond with each of the surrounding six anions. The common sharing of hydroxyl between adjacent octahedra results in the formation of sheets, that are electrically neutral and with the composition of $Al(OH)_3$, as shown in Figure 10.7.

Mica particles are composed of many layers. An individual mica *layer* is composed of two Si-O tetrahedral *sheets* with an Al-OH octahedral *sheet* sandwiched in-between. The three sheets are held together by the common sharing of oxygen to form a layer. This basic structure is shown by the models in Figure 10.8.

The left part of Figure 10.8 shows two pairs of Si-O tetrahedra with their apical oxygens facing inward and touching each other. The middle

model shows the insertion of an aluminum ion (Al^{3+}) that is surrounded by 4 oxygen (O^{2-}) from the Si-O tetrahedra. The aluminum shares half a bond with each of the surrounding 4 oxygen. This bonding holds the two pairs of tetrahedra together. The bonding from the bottom to the top of the layer is O-Si-O-Al-O-Si-O (see the middle model in Figure 10.8).

The remaining aluminum bond is neutralized by sharing a half bond with each of a pair of hydroxyls, one on the front side and the other on the back side of the model on the right. The aluminum is in the center of an octahedron, an 8-sided structure. The result is the formation of a 2:1 layer structure, consisting of two Si-O tetrahedral

FIGURE 10.9   Schematic drawing of a layer of a 2:1 mineral composed of two tetrahedral sheets and one octahedral sheet.

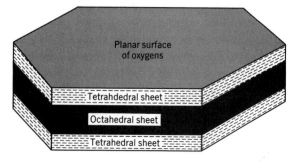

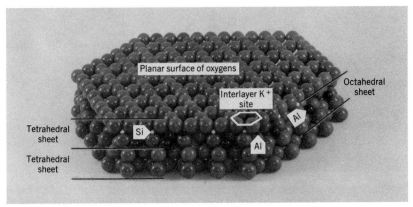

FIGURE 10.10 Model of a single layer of mica (muscovite) showing: (1) the 2:1 layer structure, (2) isomorphous substitution of Al^{3+} for Si^{4+} (can be seen in the upper tetrahedral layer below the interlayer K^+ site), and (3) the upper planar surface of oxygens forming hexagonal shaped cavities that are occupied by K^+ that neutralize the negative charge created by isomorphous substitution.

sheets with an Al-OH octahedral sheet sandwiched in the middle.

The unsatisfied charge of oxygen and hydroxyl at the edges causes this "tiny" crystal to incorporate additional silicon and aluminum and to grow outward in all directions to form a plate or layer, as illustrated in Figure 10.9. An electron micrograph of plate-shaped clay particles is shown in Figure 3.1.

During the crystallization of mica from magma, some Al^{3+} substitute for some of the Si^{4+} in the tetrahedrons (under conditions of high temperature and pressure). The substitution of one atom for another in the crystal lattice is termed *isomorphous substitution*. About one in every four silicons is substituted, and each substitution leaves the crystal with an excess negative charge because the silicon ion with a valence of 4^+ is replaced by aluminum ion with a valence of 3^+. The negative charge from isomorphous substitution is permanent and is neutralized by potassium ions (K^+) scattered over the surfaces of the clay layers, as illustrated by the model in Figure 10.10.

The potassium ions occur between adjacent clay layers as interlayer K^+. The K^+ are attracted to both adjacent layers and they hold the layers tightly together, making the mineral a *nonexpanding* mineral. The mica structure is illustrated by the model in Figure 10.11*a*. During weathering, the potassium ions along the particle edges are dislodged and released into the soil solution. The loss of potassium inward from the edges of mica particles is shown in Figure 10.11*b*. The loss of interlayer potassium from the mica structure: (1) causes an unsatisfied net negative charge to develop in the layers because of the loss of K^+, and (2) repulsion of adjacent layers, since they are both permanently and negatively charged. When the loss of potassium is complete, the mineral has attained maximum unsatisfied negative charge, and this negative charge accounts for the mineral's high cation exchange capacity (CEC). The loss of K^+ bonding in the interlayer space allows adjacent negatively charged layers to repel each other and move apart slightly or expand. This altered 2:1 mica mineral with maximimum negative charge and an expanding layer lattice is *vermiculite*.

The permanent negative charge of vermiculite is the source of most of its cation exchange capacity. The negatively charged sites in the inter-

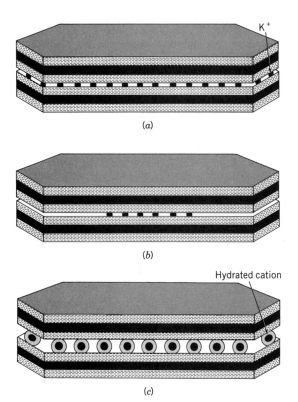

(a)

(b)

Hydrated cation

(c)

FIGURE 10.11 Schematic diagram showing changes during weathering of mica to hydrous mica and to vermiculite: 11a is unweathered mica, 11b partially weathered mica showing K lost along edges, and 11c is vermiculite. Hydrated cations move into and out of the interlayer space and neutralize the negative charge in vermiculite.

layer space (also negative charge on the external surfaces) attract hydrated cations from solution. These hydrated cations can move freely into and out of the interlayer space of vermiculute. A model of vermiculite showing hydrated cations in the interlayer space is shown in Figure 10.11c.

Of the two micas, muscovite and biotite, biotite containing octahedral Fe^{2+} and Mg^{2+} ions weathers much more readily than muscovite, containing octahedral Al ions (see Table 10.3). Both micas, however, weather into vermiculites with similar properties. In the alteration of mica to vermiculite, a 2:1 nonexpanding mineral with

minimal cation exchange capacity is converted into a 2:1 mineral with an expanding lattice and having high cation exchange capacity and surface area per gram, being 600 or more square meters per gram (see Table 10.4).

Smectites

Smectite is a group name for 2:1 layer silicates with high cation exchange capacity and highly variable interlayer spacing. The most common smectite in soils is *montmorillonite*, which is a 2:1 expanding clay with limited isomorphous substitution of magnesium for aluminum in the octahedral sheet. About one in every six aluminum, Al^{3+}, is replaced by magnesium, Mg^{2+}, resulting in high cation exchange capacity. Montmorillonite formation is favored by a weathering environment high in magnesium and silicon, which means an

FIGURE 10.12 Model of 2:1 clay mineral, smectite. The model shows: (1) the highly expanding lattice, and (2) adsorption of hydrated cations (cations surrounded by six water molecules) in the interlayer space and on the upper planar surface. The interlayer space contains water and cations, and the interlayer thickness may be greater than the thickness of the layers. Soils with high smectite content have great capacity to shrink and swell with drying and wetting.

Hydrated exchangeable-cation

TABLE 10.4 Summary of Properties of Some Silicate Clays

Mineral	Type	Interlayer Condition	Cation Exchange Capacity, cmol/kg[a]	Surface Area, meters^2/ gram
Kaolinite	1:1 nonexpanding	strong H bonding	3–15	10–20
Vermiculite	2:1 expanding	complete loss of potassium, moderate bonding and expansion	100–150	600–800
Montmorillonite	2:1 expanding	very weak bonding, great expansion	80–120	600–800

[a] Determined at pH 7.0.

environment of somewhat limited leaching. Crystallization of volcanic ash products in seawater often produces montmorillonite. The expansion of layers permits easy diffusion of hydrated cations into the interlayer space and adsorption on the planar surfaces. The dominant features of montmorillonite are shown in Figure 10.12 and some properties are given in Table 10.4.

The individual clay layers in soils tend to occur as clusters that are much like a stack of playing cards. The more expansive the clay, the more likely layers will separate, causing the clay to exist as very small-sized particles. As a consequence, smectites occur as very small particles, which give this type of clay great capacity for swelling and shrinking with wetting and drying. The small

spaces between smectite particles contribute to hydraulic impermeability. Houston Black Clay soils in Texas have a high smectite content and are characterized by contraction and development of wide cracks during the dry season and by great expansion and closure of the cracks during the wet season.

Kaolinite

Reference to the formation of kaolinite was made in regard to the weathering of albite feldspar. In that case, kaolinite formation was by crystallization from solution. Another mechanism for kaolinite formation is the partial disintegration of 2:1 clays. The stripping away of one Si–O sheet, from

FIGURE 10.13 Model of kaolinite showing two offset layers. The upper layer surfaces are composed of hydroxyls (light-colored). These hydroxyl are bonded by the hydrogen of the hydroxyl to the oxygen of adjacent layers, producing a nonexpanding clay. Cation exchange capacity is low and originates at the edges due to deprotonation of exposed hydroxyls. A hydrated cation is adsorbed to such a cation exchange site.

a 2:1 layer, converts the 2:1 layer into a 1:1 layer, typical of kaolinite. This mode of formation is favored by an acid weathering environment that is found in older soils or intensively weathered soils where silicon is being removed by leaching.

The 1:1 structure has layers that have a plane of oxygens on one side and a plane of hydroxyls on the other. Stacking of the layers results in a strong attraction between oxygen of one layer and the hydrogen of hydroxyl (hydrogen bonding) of an adjacent layer, resulting in a nonexpanding clay. Kaolinite particles tend to be large in size (many are silt sized), and to have great resistance to further weathering. The tetrahedral and octahedral sheets may retain some isomorphous substitution that can contribute to cation exchange capacity. This appears to be very minor and the cation exchange capacity is low, relative to the other clay minerals discussed thus far (see Table 10.4). The cation exchange capacity occurs mainly because of deprotonation of exposed hydroxyl at the edges. A model showing some properties of kaolinite is given in Figure 10.13.

The large size of kaolinite particles often gives soils high in kaolinite clay content good physical properties. Many of the kaolinite particles occur in the coarse clay (0.0002 to 0.002 mm diameter) and silt fractions. Smectite particles typically occur in the fine clay (less than 0.0002 mm diameter). Two Bt horizons may contain the same amount of clay but, if the clay is kaolinite, the horizon could be quite water permeable compared with a Bt horizon containing smectite. Smectite is a common clay of U.S. midwestern and Great Plains soils, in United States, and a significant number of the soils have very slowly or impermeable Bt horizons. By contrast, kaolinite is common in soils on the coastal plain of southeastern United States where soils are much older. Even so, impermeable claypan Bt horizons are not common on the southeastern coastal plain.

Allophane and Halloysite

Noncrystalline (amorphous) minerals exist where glassy ash and cinders are thrown into the air by volcanoes and are deposited on land masses before the atoms become organized into crystals. In the humid tropics, amorphous minerals weather rapidly into an amorphous clay mineral called *allophane*. Soils high in allophane have a gel-like consistency when wet and lack the grittiness associated with sand particles. There is an enormous surface area and great reaction with, or adsorption of, organic matter, giving the organic matter protection against decomposition. Characteristic features of soils high in allophane include a high content of organic matter, high water-holding capacity, high and variable cation exchange capacity, and low load-bearing strength.

Slowly, allophane becomes more organized (more crystalline) and becomes *halloysite*—a mineral with structure and composition similar to kaolinite. A weathering sequence from volcanic ash weathering is allophane-halloysite-kaolinite. Soils high in allophane are important in the mountain chain that extends from southern South America to Alaska and, also, in Hawaii.

Oxidic Clays

Kaolinite formation from 2:1 minerals is indicative of a leaching regime that is very effective in removing silicon. Under such conditions, kaolinite may slowly weather, resulting in the decomposition of the layer sheets into their component parts. The Si-O tetrahedra are transformed into Si-OH tetrahedra, H_4SiO_4, (silicic acid), that is soluble and removed by leaching. Aluminum-OH octahedra are released and slowly polymerize to form an aluminum hydrous oxide ($Al(OH)_3$), called *gibbsite*. Gibbsite particles tend to be platy and hexagonal in shape like the sheet shown in Figure 10.7. Many layers stack on top of each other and exist as clay particles in soils. Gibbsite is very insoluble and resistant to further weathering and is representative of weathering stage 11 (see Table 10.3). The mineral is common in intensively weathered soils that also contain abundant iron oxides, such as hematite (Fe_2O_3) and goethite (FeOOH). Only rare and unusual soils have a significant amount of titanium oxide (anatase), which is representative of weathering stage 13. Few parent materials contain enough titanium for a significant amount of anatase to accumulate via weathering and soil formation.

The oxide clays are abundant in many red-colored soils in the humid tropics. These soils are typically acid and, in an acidic environment, some of the exposed hydroxyls on particle surfaces of oxidic clays adsorb a H^+ (protonate) and form sites of positive charge ($Al\text{-}OH + H^+ = Al\text{-}OH_2^+$). The kaolinite clay has a net negative charge, which promotes attraction of the kaolinite and oxidic clays and formation of extremely stable aggregates or peds. The peds, which are dense and frequently the size of sand grains, are very resistant to deformation and give the soil physical characteristics similar to soils dominated by quartz sand; that is, soils with an abundance of oxidic clays have high water-infiltration rates and resist erosion. The soil can be plowed shortly after a rain. A small amount of available water is retained at field capacity. The water is held at low tension or high potential, and the water is easily and rapidly absorbed by roots. Short periods of water stress for crops may occur even though the soils are in the tropical rain forest zone.

Summary Statement

1. There is a kinship amoung the crystalline aluminum silicate clay minerals because all consist of Si-O tetrahedra sheets and Al-OH octahedra sheets.
2. The major differences in the aluminum silicate clays result from the ratio of tetrahedra and octahedra sheets and the nature and amount of isomorphous substitution. A summary of the major features of these clays is given in Table 10.4.
3. The 2:1 clays are transformed into 1:1 clay by the loss of a Si-O tetrahedral sheet. This is one formation mode of kaolinite.
4. Near the end of the weathering sequence, gibbsite formation becomes important. In the oldest and most weathered soils, the mineralogy of the clay fraction is dominated by oxidic clays, together with kaolinite.
5. Volcanic activity produces amorphous ash and cinders that weather to form amorphous allophane. Allophane slowly crystallizes to form kaolinite.
6. Over time, weathering produces changes in the mineralogical composition of soils, which is reflected in the soil's physical and chemical properties. The cation exchange capacity, for example, increases to a peak with weathering in soils dominated by 2:1 clays (moderately weathered), which is followed by a very low cation exchange capacity in soils dominated by oxidic clays (intensively weathered).

ION EXCHANGE SYSTEMS OF SOIL CLAYS

Some soils are dominated by layer silicate clays, some by oxidic clays, and some soils have both kinds of clay in abundance. This results in three unique ion-exchange systems in mineral soils: (1) layer silicate, (2) oxide, and (3) oxide-coated, layer silicate. The type and kind of ion exchange have important consequences for plant growth and soil management.

Layer Silicate System

Layer silicate system soils are overwhelmingly negatively charged with high cation exchange capacity (CEC). The cation exchange capacity originates mainly from isomorphous substitution. These negatively charged sites constitute permanent charge that is not affected by pH. The charge may be modified, however, to a small extent by minor or thin coatings of oxidic clays on the particle surfaces. Some cation exchange capacity originates from the dissociation of hydrogen (deprotonation) from edge Al-OH groups that is favored by high pH (high OH concentration):

$$[\text{clay edge}]Al\text{-}OH + OH^-$$
$$= [\text{clay edge}]Al\text{-}O^- + H_2O$$

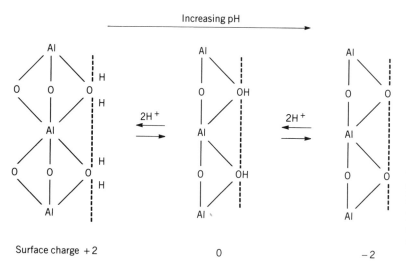

FIGURE 10.14 Protonation under acid conditions produces positive charge and deprotonation at higher pH produces negative charge where aluminum exists at the edges of clay particles. (After White, 1987.)

As soil pH increases, greater hydroxyl concentration drives the preceding reaction to the right. Thus, the amount of negative charge or cation exchange capacity increases as a result of increasing pH. This kind of cation exchange capacity is called *pH dependent* or *variable*. For soils dominated by 2:1 clays, the variable charge is a minor part of the total cation exchange capacity owing to the large amount of negative charge resulting from isomorphous substitution.

Oxidic System

Soils with an oxide system have properties dominated by oxide clays. Any kaolinite particles that are present are coated with thick and continuous layers of oxides. The oxide clay coatings give kaolinite particles the effective properties of oxidic clay. Oxidic clay particles have hydroxyls exposed on their surfaces and almost no isomorphous substitution. The charge may be both negative or positive, the amount depending on pH. It has been shown that increasing hydroxyl concentration, or pH, produces negative charge by deprotonatation of edge hydroxyls. Under quite acid conditions with abundant H^+, the edge Al-OH react with H^+ (protonates) to produce positive charge:

$$[\text{clay edge}]\text{Al-OH} + H^+ = [\text{clay edge}]\text{Al-OH}^{2+}$$

Positive charge in soil adsorbs anions, thus giving the soil an *anion exchange capacity* (AEC).

There is a pH where the negative and positive charge are equal; the net charge of the particles is zero. This is the *zero point of charge*. The development of negative charge by deprotonation and the development of positive charge by protonation are shown in Figure 10.14.

Oxidic clays (gibbsite and goethite) have no permanent CEC, have some pH dependent CEC, and have significant AEC in acid soils, where they are abundant. In theory, oxidic clay systems may be net negatively or positively charged. The organic matter in soils contributes only negative charge so that only rarely will an entire soil horizon have a net positive charge. This happens in some subsoil horizons of intensively weathered tropic soils that have a small content of organic matter.

Oxide-Coated Layer Silicate System

In the oxide-coated layer silicate system, silicate clay particles are only partially coated with a thin layer of oxidic clay. Some of the silicate clay may be 2:1 types. Soils with a partially coated oxide-silicate system have properties that are intermediate between the other two systems. Many of the red-colored tropical soils have an ion exchange system that is greatly affected by both silicate and oxidic clay.

SUMMARY

In an acid soil environment, weathering of minerals is stimulated. Minerals weather at different rates, depending on their resistance to weathering.

In general, primary minerals weather and secondary minerals are formed. Over time, the mineralogical composition of the soil changes and, consequently, soil properties change.

Weathering releases ions to the soil solution. Many of these ions are plant nutrients. Some ions precipitate to form secondary minerals and some ions are leached from the soil.

Minimally and moderately weathered soils tend to be rich in primary minerals and to be fertile with high cation exchange capacity. Intensively weathered soils are depleted of primary minerals. Few nutrient ions are released and soils are infertile. The clay minerals in intensively weathered soils have low cation exchange capacity.

Nutrient ions adsorbed to the surfaces of clay particles represent a reservoir of nutrients that is exchangeable and available to plants.

Ultimately, soils become "weathered out" and have almost no capacity to support plants. Such soils may be mined for ore. Clay edge Al-OH protonate in acid soils and form sites of anion exchange capacity and deprotonate in alkaline soils and form sites of cation exchange capacity.

REFERENCES

Birkeland, P. W. 1974. *Pedology, Weathering, and Geomorphological Research*. Oxford, New York.

Bohn. H. L., B. L. McNeal, and G. A. O'Connor. 1985. *Soil Chemistry*, 2nd ed. John Wiley, New York.

Clarke, F. W. 1924. "Data of Geochemistry." *U.S. Geo. Survey Bull*. 770. Washington, D.C.

Foth, H. D. and B. G. Ellis. 1988. *Soil Fertility*. John Wiley, New York.

Jackson, M. L. and G. D. Sherman. 1953. "Chemical Weathering of Minerals in Soils." *Advances in Agronomy*. **5:**219–318.

Jenny, H. 1980. *The Soil Resource*. Springer-Verlag, New York.

Soil Sci. Soc. Am. 1977. *Minerals in the Soil Environments*. Soil Sci. Soc. Am. Madison, Wis.

Uehara, G. and G. Gillman. 1981. *Mineralogy, Chemistry, and Physics of Tropical Soils with Variable Charge*. Westview, Boulder.

White, R. E. 1987. *Introduction to the Principles and Practice of Soil Science*. 2nd ed. Blackwell Sci. Pub. London.

CHAPTER 11

SOIL CHEMISTRY

A total chemical analysis of a soil may reveal little that is useful in predicting the ability of the soil to supply plant nutrients. Instead, it is usually a relatively small part of the total amount of an element that is available for plant growth. For these reasons, the discussion of soil chemistry focuses on ion exchange reactions, solubilities, and mineral and biochemical transformations. These reactions are illustrated in Figure 11.1.

CHEMICAL COMPOSITION OF SOILS

The average chemical composition of igneous rocks is given in Table 11.1. On an oxide basis, silicon and aluminum are first and second in abundance. This reflects the dominance of silicate and aluminosilicate minerals in igneous rocks. Iron is next in abundance in igneous rocks. Next most abundant, are four cations that are important in balancing the charge of silicon-oxygen tetrahedra in minerals—calcium, magnesium, sodium, and potassium. The remaining 1 or 2 percent includes the other elements. The aver-

age chemical composition of igneous rocks is similar to the mineralogical composition of many minimally and moderately weathered soils. Such soils contain much quartz, feldspar, and mica in the sand and silt fractions, 2:1 layer silicate clays in the clay fraction, and much of the negative charge of the clays neutralized by the adsorption of calcium, magnesium, sodium, and potassium ions.

When soils weather and the mineralogical composition changes over time, there is a corresponding change in chemical composition. During soil formation, there is a preferential loss of silicon relative to aluminum and iron. This is reflected in the chemical analysis of the intensively weathered Columbiana clay soil in Costa Rica (see Table 11.1). The soil contains only 26 percent silica, and the sum of iron and alumumium is 69 percent (on an oxide basis). The silica:alumina ratio decreased from 3.75 to 0.53 as a result of weathering and soil formation. The chemical composition of the Columbiana soil reflects a high oxidic clay content. Release and loss by leaching of calcium, magnesium, sodium, and potassium were more rapid than silica, and this is

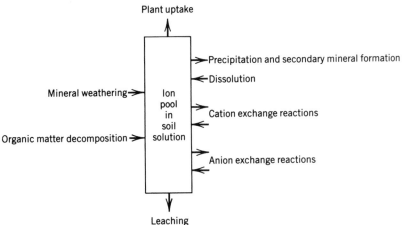

FIGURE 11.1 Major chemical processes and reactions in soils. Weathering of minerals and decomposition of organic matter supply ions to the ion pool of the soil solution. Ions are lost from the pool by plant uptake and leaching. Some ions precipitate and some ions form secondary minerals, especially clay minerals, and are subject to dissolution. Ions also sorb onto and desorb from cation and anion exchange sites.

reflected in the low content of these four cations in the intensively weathered soil relative to igneous rocks.

ION EXCHANGE

Ion exchange is of great importance in soils (see Figure 11.1). Ion exchange involves cations and anions that are adsorbed from solution onto nega-

TABLE 11.1 Chemical Composition of Average Igneous Rocks and an Intensively Weathered Soil

	Average of Igneous Rocks	Columbiana Clay (Costa Rica)
SiO_2	60	26
Al_2O_3	16	49
Fe_2O_3	7	20
TiO_2	1	3
MnO	0.1	0.4
CaO	5	0.3
MgO	4	0.7
K_2O	3	0.1
Na_2O	4	0.3
P_2O_5	0.3	0.4
SO_3	0.1	0.3
Total	100.5%	100.4%

Adapted from Bohn, McNeal, and O'Connor, 1985.

tively and postively charged surfaces, respectively. Such ions are readily replaced or exchanged by other ions in the soil solution of similar charge, and thus, are described by the term, *ion exchange*. Of the two, cation exchange is of greater abundance in soils than anion exchange.

Nature of Cation Exchange

Cation exchange is *the interchange between a cation in solution and another cation on the surface of any negatively charged material, such as clay colloid or organic colloid.* The negative charge or cation exchange capacity of most soils is dominated by the secondary clay minerals and organic matter (especially the stable or humus portion of SOM). Therefore, cation exchange reactions in soils occur mainly near the surface of clay and humus particles, called *micelles*. Each micelle may have thousands of negative charges that are neutralized by the adsorbed or exchangeable cations. Assume for purposes of illustration that X^- represents a negatively charged exchanger that has adsorbed a sodium ion (Na^+), producing NaX. When placed in a solution containing KCl, the following cation exchange reaction occurs:

$$K^+ + Cl^- + NaX = Na^+ + Cl^- + KX$$

In the reaction, K^+ in solution replaced or exchanged for adsorbed (exchangeable) Na^+, resulting in putting the adsorbed Na^+ in solution and leaving K^+ adsorbed as KX. Cations are adsorbed and exchanged on a chemically equivalent basis, that is, one K^+ replaces one Na^+, and two K^+ are required to replace or exchange for one Ca^{2+}.

The negatively charged micellar surfaces form a boundary along which the negative charge is localized. The cations concentrate near this boundary and neutralize the negative charge of the micelle. The exchangeable cations are hydrated and drag along the hydration water molecules as they constantly move and oscillate around negatively charged sites. The concentration of cations is greatest near the micellar surfaces, where the negative charge is the strongest. The charge strength decreases rapidly with increasing distance away from the micelle, and this is associated with a reduction in cation concentration away from the micelle. Conversely, the negatively charged micellar surface repels anions. This results in a decreasing concentration of cations and an increasing concentration of anions with distance away from the micellar surface. At some distance from the micellar surface, the concentration of cations and anions is equal. These features are illustrated in Figure 11.2.

An equilibrium tends to be established between the number of cations adsorbed and the number of cations in solution. The number of cations in solution is much smaller than the number adsorbed (generally 1 percent or less) unless the content of soluble salts is high. Roots absorb cations from the soil solution and upset the equilibrium. The uptake of a cation is accompanied by the excretion of H^+ from the root and this restores the charge equilibrium in both the plant and soil. Excretion of hydrogen ions tends to increase the acidity in the rhizosphere and the H^+ may be adsorbed on micelles. The H^+ is weakly adsorbed onto clay (ionic bonding); however, it is strongly adsorbed to carboxyl groups of humus (covalent bonding).

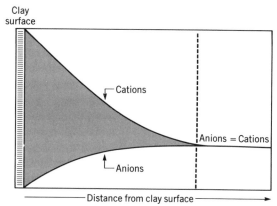

FIGURE 11.2 Distribution of cations and anions in the vicinity of a negatively charged surface.

Cation Exchange Capacity of Soils

The cation exchange capacity of soils (CEC) is defined as *the sum of positive (+) charges of the adsorbed cations that a soil can adsorb at a specific pH.* Each adsorbed K^+ contributes one + charge, and each adsorbed Ca^{2+} contributes two + charges to the CEC. The CEC is the sum of the + charges of all of the adsorbed cations. (Conversely, the CEC is equivalent to the sum of the − charges of the cation exchange sites.)

The CEC is commonly expressed as centimoles of positive charge per kilogram [cmol(+)/kg], also written as cmol/kg, of oven dry soil. A mole of positive charge (+) is equal to 6.02×10^{23} charges. For each cmol of CEC per kg there are 6.02×10^{21} + charges on the adsorbed cations.

For many years the CEC of soils was expressed as milliequivalents of cations adsorbed per 100 grams of oven dry soil (meq/100 g). Many agricultural publications still use meq/100 g, which means that many persons need to be familiar with both systems. The CEC values are the same in both systems; that is, a CEC of 10 cmol/kg is equal to a CEC of 10 meq/100 g

The CEC of a soil is equal to the CEC of both the mineral and organic fractions. The major source of CEC in the mineral fraction comes from the clay. The negative charge of organic matter is due mainly to dissociation of H^+ (deprotonation) from the $-OH$ of carboxyl and phenolic groups. The CEC of soil organic matter (SOM) ranges from 100 to 400 cmol/kg, depending on degree of decomposition. There is an increase in the groups contributing to the CEC as organic matter becomes more decomposed or more humified. As mentioned in the previous chapter, the clays have great variation in CEC, ranging from less than 10 cmol/kg for oxidic clay, to a 100 cmol/kg or more for some 2:1 clay. The sand and silt fractions contribute little to the CEC of soils. As a consequence, the CEC of soils is affected mainly by the amount and type of clay and the amount and degree of decomposition of the organic matter.

In the A horizons of mineral soils, the organic matter and clay frequently make similar contributions to the CEC. In subsoils, such as Bt horizons where clay has accumulated, the clay commonly contributes more to the CEC than the organic matter. The accumulation of humus in Bhs or Bh horizons contributes much CEC to these subsoil horizons.

Cation Exchange Capacity versus Soil pH

The definition of CEC specifies that the CEC applies to a specific pH because the CEC is pH dependent. To make valid comparisons of CEC between soils and various materials, it is necessary to make the determination of CEC at a common pH. The CEC is positively correlated with pH; therefore, acid soils have a CEC less than the maximum potential CEC.

Most of the CEC of 2:1 clays is permanent or non-pH dependent. By contrast, the CEC of SOM is entirely pH-dependent. Soils high in organic matter content, kaolinite (1:1 clay), and oxidic clays have relatively large changes in CEC with changes in pH owing to the predominance of pH-dependent charge. The changes in the CEC with changes in pH of the horizons in a very sandy soil that has a Bh horizon is shown in Figure 11.3.

Changes in CEC with changes in soil pH are

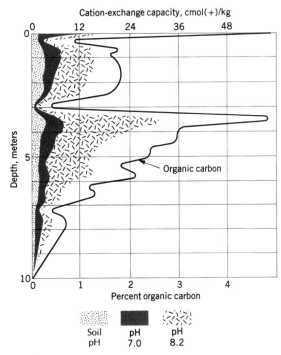

FIGURE 11.3 The cation exchange capacities in a sand-textured soil at the soils natural pH and determined at pH 7.0 and 8.2. The increases in cation exchange capacity with increasing pH are closely related to the content of organic carbon (organic matter) in this soil which has a Bh horizon and contains very little clay. (Adapted from Holzhey, Daniels, and Gamble, 1975).

important in the management of intensively weathered and acid soils in tropical regions because of the generally low CEC and the highly pH-dependent nature of the CEC. For minimally and moderately weathered mineral soils, where much of the CEC originates from the permanent charge of 2:1 clays, changes in CEC resulting from changes in soil pH have much less importance. For plants, the current soil pH is the relevant pH. Values given for the CEC are for soils at their natural, or current pH, unless otherwise noted. The CEC at the soils' current pH is called the effective cation exchange capacity (ECEC).

Kinds and Amounts of Exchangeable Cations

The major sources of cations are mineral weathering, mineralization of organic matter, and soil amendments, particularly lime and fertilizers. The amounts and kinds of cations adsorbed are the result of the interaction of the concentration of cations in solution and the energy of adsorption of the cations for the exchange surface. The cations compete for adsorption. As the concentration of a cation in the soil solution increases, there is an increased chance for adsorption. The more strongly a cation is attracted to the exchange surface (energy of adsorption); the greater is the chance for adsorption.

The energy of adsorption of specific ions is related to valence and degree of hydration. The energy of adsorption of divalent cations is about twice that of monovalent cations. For cations of equal valence, the cation with the smallest hydrated radius is most strongly adsorbed because it can move closer to the site of charge. The dehydrated and hydrated radi of four monovalent cations are given in Table 11.2. Rubidium (Rb) is the most strongly adsorbed, whereas lithium (Li) is the most weakly adsorbed. Note that the hydrated radius is inversely related to the nonhydrated radius. This is because the water molecules can move closer to the charge that is located at the center of a small ion and, therefore, the water molecules are more strongly attracted to smaller ions. The more strongly that water mole-

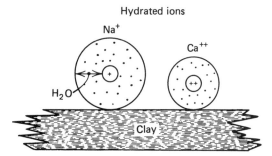

FIGURE 11.4 Calcium ions are more strongly adsorbed by clay than are sodium ions, because calcium is divalent and has a smaller hydrated radius.

cules are attracted to ions, the greater is the number of hydrated molecules and the hydrated radius of the ions.

The differences in the hydrated radius and valence of Ca^{2+} and Na^+ are illustrated in Figure 11.4. Calcium is adsorbed more strongly than sodium because: (1) it is divalent, and (2) it has the smallest hydrated radius. As a result of the small energy of adsorption of sodium, it is more likely to exist in the soil solution and be removed from the soil by leaching. As a result of the strong energy of adsorption, calcium is typically more abundant as an exchangeable cation than are magnesium, potassium, or sodium. The energy of adsorption sequence is: $Ca > Mg > K > Na$. The cmol/kg (meq/100 g) of these four cations for a moderately weathered soil is given in Table 11.3. Note that their abundance follows the energy of adsorption sequence and the dominance of calcium.

The symbol, X, has already been used in a cation exchange reaction, as in NaX. In NaX, a Na^+ is adsorbed onto a cation exchange site. By transposing the X to form XNa, exchangeable sodium is formed. *XNa* is an abbreviation for *exchangeable Na*. Pronounce XNa, as "X Na," or as exchangeable Na, and write it as *XNa*. Exchangeable calcium is XCa. This nomenclature will be used in reference to the exchangeable cations.

TABLE 11.2 Ionic Radii and Exchange Efficiency of Several Monovalent Ions

| Ion | Radii of Ions, nanometers | | Order of Energy of Adsorption |
	Dehydrated	Hydrated	
Li	0.078	1.003	4th
Na	0.098	0.790	3rd
K	0.133	0.532	2nd
Rb	0.149	0.509	1st

TABLE 11.3 Exchangeable Calcium, Magnesium, Potassium and Sodium in Tama Silt Loam

Depth, cm	Horizon	Exchangeable Cations, cmol (+) /kg			
		Ca	Mg	K	Na
0–5	Ap	14.0	3.4	0.5	0.1
15–28	A	13.8	4.2	0.4	0.1
28–50	AB	14.5	6.1	0.4	0.1
50–89	Bt	14.7	6.7	0.3	0.1
89–130	BC	14.8	5.7	0.3	0.1
130–155	C	16.1	5.6	0.4	0.2

Adapted from *Soil Survey Investigations Report,* No. 3, Iowa, SCS, USDA, 1966.

Exchangeable Cations as a Source of Plant Nutrients

Of the four cations referred to in Table 11.3, three are essential for plants. The sodium may be absorbed by plants, may substitute for some potassium, but sodium is not considered an essential element. It is, however, needed by animals, and its uptake by plants and its presence in food is important. It is paradoxical that, in regard to XCa and XK, the XCa is much more abundant in soils than is XK. On the other hand, plants generally absorb much more potassium than calcium. The result is that calcium is rarely deficient in soils for plant growth; by contrast, potassium frequently limits crop yields in the humid regions and is a common element in fertilizers. This is reflected in the data of Table 11.4, which are based on the data in Table 11.3. A plow layer would contain enough XCa to last a high-yielding corn crop 112 years, the XMg would last for 27 years, whereas the XK would provide only a 2-year supply.

When the exchangeable cations in the underlying horizons are considered, it can be seen that the exchangeable cations are an important supply of available nutrients. Small, but important, quantities of other cations, including iron, manganese, copper, and zinc also exist in the soil as exchangeable cations.

Anion Exchange

Anion exchange sites arise from the protonation of hydroxyls on the edges of silicate clays and on the surfaces of oxidic clays. The anion exchange capacity is inversely related to soil pH and is, perhaps, of greatest importance in acid soils dominated by oxidic clays. Nitrate (NO_3^-) is very weakly adsorbed in soils and, as a consequence, remains in the soil solution where it is very susceptible to leaching and removal from soils. Anion exchange reactions with phosphate may result in such strong adsorption as to render the phosphate unavailable to plants. Uptake of anions by roots is accompanied by the excretion of OH^- or HCO_3^-.

In general, plants absorb as many anions as cations. The availability to plants of the anions nitrate, phosphate, and sulfate is related to mineralization from organic matter, as well as anion exchange.

TABLE 11.4 Amounts of Exchangeable Calcium, Magnesium and Potassium in 2 Million Pounds in the Ap horizon of Tama Silt Loam and Adequacy for Plant Growth

Exchangeable cation	cmol (+) /kg	Pounds, pp2m	Plant Supply, years
Calcium	14.0	5600	112
Magnesium	3.4	816	27
Potassium	0.5	390	2

SOIL pH

The pH is *the negative logarithm of the hydrogen ion activity*. The pH of a soil is one of the most important properties involved in plant growth. There are many soil pH relationships, including those of ion exchange capacity and nutrient availability. For example, iron compounds decrease in solubility with increasing pH, resulting in many instances where a high soil pH (soil alkalinity) causes iron deficiency for plant growth.

Determination of Soil pH

Soil pH is commonly determined by: (1) mixing one part of soil with two parts of distilled water or neutral salt solution, (2) occasional mixing over a period of 30 minutes to allow soil and water to approach an equilibrium condition, and, (3) measuring the pH of the soil-water suspension using a pH meter. A rapid method using pH indicator dye is illustrated in Figure 11.5. The common range of soil pH is 4 to 10.

Sources of Alkalinity

Parent materials have a wide ranging mineralogical composition and pH, and young soils in-

herit these properties from the parent material. Throughout the world there are parent materials and soils that contain several percent or more of calcium carbonate; they are *calcareous*. When calcareous soil is treated with dilute HCl, carbon dioxide gas is produced, and the soil is said to *effervesce*.

Carbonate Hydrolysis The hydrolysis of calcium carbonate produces OH^-, which contributes to alkalinity in soils:

$$CaCO_3 + H_2O = Ca^{2+} + HCO_3^- + OH^-$$

Calcium carbonate is only slightly soluble, and this reaction can produce a soil pH as high as 8.3, assuming equilibrium with atmospheric carbon dioxide. In calcareous soil, carbonate hydrolysis controls soil pH. When a soil contains Na_2CO_3, the pH may be as high as 10 or more, which is caused by the greater solubility of Na_2CO_3 and greater production of OH^- by hydrolysis in a similar manner. Sodium-affected soils with 15 percent or more of the CEC saturated with Na^+ are called sodic and are also highly alkaline.

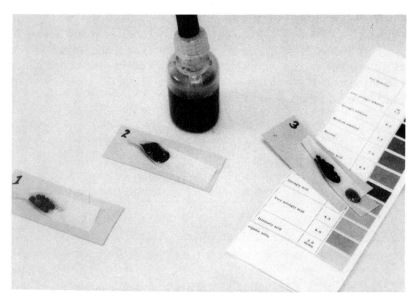

FIGURE 11.5 To determine soil pH with color indicator dye solution: (1) Place a small amount of soil in the folded crease of a strip of wax paper; (2) add 1 or 2 drops more indicator than the soil can absorb, tilt up and down to allow indicator to equilibrate with soil, and (3) separate excess dye from soil with tip of a pencil and match dye color with the color chart.

Mineral Weathering Some rocks and minerals weather to produce an acidic effect. The weathering of many primary minerals, however, contributes to alkalinity. This is the result of the consumption of H^+ and the production of OH^-. For example, the hydrolysis of anorthite, (calcium feldspar), produces a moderately strong base:

$$3CaAl_2Si_2O_8 + 6H_2O =$$
(anthorite)

$$2HAl_4Si_6O_{10}(OH)_2 + 3Ca(OH)_2$$
(aluminosilicate)

In the reaction, an aluminosilicate clay mineral was formed together with the moderately strong base, calcium hydroxide. The net effect is basic. The generalized weathering reaction, in which M represents metal ions such as calcium, magnesium, potassium, or sodium, is

M-silicate mineral + H_2O =
H-silicate mineral + M^+ + OH^-

As long as a soil system remains calcareous, carbonate hydrolysis dominates the system and maintains a pH that ranges from 7.5 to 8.3 or more. Mineral weathering is minor under these conditions, and the mineralogy and pH of the soils change little, if at all, with time. The development of soil acidity requires the removal of the carbonates by leaching. When acidity develops, the weathering of primary minerals is greatly increased, causing an increase in the release of cations—calcium, magnesium, potassium and sodium. Leaching of these cations in humid regions, however, eventually results in the development of soil acidity.

Calcium and magnesium are alkaline earth metals or cations, and potassium and sodium are alkali metals. These cations are called *basic cations* because soils tend to be alkaline (basic) or neutral when the CEC is entirely saturated with them. Leaching of the basic cations (Ca, Mg, K, and Na) results in their replacement by Al^{3+} and H^+, and soils then become acid. The cations Al^{3+} and H^+ are called the *acidic cations*.

Sources of Acidity

Three important processes, or reactions, contribute to a more or less continuous addition of H^+ to soils, and these processes tend to make soils acid. In all soils respiration by roots and other soil organisms produces carbon dioxide that reacts with water to form carbonic acid (H_2CO_3). This is a weak acid, which contributes H^+ to the soil solution. Second, acidity is produced when organic matter is mineralized, because organic acids are formed and the mineralized nitrogen and sulfur are oxidized to nitric and sulfuric acid, respectively. Third, natural or normal precipitation reacts with carbon dioxide of the atmosphere and the carbonic acid formed gives natural precipitation a pH of about 5.6. The results of these reactions add acidity continuously to soils.

In desert regions there is little precipitation and, thus, these reactions contribute little acidic input to soils. As precipitation increases there is an increase in all three of the previously mentioned processes. A larger amount of precipitation contains more acidity. Greater precipitation results in more plant growth, causing more respiration and organic matter mineralization. Many studies have shown a relationship between annual precipitation, the leaching of carbonates, and soil pH, as given in Figure 11.6.

In soils with a pH of 7, there is a balance in the processes or reactions that produce alkalinity and acidity. On the central U.S. Great Plains this occurs with about 26 inches (65 cm) of annual precipitation (near the Iowa-Nebraska border). The dominant upland soils have developed under a grass vegetation and are quite fertile for agricultural crops.

Development and Properties of Acid Soils

In a calcareous soil, OH^- from carbonate hydrolysis reacts with any H^+ that is produced to form water; thus, calcareous soils remain alkaline in spite of the near constant production or addition of H^+. As long as soils remain calcareous, they

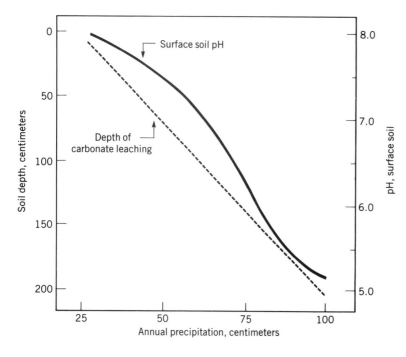

FIGURE 11.6 General relationship of upland soils between annual precipitation, depth of leaching of carbonates, and pH of the surface soil in the central United States. (Adapted from Jenny, 1941.)

remain alkaline. Development of neutrality, and eventually acidity, is dependent on the removal of the carbonates. This requires sufficient precipitation to cause effective leaching. This amount of precipitation also results in considerable acidic input. The weathering of primary minerals and the release of basic cations tend to repress the acidifying effects of respiration, precipitation, and mineralization of organic matter. Any H^+ in the soil solution neutralizes the OH^- produced by mineral weathering and shifts the weathering reactions to the right, or encourages further weathering of primary minerals and formation of OH^-. Despite the neutralizing effect of mineral weathering, soils in humid regions become increasingly acid with time as leaching continues.

The effects of the acidifying processes are progressive; the effects producing acid A horizons first, then acid B horizons, and, lastly, acid C horizons. Many minimally weathered and leached soils have a neutral or alkaline reaction in all horizons and in intensively weathered and

leached soils, all of the horizons are acid. By contrast, many moderately weathered and leached soils have acid A and B horizons and nuetral or alkaline (many being calcareous) C horizons.

Role of Aluminum After leaching has removed the carbonates, a progressive loss of XCa, XMg, XK, and XNa occurs as leaching continues. For example, the removal of an exchangeable potassium ion, K^+, is accompanied by the removal of an acid anion, like nitrate. The H^+ associated with the nitrate remains behind and competes for adsorption onto the exchange position vacated by the potassium. Adsorption of H^+ at the edge of silicate clay minerals makes aluminum unstable, and it exits from the edge of the clay lattice. Subsequent hydrolysis of the released Al^{3+} produces H^+:

$$Al^{3+} + H_2O = Al(OH)^{2+} + H^+$$

(lattice Al)   (hydroxy-aluminum)

Hydroxy-aluminum from the preceding reaction hydrolyzes to produce additional H^+:

$$\underset{\text{(hydroxy-aluminum)}}{Al(OH)^{2+}} + 2H_2O = \underset{\text{(Al hydroxide)}}{Al(OH)_3} + 2H^+$$

Consequently, acidity in soil stimulates the development of additional acidity through aluminum hydrolysis, and aluminum hydrolysis becomes a very important source of H^+ when soils become acid.

The hydroxy-aluminum (hydroxy-Al) formed in the preceding reactions consists of Al-OH octahedra that polymerize and produce large cations with a high valence. These large cations are very strongly adsorbed onto cation exchange sites because of their high valence. In fact, the adsorption is so strong that they are not exchangeable in the sense that XCa and XNa are exchangeable. Sites that adsorb hydroxy-aluminum cease to function for cation exchange, and the adsorption of hydroxy-aluminum reduces the CEC. This is one more reason for the soil's pH-dependent cation exchange capacity. The adsorption of hydroxy-aluminum in the interlayer space of expanding clays renders them nonexpandable. The adsorption of large hydroxy-aluminum cations in the interlayer space of 2:1 clay is illustrated in Figure 11.7.

When soil pH declines below 5.5, Al^{3+} begins to occupy cation exchange sites and exist as an exchangeable cation (XAl). The aluminum ion has a valence of 3^+, and it is adsorbed much more strongly than divalent and monovalent cations. The amount of XAl increases over time as the soil becomes more leached and more acid. The increase in XAL is accompanied by a decrease in XCa, XMg, XK, and XNa.

Moderately versus Intensively Weathered Soils As long as soils remain only moderately weathered, the weathering of primary minerals and the release of calcium, magnesium, potassium, and sodium contribute to the maintenance of a good supply of these elements. These elements, or basic cations, occupy most of the cation exchange sites, and there is little exchangeable aluminum. As moderately weathered soils evolve into intensively weathered soil, however, a considerable change in mineralogical composition occurs. The high CEC 2:1 clays disappear, and low CEC clays, such as kaolinite and oxidic clay, accumulate. There is a large decrease in the CEC, and soils have a small capacity for adsorption of nutrient cations. In addition, with the disappearance of most of the primary minerals, there is a very limited release of calcium, magnesium, potassium, and sodium by weathering. The aluminum saturation of the CEC ranges from about 0 percent aluminum saturation at pH 5.5 to nearly 100 percent aluminum (XAl) saturation at pH 4. The saturation of the CEC with basic cations (calcium, magnesium, potassium, and potassium) decreases from nearly 100 percent at pH about 5.5 or 6 to about 0 percent at pH 4.0, causing a great decline in the amount of exchangeable nutrient cations. In essence, intensive weathering results in acid soils that are naturally infertile.

The data in Table 11.5 are for three intensively weathered soils on the coastal plain of the southeastern United States. The pH is below 5, and the CEC averages only 2.68 cmol/kg (2.68 meq/100 g). These soils are characterized by very low fertility as suggested by the small amount of XCa

FIGURE 11.7 Drawing of two layers of a 2:1 expanding clay with hydroxy-Al adsorbed in the interlayer space. This adsorption is very strong and reduces the cation exchange capacity. The mutal attraction of the two clay layers to the positively charged hydroxy-Al causes the clay to become a nonexpanding, hydroxy-interlayered clay.

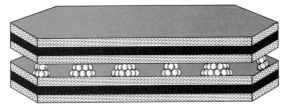

TABLE 11.5 The pH, Exchangeable Cations (KCl Extractable), Cation Exchange Capacity, and Aluminum Saturation in Three Intensively Weathered Soils

Soil	pH	Exchangeable Cations cmol (+) /kg			Cation Exchange Capacity, cmol (+) /kg	Percentage Aluminum Saturation
		Ca	Mg	Al		
Norfolk	4.9	0.06	0.02	1.43	1.51	95
Rains	4.7	0.11	0.02	1.62	1.75	93
Dunbar	4.6	1.00	0.11	3.67	4.78	77
Average		0.39	0.05	2.24	2.68	88

Adapted from Evans and Kamprath, 1970.

and XMg (XK and XNa are also low). The average aluminum saturation of the cation exchange capacity (cmol of XAl divided by the CEC and times 100) is 88 percent. When the aluminum saturation of soils exceeds 60 percent, there is frequently enough aluminum (Al^{3+}) in solution to be toxic to plant roots. The relationship between the amounts of exchangeable aluminum and calcium and the growth of corn roots is shown in Figure 11.8.

Whereas calcium and magnesium deficiences for plants are rather uncommon on acid soils that are moderately weathered, low XMg and XCa commonly result in plant deficiencies on acid and intensively weathered soils. A comparison of the data in Table 11.5 with the data in Table 11.3 shows these large differences in exchangeable cations.

Role of Strong Acids A soil pH below about 3.5 and down to 2 is indicative of strong acid in the soil. Pyrite (FeS_2) occurs in some soils and the sulfur is oxidized to sulfuric acid. Soils containing strong acid may produce a toxicity of both aluminum and iron; the soil may be too acid for plant growth. These soils are called *acid sulfate* soils or *cat clays,* and they tend to form in mangrove swamps along saltwater beaches (see Figure 11.9). Sometimes the tailings from mining operations contain pyrite, and the formation of sulfuric acid and low pH make reclamation and revegetation of such materials difficult.

Acid Rain Effects The water in precipitation reacts with carbon dioxide in the atmosphere to form carbonic acid. This causes natural, or ordinary, precipitation to be acidic and have a pH of about 5.6. Industrial gases contain nitrogen and sulfur that result in the formation of acids and cause rainfall to have a pH less than 5.6. Such rain is called *acid rain.* Studies have shown that the amount of acidity contributed to soils by acid rain is usually much less than that normally contributed by respiration, organic matter mineralization, and the acidity in normal rainfall.

In soils that are underlain by calcareous layers, any acid water that percolates through the soil will be neutralized and will not contribute to the acidity of groundwater or nearby lakes. On the Laurentian Shield in northeastern Canada, soils generally are underlain by acid parent materials and rocks. Here, additions of acidity from acid rain to acid soils contribute to the acidification of groundwater and lakes.

Soil Buffer Capacity

The *buffer capacity* is the ability of ions associated with the solid phase to buffer changes in ion

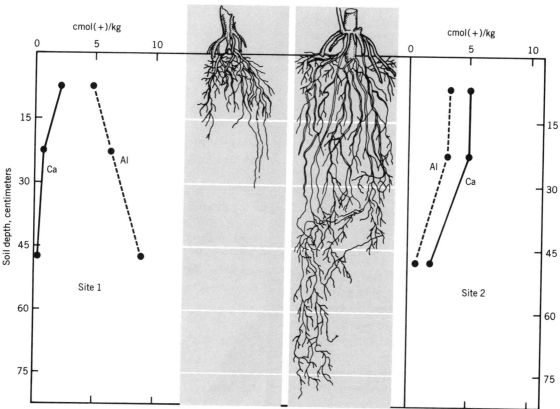

FIGURE 11.8   On left, poor corn root growth in a Hawaiian soil associated with low exchangeable calcium and high exchangeable aluminum. On right, good corn root growth associated with low exchangeable aluminum and high exchangeable calcium. (Drawn from Green, R. E., Fox, R. L., and D. D. F. Williams, 1965.)

FIGURE 11.9   Mangrove swamps are one of the typical environments where acid sulfate soils develop.

concentration in the solution phase. In acid soils, buffering refers to the ability of the XAl, XH, and hydroxy-aluminum to maintain a certain concentration of H^+ in solution. The amount of H^+ in the soil solution of a soil with a pH of 6.0, for example, is extremely small compared with the nondissociated H^+ adsorbed and the amount of aluminum that can hydrolyze to produce H^+. Neutralization of the active or solution H^+ results in rapid replacement of H^+ from the relatively large amount of H^+ associated with the solid phase. Thus, the soil exhibits great resistance to undergo a pH change.

Summary Statement

With a long period of time, soil evolution in humid regions causes changes in both the mineralogical and chemical properties and changes in soil pH. Minimally weathered soils tend to be alkaline or slightly acid and fertile. By contrast, intensively weathered soils are very acid and infertile. A summary of the major processes and changes that produce various ranges in soil pH and accompanying properties is as follows.

1. In the pH range 7.5 to 8.3 or above, soil pH is controlled mainly by carbonate hydrolysis.
2. If a soil is calcareous, removal of carbonates by leaching is required before a neutral or acid soil can develop.
3. Soils naturally at or near neutral tend to occur in regions of limited precipitation where there is a balance between the production of H^+ and OH^-. The acidic inputs from biological respiration, organic matter mineralization, and precipitation are balanced by the basic inputs of mineral weathering.
4. In humid regions, soils tend to become acid over time as basic cations decrease and acidic cations increase. In the range of about 7 to 5.5, H^+ for the soil solution is supplied mainly by hydroxy-aluminum hydrolysis, and in the range 5.5 to about 4 the H^+ for the soil solution comes mainly from exchangeable aluminum hydrolysis. Soils high in organic matter may receive a significant amount of H^+ from exchangeable hydrogen.
5. Intensively weathered soils in the advanced stages of weathering have: (1) low CEC, (2) high saturation of the cation exchange capacity with aluminum, and (3) very small amounts of exchangeable calcium, magnesium, potassium, and sodium. Intensively weathered soils tend to produce aluminum toxicity and deficiencies of magnesium and calcium for plant growth.
6. In the most acid soils, pH < 4, free strong acid exists.

SIGNIFICANCE OF SOIL pH

Studies have shown that the actual concentrations of H^+ or OH^- are not very important, except under the most extreme circumstance. The associated chemical or biological environment of a certain pH is the most important factor. Some soil organisms have a rather limited tolerance to varia-

FIGURE 11.10. Red pine was planted along the edges of this field to serve as a windbreak. The red pine in the foreground and background failed to survive because these areas were part of an old limestone road and the soil was alkaline. White cedar was later planted and is growing very well on the old road.

Old limestone gravel road

tions in pH, but other organisms can tolerate a wide pH range. The effects of limestone gravel from an old roadway on the growth of red pine and white cedar trees is shown in Figure 11.10.

Nutrient Availability and pH

Perhaps the greatest general influence of pH on plant growth is its effect on the availability of nutrients for plants, as shown in Figure 11.11. Nitrogen availability is maximum between pH 6 and 8, because this is the most favorable range for the soil microbes that mineralize the nitrogen in organic matter and those organisms that fix nitrogen symbiotically. High phosphorus availability at high pH—above 8.5—is due to sodium phosphates that have high solubility. In calcareous soil, pH 7.5 to 8.3, phosphorus availability is re-

duced by the presence of calcium carbonate that represses the dissolution of calcium phosphates. Maximum phosphorus availablity is in the range 7.5 to 6.5. Below pH 6.5, increasing acidity is associated with increasing iron and aluminum in solution and the formation of relatively insoluble iron and aluminum phosphates.

Note that potassium, calcium, and magnesium are widely available in alkaline soils. As soil acidity increases, these nutrients show less availability as a result of the decreasing CEC and decreased amounts of exchangeable nutrient cations: decreased amounts of XCa, XMg, and XK.

Iron and manganese availability increase with increasing acidity because of their increased solubility. These two nutrients are frequently deficient in plants growing in alkaline soils because of the insolubility of their compounds (see Figure

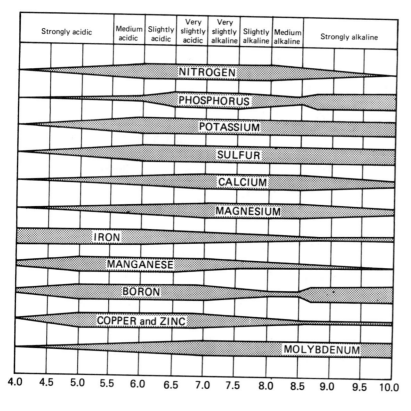

FIGURE 11.11 General relationship between soil pH and availability of plant nutrients in minimally and moderately weathered soils; the wider the bar, the more availability.

11.11). Boron, copper, and zinc are leachable and can be deficient in leached, acid soils. Conversely, they can become insoluble (fixed) and unavailable in alkaline soils. In acid soils, molybdenum is commonly deficient owing to its reaction with iron to form an insoluble compound.

The relationships shown in Figure 11.11 are for minimally and moderately weathered soils. For plant nutrients as a whole, good overall nutrient availability occurs near pH 6.5. In intensively weathered soils, some nutrients such as boron and zinc may be in very short supply and become deficient at lower pH than for minimally and moderately weathered soils. As a consequence, a desirable pH is more typically 5.5 in intensively weathered soils.

Effect of pH on Soil Organisms

The greater capacity of fungi, compared with bacteria, to thrive in highly acid soils was cited in Chapter 8. The pH requirement of some disease organisms is used as a management practice to control disease. One of the best known examples is that of the maintenance of acid soil to control potato scab. Potato varieties have now been developed that resist scab organisms in neutral and alkaline soils. Damping-off disease in nurseries is controlled by maintaining soil pH at 5.5 or less.

Earthworms are inhibited by high soil acidity. Peter Farb (1959) relates an interesting case in which soil pH influenced both earthworm and mole populations. On a certain tennis court in England, each spring after rain and snow had washed away the court markings, the marking lines were still visible because moles had unerringly followed them. This intriguing mystery was investigated and produced the following solution. The tennis court had been built on acid soil and chalk (lime) was used to mark the lines. As a result, the pH in the soil below the lines was increased. Earthworms then inhabited the soil under the markings, and the soil remaining on the tennis court remained acid and uninhabitated by

earthworms. Since earthworms are the primary food of moles, the moles restricted their activity only to the soil under the marking lines.

Toxicities in Acid Soils

Figure 11.11 shows that iron and manganese are most available in highly acid soils. Manganese toxicity may occur when soil pH is about 4.5 or less. High levels of exchangeable aluminum in many acid subsoils of the southeastern United States cause high levels of solution aluminum and restrict root growth. Plants, and even varieties of the same species, exhibit differences in tolerance to high levels of aluminum, iron, and manganese in solution and to other factors associated with soil pH. This gives rise to the pH preferences of plants.

pH Preferences of Plants

Blueberries are well known for their acid soil requirement. In Table 11.6, the optimum pH range for blueberries is 4 to 5. Other plants that prefer a pH of 5 or less include: azaleas, orchids, sphagnum moss, jack pine, black spruce, and cranberry. In Table 11.6 the names of plants that prefer a pH of 6 or less are italicized. Most field and garden crops prefer a pH of about 6 or higher. Because plants have specific soil pH requirements, there is a need to alter or manage soil pH for their successful growth.

MANAGEMENT OF SOIL pH

There are two approaches to assure that plants will grow without serious inhibition from unfavorable soil pH: (1) plants can be selected that grow well at the existing soil pH, or (2) the pH of the soil can be altered to suit the needs of the plants. Most soil pH changes are directed toward reduced soil acidity and increased pH by liming.

TABLE 11.6 Optimum pH Ranges of Selected Plants Growing on Minimally and Moderately Weathered Soils

Field Crops		Alyssum	6.0–7.5	Oak, Black	6.0–7.0
Alfalfa	6.2–7.8	*Azalea*	*4.5–5.0*	Oak, Pin	5.0–6.5
Barley	6.5–7.8	Barberry, Japanese	6.0–7.5	Oak, White	5.0–6.5
Bean, field	6.0–7.5	Begonia	5.5–7.0	*Pine, Jack*	*4.5–5.0*
Beets, sugar	6.5–8.0	Burning bush	5.5–7.5	*Pine, Loblolly*	*5.0–6.0*
Bluegrass, Ky.	5.5–7.5	Calendula	5.5–7.0	*Pine, Red*	*5.0–6.0*
Clover, red	6.0–7.5	Carnation	6.0–7.5	*Pine, White*	*4.5–6.0*
Clover, sweet	6.5–7.5	Chrysanthemum	6.0–7.5	*Spruce, Black*	*4.0–5.0*
Clover, white	5.6–7.0	*Gardenia*	*5.0–6.0*	Spruce, Colorado	6.0–7.0
Corn	5.5–7.5	Geranium	6.0–8.0	*Spruce, White*	*5.0–6.0*
Flax	5.0–7.0	*Holly, American*	*5.0–6.0*	*Sycamore*	*6.0–7.5*
Oats	5.0–7.5	Ivy, Boston	6.0–8.0	Tamarack	6.0–7.5
Pea, field	6.0–7.5	Lilac	6.0–7.5	Walnut, Black	6.0–8.0
Peanut	5.3–6.6	Lily, Easter	6.0–7.0	Yew, Japanese	6.0–7.0
Rice	5.0–6.5	*Magnolia*	*5.0–6.0*	**Weeds**	
Rye	5.0–7.0	*Orchid*	*4.0–5.0*	Dandelion	5.5–7.0
Sorghum	5.5–7.5	*Phlox*	*5.0–6.0*	Dodder	5.5–7.0
Soybean	6.0–7.0	Poinsettia	6.0–7.0	Foxtail	6.0–7.5
Sugar Cane	6.0–8.0	Quince, flowering	6.0–7.0	Goldenrod	5.0–7.5
Tobacco	5.5–7.5	*Rhododendron*	*4.5–6.0*	Grass, Crab	6.0–7.0
Wheat	5.5–7.5	Rose, hybrid tea	5.5–7.0	Grass, Quack	5.5–6.5
		Snapdragon	6.0–7.5	*Horse Tail*	*4.5–6.0*
Vegetable Crops		Snowball	6.5–7.5	*Milkweed*	*4.0–5.0*
Asparagus	6.0–8.0	Sweet William	6.0–7.5	Mustard, Wild	6.0–8.0
Beets, table	6.0–7.5	Zinnia	5.5–7.5	Thistle, Canada	5.0–7.5
Broccoli	6.0–7.0				
Cabbage	6.0–7.5	**Forest Plants**		**Fruits**	
Carrot	5.5–7.0	Arbor Vitae	6.0–7.5	Apple	5.0–6.5
Cauliflower	5.5–7.5	Ash, White	6.0–7.5	Apricot	6.0–7.0
Celery	5.8–7.0	*Aspen, American*	*3.8–5.5*	Blueberry, High	
Cucumber	5.5–7.0	Beech	5.0–6.7	Bush	4.0–5.0
Lettuce	6.0–7.0	*Birch, European*		Cherry, sour	6.0–7.0
Muskmelon	6.0–7.0	*(white)*	*4.5–6.0*	Cherry, sweet	6.0–7.5
Onion	5.8–7.0	*Cedar, White*	*4.5–5.0*	Crab apple	6.0–7.5
Potato	4.8–6.5	Club Moss	4.5–5.0	*Cranberry, large*	*4.2–5.0*
Rhubarb	5.5–7.0	*Fir, Balsam*	*5.0–6.0*	Peach	6.0–7.5
Spinach	6.0–7.5	*Fir, Douglas*	*6.0–7.0*	*Pineapple*	*5.0–6.0*
Tomato	5.5–7.5	*Heather*	*4.5–6.0*	Raspberry, Red	5.5–7.0
		Hemlock	*5.0–6.0*	Strawberry	5.0–6.5
Flowers and Shrubs		Larch, European	5.0–6.5		
African violet	6.0–7.0	Maple, Sugar	6.0–7.5		
Almond, flowering	6.0–7.0	*Moss, Sphagnum*	*3.5–5.0*		

Data from Spurway, 1941. Plants with an optimum pH range of 6.0 and below are italicized.

The Lime Requirement

Agricultural lime is a soil amendment containing calcium carbonate, magnesium carbonate, and other materials, which are used to neutralize soil acidity and to furnish calcium and magnesium for plant growth. Most frequently, agricultural lime is ground limestone that is mainly $CaCO_3$ and is calcitic lime. If the lime also contains a significant amount of $MgCO_3$, the lime is dolomitic. The amount of lime needed or the *lime requirement* is the amount of liming material required to produce a specific pH or to reduce exchangeable and soil solution aluminum.

Lime Requirement of Intensively Weathered Soils Intensively weathered acid soils are characterized by low CEC that is highly aluminum saturated and with low calcium and magnesium saturation. The main benefits from liming are: (1) reduced exchangeable aluminum and reduced aluminum in solution, (2) an increase in the amount of XCa and/or XMg, and (3) increased availability of molybdenum.

The lime requirement of intensively weathered soils is the amount of liming material needed to change the soil to a specific soluble aluminum content. This is commonly considered to be equal to 1.5 times the amount of exchangeable aluminum. Thus, the lime requirement of the Rains soil (see Table 11.5) is equal to

$$1.5 \times 1.62 \text{ cmol/kg} = 2.43 \text{ cmol/kg}$$

The lime requirement for the Rains example is the amount of lime equal to 2.43 cmol of + charge per kilogram of soil. A mole of $CaCO_3$ weighs 100 grams. Because of the divalent nature of the calcium, a half mole of $CaCO_3$, 50 g, will neutralize a mole of + charge. The amount of $CaCO_3$ equal to 1 cmol of + charge is 0.5 g (50 g/100). Therefore, 1.21 g/kg (2.43 cmol/kg × 0.5 g/kg) of $CaCO_3$ is the lime requirement. Assuming a 2 million kg furrow slice of soil per hectare, the lime needed is 2,420 kg/ha.

The milliequivalent weight of $CaCO_3$ is 0.05 g. If $CaCO_3$ is used, the lime requirement is 0.12 g (2.43 meq × 0.05 g per meq) of $CaCO_3$ per 100 g of soil. Assuming an acre furrow slice weighs 2 million pounds, 2,400 pounds of $CaCO_3$ are required per acre. The rate must be adjusted to account for different thicknesses or weights of plow layers.

In many soils, roots are susceptible to alumi-num toxicity in the subsoil that restricts rooting depth and water uptake. Soils are droughty, which is a difficult problem to solve, because the lime has low solubility and moves downward slowly. Overliming the plow layer, to promote the downard movement of lime, may increase pH enough to cause micronutrient deficiences or toxicities.

Lime Requirement of Minimally and Moderately Weathered Soils Moderately and minimally weathered acid soils typically have a large CEC that has low aluminum saturation. Aluminum toxicity and calcium deficiency are rare and, only occasionally, is there a magnesium deficiency. The lime requirement is the amount needed to increase or adjust pH to a more favorable value, as shown in Figure 11.12 for production of soybeans.

Consider a moderately weathered soil with a pH of 5.0 for the growth of soybeans, which have a pH preference of 6 to 7. The lime requirement is the amount of lime required to adjust the pH to a more favorable value for the crop being grown; for soybeans it would be about 6.5. The same is true for many other legumes, including alfalfa and clover. Farmers who continuously grow corn can maintain a lower soil pH than those producing soybeans and alfalfa, because corn is less respon-

FIGURE 11.12 Growth of soybeans versus soil pH. A pH of 6.5 is good for soybean production.

sive to pH adjustment than are soybeans and alfalfa.

The amount of lime required to adjust the pH of soils depends on the amount of pH adjustment needed and on the amount of buffer acidity. The amount of buffer acidity is related to the CEC. As shown in Figure 11.13, much less base (or lime), is required to increase the pH of Plainfield sand than of Granby loam.

It is a common practice in soil testing laboratories to make a direct measurement of the lime requirement by using a buffer solution. A buffer solution with a pH of 7.5, for example, is mixed with a known quantity of soil. After a standard period of stirring, the pH of the buffer-soil suspension is made. The amount of depression in pH from that of the buffer solution is directly related to the amount of lime needed to adjust soil pH to a particular value.

A less accurate but useful method, especially for the gardener or homeowner, is to determine soil pH with inexpensive color indicator dyes. The lime requirement is obtained from a chart that relates lime need to soil pH and texture, essentially CEC, as shown in Table 11.7. The data in Table 11.7 show lower lime requirements in the southern coastal states, because the soils contain clays (such as kaolinite) with lower CEC.

The Liming Equation and Soil Buffering

In acid soils, H^+ in the soil solution comes mainly from aluminum hydrolysis. Generally, a minor amount comes from XH unless there is a high organic matter content. Below pH 5.5, the hydrolysis is of exchangeable aluminum and, between 5.5 and 7, mainly from hydroxy-aluminum hydrolysis. The dominant aluminum form at pH 7 is $Al(OH)_3$, which has very low solubility and, for all practical purposes, is inert. Lime hydrolyzes in the soil to form OH^-:

$$CaCO^3 + H_2O = Ca^{2+} + HCO_3^- + OH^-$$

The hydroxyls neutralize the H^+ that are produced from aluminum hydrolysis, gradually the exchangeable aluminum and hydroxy-aluminum are converted to $Al(OH)_3$, and the pH increases.

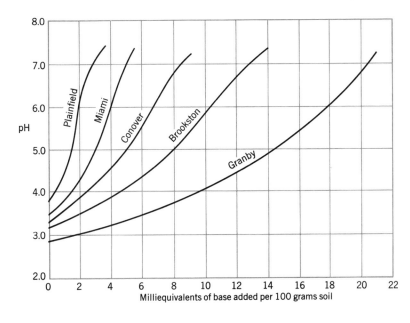

FIGURE 11.13 Titration curves showing the different amounts of base needed to increase the pH of soils having increasing cation exchange capacities from left to right.

TABLE 11.7   Suggested Applications of Finely Ground Limestone to Raise the pH of a 7-inch Layer of Several Textural Classes of Acidic Soils, in Pounds per 1,000 ft^2

Textural Class	pH 4.5 to 5.5		pH 5.5 to 6.5	
	Northern and Central States	Southern Coastal States	Northern and Central States	Southern Coastal States
Sands and loamy sands	25	15	30	20
Sandy loams	45	25	55	35
Loams	60	40	85	50
Slit loams	80	60	105	75
Clay loams	100	80	120	100
Muck	200	175	225	200

From Kellogg, 1957.

The overall reaction representing the neutralization of Al-derived soil acidity is

$$2AlX + 3CaCO_3 + 3H_2O = 3CaX + 2Al(OH)_3 + 3CO_2$$

Exchangeable and hydroxy aluminum are precipitated as insoluble aluminum hydroxide, and the amount of XCa is increased. The reduction in acidic cations and increase in basic cations results in an increase in soil pH. Liming can be viewed as reversing the processes that produced the soil acidity.

Some Considerations in Lime Use

In agriculture, liming is usually done by applying ground limestone with spreaders, as shown in Figure 11.14. Major considerations are the purity, or neutralizing value, and fineness of the lime. Calcium carbonate is quite insoluble, and particles that will not pass a 20-mesh screen are quite ineffective. It takes at least 6 months for lime to appreciably increase soil pH, and lime may be applied whenever it is convenient to do so. Calcium hydroxide ($Ca(OH)_2$) is also used as a liming material. It has greater neutralizing value per pound and is more soluble than the carbonate form. Wood ashes can also be used to increase soil pH.

Acid soils that have been limed are still subjected to the natural acidifying effects of biological respiration, organic matter mineralization, and precipitation. Thus, soils again become acid and must be relimed about every 2 to 5 years, depending on variables such as lime application rate, CEC, rainfall, crop removal of calcium and magnesium, and fertilization practices.

Management of Calcareous Soils

Millions of acres of soil are calcareous in arid regions. In humid regions, many calcareous soils occur on floodplains and on recently drained lake plains. The pH ranges from about 7.5 to 8.3, and many of these calcareous soils are excellent for

FIGURE 11.14   Broadcast application of lime on acid soil.

agricultural use. Plants growing on calcareous soils are sometimes deficient in iron, manganese, zinc, and/or boron. To lower soil pH would require the leaching of the carbonates, which is impractical. As a result, crops are fertilized with appropriate nutrients to overcome the deficiencies, or crops that grow well on calcareous soil are selected.

Soil Acidulation

The addition of acid sphagnum peat to soil may have some acidifying effect. However, significant and dependable increases in soil acidity are probably best achieved through the use of sulfur. Sulfur is slowly converted to sulfuric acid by soil microbes, and the soil slowly becomes more acid over a period of several months or a year. As with lime, the amount of sulfur required varies with the pH change desired and the soil CEC. Sulfur is used to change soil pH in large agricultural fields in arid regions, but it is less common a practice than the use of lime in humid regions. Sulfur is commonly used in nurseries and gardens. Recommendations for small areas are given in Table 11.8.

EFFECTS OF FLOODING ON CHEMICAL PROPERTIES

In well-aerated soils, the soil atmosphere contains an abundant supply of oxygen for microbial respiration and organic matter decomposition.

Oxygen serves as the dominant electron acceptor in the transformation of organic substrates to carbon dioxide and water. When soils are flooded naturally, or artificially as in rice production, a unique oxidation-reduction profile is created. A thin, oxidized surface soil layer results from the diffusion of oxygen from the water into the soil surface. This layer is underlain by a much thicker layer that is reduced (see Figure 4.11). Oxygen is deficient in the reduced zone, which initiates a series of reactions that drastically alter the soil's chemical environment.

Dominant Oxidation and Reduction Reactions

Flooded soils with a good supply of readily decomposable organic matter may become depleted of oxygen within a day. Then, anaerobic and facultative microbes multiply and continue the decomposition process. In the absence of oxygen, other electron acceptors begin to function, depending on their tendency to accept electrons. Nitrate is reduced first, followed by manganic compounds, ferric compounds, sulfate and, finally, sulfite. The reactions are as follows (reduction to the right and oxidation to the left):

$$\textbf{(1)} \quad 2NO_3{}_- + 12H^+ + 10e^- = N_2 + 6H_2O$$
$$\textbf{(2)} \quad MnO_2 + 4H^+ + 2e^- = Mn^{2+} + 2H_2O$$
$$\textbf{(3)} \quad Fe(OH)_3 + e^- = Fe(OH)_2 + OH^-$$
$$\textbf{(4)} \quad SO_4{}^{2-} + H_2O + 2e^- = SO_3{}^{2-} + 2OH^-$$
$$\textbf{(5)} \quad SO_3{}^{2-} + 3H_2O + 6e^- = S^{2-} + 6OH^-$$

TABLE 11.8 Suggested Applications of Ordinary Powdered Sulfur to Reduce the pH of 100 ft^2 of a 20 Centimeter Layer of Sand or Loam Soil

| Original pH | Pints of Powdered Sulfur, for Desired pH | | | | | | | | | |
| | 4.5 | | 5.0 | | 5.5 | | 6.0 | | 6.0 | |
	Sand	Loam	Sand	Loam	Sand	Loam	Sand	Loam	Sand	Loam
5.0	$\frac{2}{3}$	2								
5.5	$1\frac{1}{3}$	4	$\frac{2}{3}$	2						
6.0	2	$5\frac{1}{2}$	$1\frac{1}{3}$	4	$\frac{2}{3}$	2				
6.5	$2\frac{1}{2}$	8	2	$5\frac{1}{2}$	1	4	$\frac{2}{3}$	2		
7.0	3	10	$2\frac{1}{2}$	8	2	$5\frac{1}{2}$	1	4	$\frac{2}{3}$	2

From Kellogg, 1957.

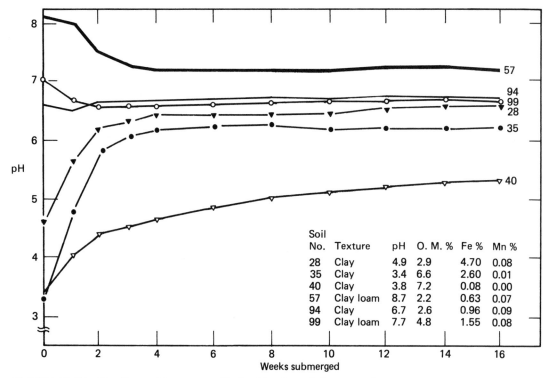

FIGURE 11.15 Flooding soil increases pH of acid soils and decreases the pH of alkaline soils. The pH changes are related to the soil's content of iron and organic matter. (Data from F. N. Ponnamperuma, 1976.)

Soils have various levels of reducing conditions, depending on the quantity of decomposable organic matter, the amount of oxygen brought in by water, and the temperature. The formation of S^{2-} (sulfide) occurs in a strongly reduced environment, resulting in the formation of FeS_2 and H_2S. In some instances, toxic amounts of H_2S are produced for rice production. About two weeks are needed after flooding to create rather stable conditions.

Changes in Soil pH

From the reactions just cited, there is a consumption of H^+ and an increase in hydroxyls. In most flooded rice fields, iron is the most abundant electron acceptor (equation 3), and iron reduction tends to increase soil pH. The production of carbon dioxide, and the accompanying formation of carbonic acid, have an acidifying effect. Alkaline soils tend to be low in available iron, minimally weathered, and the production of carbonic acid lowers soil pH. In acid soils the abundance of soluble or available iron results in a net increase in pH due to hydroxyl production. As a consequence, flooding of most rice fields causes the pH to move toward neutrality, as shown in Figure 11.15. There is an increase in nutrient availability, and soils generally do not need lime if they are flooded and puddled for rice production. The increase in pH of highly acid soils is sufficient to reduce the threat of aluminum toxicity.

SUMMARY

The kinds and amount of clay minerals and organic matter determine the cation exchange capacity.

The abundance of adsorbed cations in neutral and slightly acid soils is Ca > Mg > K > Na. These cations are held against leaching and are available for plant uptake.

The anion exchange capacity is typically much less than the cation exchange capacity, and it results from protonation of hydroxyls.

Calcareous soils are alkaline as a result of hydrolysis of calcium carbonate. Leaching of the carbonates in calcareous soils produces a pH of about 7. Continued leaching results in decreased amounts of exchangeable calcium, magnesium, potassium, and sodium. There is an increase in exchangeable H and Al with increasing acidity.

In acid mineral soils, H^+ for the soil solution comes mainly from hydrolysis of aluminum, both hydroxy and exchangeable forms. Exchangeable H^+ increases as organic matter content increases in acid soils.

The lime requirement of intensively acid soils is equal to about 1.5 times the amount of exchangeable aluminum. In minimally and moderately weathered soils, the lime requirement is the amount of lime needed to increase soil pH to a desired value.

Sulfur is used to increase soil acidity.

Flooding soils results in a decrease in the pH of alkaline soils and an increase in the pH of acid soils.

REFERENCES

Bohn, H. L., B. L. McNeal, and G. A. O'Connor. 1985. *Soil Chemistry*, 2nd ed. Wiley, New York.

Binkley, D. 1987. "Use of the Terms 'Base Cation' and 'Base Saturation' Should Be Discouraged." *Soil Sci. Soc. Am. J.* **51**:1089–1090. Madison, Wis.

Clarke, F. W. 1924. "Data of Geochemistry." *U.S. Geo. Sur. Bul. 770.* Washington, D.C.

Evans, C. E. and E. J. Kamprath. 1970. "Lime Response as Related to Percent Aluminum Saturation, Solution Aluminum, and Organic Matter Content." Soil Sci. Soc. Am. Proc. **34**:893–896.

Farb, P. 1959. *Living Earth.* Harper & Row, New York.

Foth, H. D. 1987. "Proposed Designation Change for Exchangeable Bases." *Agronomy Abstracts,* Madison, Wis., pp. 1.

Foth, H. D. and B. G. Ellis. 1988. *Soil Fertility.* Wiley, New York.

Foy, C. D. and G. B. Burns. 1964. "Toxic Factors in Acid Soils." *Plant Food Review.* Vol. 10.

Green, R. E., R. L., Fox, and D.D.F. Williams. 1965. "Soil Properties Determine Water Availability to Crops." *Hawaii Farm Science.* **14, (3)**:6–9. Hawaii Agr. Exp. Sta., Honolulu.

Holzhey, C. S., R. B. Daniels, and E. E. Gamble. 1975. "Thick Bh Horizons in the North Carolina Coastal Plain: II Physical and Chemical Properties and Rates of Organic Additions from Surface Sources." *Soil Sci. Soc. Am. Proc.* **39**:1182–1187.

Krug, E. C. and C. R. Frink. 1983. "Acid Rain on Acid Soils." *Science.* **221**:520–525.

Jenny, Hans. 1941. *Factors of Soil Formation.* McGraw-Hill, New York.

Kellogg, C. E. 1957. "Home Gardens and Lawns." *Soil.* USDA Yearbook. Washington, D.C.

Kamprath, E. J. and C. D. Foy. 1985. "Lime and Fertilizer Interactions in Acid Soils. *Fertilizer Technology and Use.* Am. Soc. Agron. Madison, Wis.

Ponnamperuma, F. N. 1976. "Specific Soil Chemical Characteristics for Rice Production in Asia." *IRRI Res. Papers Series,* No. 2. Manila.

Spurway. C. H. 1941. "Soil Reaction Preferences of Plants." *Mich. Agr. Exp. Sta. Spec. Bull.* 306. East Lansing.

Thomas, G. W. and W. L. Hargrove. 1984. "The Chemistry of Soil Acidity." In *Soil Acidity and Liming.* Agronomy **12**:1–58. Am. Soc. Agron. Madison, Wis.

CHAPTER 12

PLANT-SOIL MACRONUTRIENT RELATIONS

For many centuries, people knew that manure, ashes, blood, and other substances had a stimulating effect on plant growth. The major effect was found to result from the essential elements contained in these materials. Six of the essential nutrients are used in relatively large amounts and are the macronutrients—usually over 500 parts per million (ppm) in the plant. Macronutrients include nitrogen, phosphorus, potassium, calcium, magnesium, and sulfur. A list of the macronutrients and their major roles in plant growth are given in Table 12.1. The quantities of the macronutrients contained in harvested crops are given in Table 12.2.

DEFICIENCY SYMPTOMS

Translocation of nutrients within a plant is an ongoing process. Early in the gowing season, a corn plant consists mainly of leaves, and the nitrogen exists mostly in the leaves, as shown in Figure 12.1. As the growing season continues, considerable nitrogen accumulates in the stalks and cobs. The grain (fruit) develops last and has a high priority for the nutrients within the plant. In the latter part of the growing season, mobile nutrients are translocated from leaves, stalks, and cobs and are used for the development of the grain (see Figure 12.1 for nitrogen). When the nutrient content of leaves decreases below critical values during the growing season, owing to translocation to other plant parts, specific symptoms develop.

There is a considerable difference in the mobility of various nutrients within plants. A shortage of a mobile nutrient, such as nitrogen, results in the translocation of the nutrient out of the older and first formed tissue into the younger tissues. This causes the deficiency symptoms to appear first on the older or lower leaves. This is true for nitrogen, phosphorus, potassium, and magnesium. Immobile plant nutrients cannot be remobilized when a shortage of the nutrient occurs, resulting in a deficiency in the most recent or newly formed tissues. Macronutrients with limited mobility in plants are calcium and sulfur.

NITROGEN

The atmosphere is made up of 79 percent nitrogen, by volume, as inert N_2 gas. Although a large quantity of nitrogen exists in the atmosphere, the nutrient that is absorbed from the soil in the great-

TABLE 12.1 Essential Macronutrient Elements and Role in Plants

Element	Role in Plants
Nitrogen (N)	Constituent of all proteins, chlorophyll, and in coenzymes, and nucleic acids.
Phosphorus (P)	Important in energy transfer as part of adenosine triphosphate. Constituent of many proteins, coenzymes, nucleic acids, and metabolic substrates.
Potassium (K)	Little if any role as constituent of plant compounds. Functions in regulatory mechanisms as photosynthesis, carbohydrate translocation, protein synthesis, etc.
Calcium (Ca)	Cell wall component. Plays role in the structure and permeability of membranes.
Magnesium (Mg)	Constituent of chlorophyll and enzyme activator.
Sulfur (S)	Important constituent of plant proteins.

Complied from many sources.

est quantity and is the most limiting nutrient for food production is nitrogen.

The Soil Nitrogen Cycle

The *nitrogen cycle* consists of a sequence of biochemical changes wherein nitrogen is used by living organisms, transformed upon death and decomposition of the organisms, and converted ultimately to its original state of oxidation. The major segments of the soil nitrogen cycle are shown as five steps or processes in Figure 12.2.

The reservoir of nitrogen for plant use is essentially that in the atmosphere, N_2. There is virtually an inexhaustible supply of nitrogen in the atmosphere, since (at sea level) there are about 77,350 metric tons of N_2 in the air over 1 hectare (34,500, tons per acre). It takes about a million years for the nitrogen in the atmosphere to move through one cycle.

The N_2 is characterized by both an extremely strong triple bonding between the two nitrogen atoms and a great resistance to react with other elements. The process of converting N_2 into forms that vascular plants can use is termed *nitrogen fixation*, (step 1 of Figure 12.2). The other major steps in the soil nitrogen cycle are: (2) mineralization, (3) nitrification, (4) immobilization, and (5) denitrification.

Dinitrogen Fixation

Dinitrogen fixation is the conversion of molecular nitrogen (N_2) to ammonia and subsequently into organic forms utilizable in biological processes. Some N_2 is fixed by lightning and other ionizing phenomena of the upper atmosphere. This fixed nitrogen is added to soils in precipitation.

Most nitrogen fixation is biological, being either symbiotic or nonsymbiotic. Nitrogen-fixing organisms contain an enzyme, nitrogenase, which combines with a dinitrogen molecule (N_2) and fixation occurs in a series of steps that reduces N_2 to NH_3, as shown in Figure 12.3. Molybdenum is a part of nitrogenase and is essential for biological nitrogen fixation. The nitrogen-fixing organisms also require cobalt, which is an example of plants need for cobalt.

Symbiotic Legume Fixation Legume plants are dicots that form a symbiotic nitrogen fixation relationship with heterotrophic bacteria of the genus *Rhizobium*. When the root of the host plant invades the site of a dormant bacterial cell, a so-called "recognition event" occurs. A root hair in contact with the bacterial cell curls around the cell. The dormant bacterial cell then germinates and forms an infection thread that penetrates the root and migrates to the center of the root (see Figure 12.4). Once inside the root, bacteria multiply and are transformed into swollen, irregular-shaped bodies called *bacteroids*. An enlargement of the root occurs and, eventually, a gall or nodule is formed (see Figure 8.12). The bacteroids re-

TABLE 12.2 Approximate Pounds per Acre of Macronutrients Contained in Crops

Crop	Acre Yield	Nitrogen	Phosphorus Potassium as P	as K	Calcium	Magnesium	Sulfur
Grains							
Barley (grain)	40 bu.	35	7	8	1	2	3
Barley (straw)	1 ton	15	3	25	8	2	4
Corn (grain)	150 bu.	135	23	33	16	20	14
Corn (stover)	4.5 tons	100	16	120	28	17	10
Oats (grain)	80 bu.	50	9	13	2	3	5
Oats (straw)	2 tons	25	7	66	8	8	9
Rice (grain)	80 bu.	50	9	8	3	4	3
Rice (straw)	2.5 tons	30	5	58	9	5	—
Rye (grain)	30 bu.	35	5	8	2	3	7
Rye (straw)	1.5 tons	15	4	21	8	2	3
Sorghum (grain)	60 bu.	50	11	13	4	5	5
Sorghum (stover)	3 tons	65	9	79	29	18	—
Wheat (grain)	40 bu.	50	11	13	1	6	3
Wheat (straw)	1.5 tons	20	3	29	6	3	5
Hay							
Alfalfa	4 tons	180	18	150	112	21	19
Bluegrass	2 tons	60	9	50	16	7	5
Coastal Bermuda	8 tons	185	31	224	59	24	—
Cowpea	2 tons	120	11	66	55	15	13
Peanut	2.25 tons	105	11	79	45	17	16
Red clover	2.5 tons	100	11	83	69	17	7
Soybean	2 tons	90	9	42	40	18	10
Timothy	2.5 tons	60	11	79	18	6	5
Fruits and vegetables							
Apples	500 bu.	30	5	37	8	5	10
Beans, dry	30 bu.	75	11	21	2	2	5
Cabbage	20 tons	130	16	108	20	8	44
Onions	7.5 tons	45	9	33	11	2	18
Oranges (70 pound boxes)	800 boxes	85	13	116	33	12	9
Peaches	600 bu.	35	9	54	4	8	2
Potatoes (tubers)	400 bu.	80	13	125	3	6	6
Spinach	5 tons	50	7	25	12	5	4
Sweet potatoes (roots)	300 bu.	45	7	62	4	9	6
Tomatoes (fruit)	20 tons	120	18	133	7	11	14
Turnips (roots)	10 tons	45	9	75	12	6	—
Other crops							
Cotton (seed and lint)	1,500 lb	40	9	13	2	4	2
Cotton (stalks, leaves, and burs)	2,000 lb	35	5	29	28	8	—

TABLE 12.2   *(continued)*

Crop	Acre Yield	Nitrogen	Phosphorus as P	Potassium as K	Calcium	Magnesium	Sulfur
Peanuts (nuts)	1.25 tons	90	5	13	1	3	6
Soybeans (grain)	40 bu.	150	16	46	7	7	4
Sugar beets (roots)	15 tons	60	9	42	33	24	10
Sugarcane	30 tons	96	24	224	28	24	24
Tobacco (leaves)	2,000 lb	75	7	100	75	18	14
Tobacco (stalks)	—	35	7	42	—	—	—

Data from *Plant Food Review,* 1962, pp. 22–25, publication of The Fertilizer Institute.
[a] These values are about the same as kilograms per hectare (multiply pounds per acre by 1.121).

ceive food, nutrients, and probably certain growth compounds from the host plant. The fixed nitrogen in the nodule is transported from the nodule to other parts of the host plant, which is beneficial for the host.

One method used to study the amount of nitrogen that is fixed is to compare the amount of nitrogen in nodulated and nonnodulated legume plants. Weber (1966) used this method and found that as much as about 160 kilograms of nitrogen per hectare (140 lb per acre) was fixed when soybeans were grown. This is shown in Figure 12.5 for the zero amount of nitrogen applied per hectare. In addition, about 75 percent of the nitrogen in the tops of the soybean plants was derived from nitrogen that was fixed, and 25 percent was

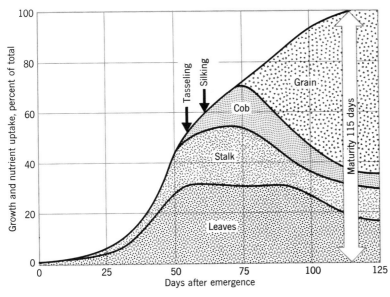

FIGURE 12.1 Growth and nitrogen in corn plants. Part of the nitrogen taken up and present in the leaves, stalks, and cobs early in the growing season is later moved to other organs. The grain has high priority for the nitrogen as shown by the large amount in the grain at harvest time. (From J. Hanway, 1960.)

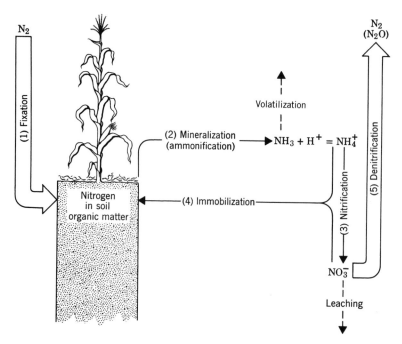

FIGURE 12.2 The major processes of the soil nitrogen cycle are fixation, mineralization, nitrification, immobilization, and denitrification. Nitrogen can also be lost from the soil by leaching and volatilization.

FIGURE 12.3 A simplified series of steps in nitrogen, fixation. (Adapted from Delwiche, in *The Science Teacher*, Vol. 36, p. 19, March 1969.)

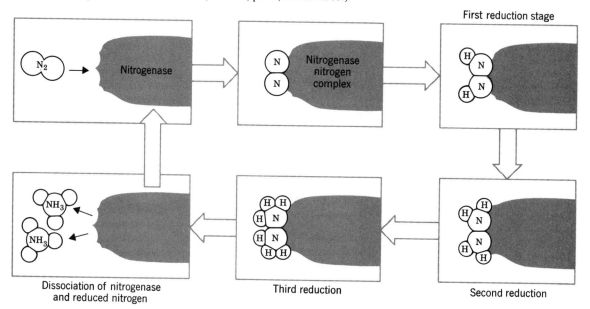

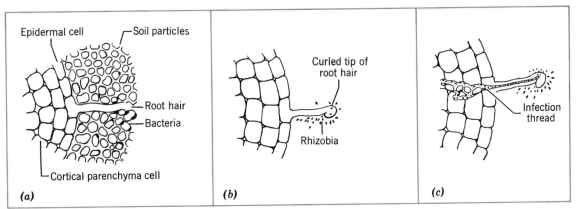

FIGURE 12.4 Early stages in nodule formation. (*a*) Recognition event between the root hair and bacteria. (*b*) Curling of root hair. (*c*) Early penetration of the infection thread. (From P. W. Wilson, *The Biochemistry of Symbiotic Nitrogen Fixation,* The University of Wisconsin Press, 1940, p. 74. Used by permission of P. W. Wilson and the University of Wisconsin Press.)

derived from the soil. Also note in Figure 12.5 that the amount of fixed nitrogen decreased as the amount of nitrogen applied to the soil as fertilizer was increased. At the highest nitrogen fertilizer rate, little nitrogen fixation occurred, and nearly all of the nitrogen in the plant was absorbed from

FIGURE 12.5 Amount of total nitrogen fixed and percent of nitrogen in soybean plants from symbiotic fixation in relation to the application of nitrogen fertilizer. (Data from C. R. Weber 1966.)

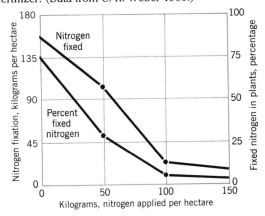

the soil. From these and other data, it is reasonable to conclude that 50 to 200 kilograms per hectare (or pounds per acre) are commonly fixed annually when legumes such as soybeans, clover, or alfalfa are grown. Peas and navy beans fix less nitrogen, so it is a standard practice to use nitrogen fertilizers to increase pea and bean yields.

Many trees are legumes and fix nitrogen. Black locust is an example of a nitrogen-fixing legume tree. In the tropics, *Leucaena* trees have been used to supply nitrogen for food crops in mixed cropping systems. The trees are also effective for erosion control. Other additional benefits of the *Leucaena* are that the leaves can be used as forage for animals and the wood can be used for fuel. The same benefits are obtained from the growth of cowpeas, which forms a small shrub.

In actual farming practice, the amount of nitrogen added to soils by the growth of legume crops depends on the amount of the crop that is harvested and removed from the soil. If an entire crop of legumes is plowed under, all the fixed nitrogen is added to the soil. If all of the legume crop is harvested and removed from the field, there may actually be a net loss of soil nitrogen by the production of the legume crop. This occurs when soybeans are grown and only the beans are harvested. Legume residues—roots, stems and leaves—are relatively rich in nitrogen and are decomposable (mostly labile). Even though the production of a legume crop may not increase total soil nitrogen, the decomposition of the plant residues releases considerable available nitrogen during the following months.

Nonlegume Symbiotic Fixation Many nonlegume plant species have root nodules and also fix nitrogen symbiotically. This means that symbiotic fixation of nitrogen is important in natural ecosystems by both legume and nonleguminous plants. Red alder is an example of a nonlegume capable of symbiotic nitrogen fixation. This feature makes red alder (*Alnus*) a good pioneer species for invading freshly exposed parent materials and lands that are burned over, where soils have low nitrogen-supplying capacity because of low organic matter content. *Ceanothus* species (a flowering shrub), in the forests of the U.S. Pacific Northwest, fixes a large amount of nitrogen. The nitrogen-fixing organisms in this nonlegume are actinomycetes. The nonlegume symbiotic nitrogen-fixing systems result in more fixation of nitrogen in natural ecosystems than in the symbiotic herbaceous legumes found in agricultural fields.

In water, the fern, *Azolla*, forms an association with blue-green algae. The blue-green algae live in the fronds of the fern and fix nitrogen. The algal partner can fix all the nitrogen that the association can use. Prolific growth of *Azolla* forms a floating mat of biomass on the surface of the water in rice paddies and other bodies of water. In Asia, *Azolla* is cultivated, and the biomass is harvested and used as a manure for crop production and as animal feed.

Nonsymbiotic Nitrogen Fixation Certain groups of bacteria living independently in the soil can fix nitrogen. These bacteria fix nitrogen non-symbiotically, and a dozen or more have been found. The two organisms that have been studied most belong to the genus *Azotobacter* and the genus *Clostridium*.

Azotobacter are widely distributed in nature. They have been found in soils with pH 6 or above in almost every instance where examinations have been made. The greatest limiting factor affecting their distribution in soil appears to be pH. These organisms are favored by pH 6 or more, good aeration, abundant organic matter, and suitable moisture. *Clostridium* is also a heterotrophic bacteria, but is an anaerobe that can thrive in soils too acid for *Azotobacter*. It exists mostly as a spore and becomes vegetative for brief periods after rains when anaerobic conditions exist. Both of these heterotrophs depend on decomposable organic matter for their energy supply and thus, have limited activity in soils and contribute little nitrogen for plant growth in cultivated fields.

Summary Statement In every ecological niche there are nitrogen fixing organisms and systems that transfer N_2 from the atmosphere to the soil. Plants that fix nitrogen are good pioneers for the colonization of parent materials and soils that are low in organic matter and nitrogen. In cultivated fields, legume-*Rhizobia* symbiotic fixation of nitrogen has traditionally been a primary source of nitrogen in agriculture. The amount of nitrogen fixed symbiotically in most cultivated soils is related to the legume species, the nitrogen-fixing *Rhizobia* species, and the amount of nitrogen available in the soil. Whether or not the production of a legume crop will increase the total content of soil nitrogen is dependent on the amount of the crop that is harvested and removed from the field. *Azolla* fixes nitrogen symbiotically and is cultivated in Asia. The biomass is harvested and is used as manure in crop production and is used as animal feed.

Mineralization

At any given time, about 99 percent of soil nitrogen is organic nitrogen in organic matter. About 1 to 2 percent of the total organic nitrogen is decomposed and the nitrogen mineralized each year. Many different kinds of heterotrophic organisms are involved. The first inorganic or mineral form of nitrogen produced by decomposition (mineralization) is ammonia (NH_3). For this reason, the nitrogen mineralization process has also been called *ammonification*; (see step 2 in Figure 12.2).

Some NH_3 volatilizes from animal droppings or manure, and some NH_3 produced at or near the soil surface escapes into the atmosphere by volatilization, especially when soil pH is 8 or more. Ammonia is a polar gas and, in the soil, NH_3 reacts with water and then combines with a H^+ to form ammonium, NH_4^+. Ammonmium is adsorbed onto the cation exchange sites and is retained in soils as an exchangeable cation. The ammonium ion is about the same size as the potassium ion, and some NH_4^+ becomes entrapped in the interlayer space of micaeous minerals, somewhat like the entrappment of K^+. This entrappment of ammonium nitrogen is called *ammonium fixation*. Fixed NH_4^+ has low plant availability.

It appears that all plants obtain some of their nitrogen from the soil. Plants that participate in symbiotic nitrogen fixation may obtain the great bulk of their nitrogen through fixation. Plants that do not have a symbiotic nitrogen-fixing relationship must meet all of their nitrogen needs from the soil. How much nitrogen the soil can supply plants depends on the amount of nitrogen mineralized, which is dependent on the kinds and amounts of soil organic matter and the rate of nitrogen mineralization.

The total nitrogen in a grassland soil in Canada was divided into four pools, depending on decomposability, as shown in Table 12.3. The living biomass accounted for only 4 to 6 percent of the soil nitrogen, but contributed 30 percent to the amount of nitrogen mineralized in a 12-week period. By contrast, the oldest and most stable nitrogen compounds accounted for 50 percent of the total soil nitrogen and contributed a mere 1 percent to the mineralized nitrogen (see Table 12.3).

Each percent of soil organic matter represents about 1,000 pounds of nitrogen per acre-furrow-slice, assuming soil organic matter is 5 percent nitrogen. If 1 percent of the total organic nitrogen is mineralized each year, the nitrogen mineralizing potential of the plow layer is 10 pounds per acre annually for each percent of organic matter. A plow layer containing 3 percent organic matter would annually mineralize about 30 pounds of nitrogen per acre.

Nitrification

Nitrification is the biological oxidation of ammonium (NH_4^+) to nitrite (NO_2^-) and to nitrate (NO_3^-). This process is shown as step 3 in Figure 12.2. Nitrification is a two-step process, with nitrite (NO_2^-) as the intermediate product. Specific

TABLE 12.3 Nitrogen Pools and Their Contribution to Mineralized Nitrogen in a Chernozemic (grassland) Soil

Pool	Percentage of Soil N	Half-life, years[a]	Relative Contribution of Mineralized N, %
Biomass	4–6	0.5	30
Active nonbiomass	6–10	1.5	34
Stabilized	36	22	35
Old	50	600	1

Data from Paul, 1984.
[a] Time for one-half of the material to decompose.

autotrophic bacteria are involved. The reactions and organisms are as follows.

$$NH_4^+ + 1.5O_2 \text{ (Nitrosomonas)} = NO_2^- + 2H^+ + H_2O + \text{energy}$$

$$NO_2^- + 0.5O_2 \text{ (Nitrobacter)} = NO_3^- + \text{energy}$$

The nitrifying bacteria obtain energy from these oxidizing reactions. Note also that the first step in nitrification produces H^+, which contributes to soil acidity. Nitrification changes the form of available nitrogen from a cation to an anion. This is of no great concern in plant nutrition, since most plants readily use both nitrate and ammonium.

Nitrification is affected by oxygen supply and soil pH. Nitrification is inhibited in water-saturated soils. Plants that are native to water-saturated environments, where the lack of O_2 does not permit nitrification to occur, tend to prefer ammonium. Paddy rice is more productive when an ammonium fertilizer rather than a nitrate fertilizer is applied.

When soils have a pH of 6 or more and are well aerated, there is an abundance of nitrifying bacteria—bacteria that quickly oxidize NH_4^+ to $NO3^-$. Well-aerated soils with pH of 6 or more also provide an environment that is favorable for production of most economic crops. Thus, conditions favorable for crop production are favorable for nitrification, resulting in NO_3^- being a common form of nitrogen absorbed by plants and microorganisms.

In most well-aerated soils the dominant form of available nitrogen is nitrate. It remains soluble in the soil solution and moves readily to roots in water by mass flow. In essence, plants can rapidly absorb both the water and nitrate from a soil if both are in good supply (available). This has two important consequences. First, mass flow is the major means by which nitrogen is moved to root surfaces for absorption. It accounts for about 80 percent of the nitrogen absorbed by corn. Second, the nitrogen from fertilizer is quickly used, and more than one application of nitrogen fertilizer

per year is usually necessary for lawns and other long-season crops to maintain vigorous growth.

Nitrate is subject to leaching, whereas, NH_4^+ is adsorbed onto the cation exchange sites and resists leaching. This has important implications for the economy of use of nitrogen fertilizer and nitrate pollution of groundwater. During the winter and spring, plants may be dormant and not be active in nitrogen uptake. This is the time of year when surplus water and groundwater recharge are likely to occur and nitrate leaching is most likely. Nitrogen fertilizer applied in the fall is usually as ammonium, commonly with a nitrification inhibitor. The nitrification inhibitor is used to maintain the nitrogen in the ammonium form and reduce possible leaching loss of nitrogen as nitrate, and this reduces the danger of groundwater pollution.

Immobilization

Both ammonium and nitrate are available forms of nitrogen for roots and microorganisms. Their uptake and use result in the conversion of inorganic (mineral) forms of nitrogen into organic form. The process is *immobilization* (shown as step 4 in Figure 12.2). Immobilized nitrogen is stable in the soil, in the sense that is is not readily lost from the soil by leaching, volitalization, or denitrification, and is not taken up by roots and microorganisms. Soil nitrogen is subject to repeated cycling within the soil through mineralization, nitrification, and immobilization.

Mineralization and immobilization are two opposing processes that generally control the level of available nitrogen that exists in the soil. As noted earlier, a soil with 3 percent organic matter might mineralize 30 pounds of nitrogen per acre per year. Part of this nitrogen is used by the mineralizing organisms. Perhaps, only one half of the amount of nitrogen mineralized might be in excess of that needed by the microorganisms. This would result in only 15 pounds of nitrogen available for uptake by roots. Large crop yields may contain several hundred pounds of nitrogen per acre, which means that large amounts of nitrogen fertilizer are commonly applied to produce non-leguminous crops.

Carbon-Nitrogen Relationships Some plant-residue materials such as wheat straw, corn stover, and sawdust have a very small nitrogen content relative to their carbon content. As a consequence, the ratio of carbon to nitrogen (C : N ratio) is relatively large compared with soil humus and the leguminous plants, which tend to be rich in nitrogen (see Table 12.4). Bacteria require about 5 grams of carbon for for each gram of nitrogen assimilated or use carbon and nitrogen in a ratio of 5:1. Consequently, when straw is added to the soil, there is insufficient nitrogen to meet the needs of the decomposing organisms. Under these conditions, decomposition of the straw is limited by an insufficient supply of nitrogen; unless, nitrogen is available from some other source. This other source could be nitrogen mineralized within the soil from SOM. The decomposing microorganisms have first priority for any mineralized nitrogen (nitrogen released in decomposition). When there is insufficient nitrogen in the straw to supply the needs of the decomposers, they will use any nitrogen that is available. This results in a deprivation of nitrogen for growing plants, and means that plants may be starved for nitrogen for a period of several weeks. Eventually, the straw material will be consumed and the carbon and nitrogen assimilated end up as live and/or dead microbial tissue. This tissue, along with other SOM (with relatively small C : N ratios) become the major substrate for the decomposers, and the temporary period of nitrogen starvation ceases. These changes are illustrated in Figure 12.6.

In general, materials with C : N ratios between 20 and 30 have just about enough nitrogen to satisfy the needs of the decomposers. Materials with higher ratios produce a temporary period of nitrogen deficiency for plant growth. Humus, animal manure, and legume residues are relatively rich in nitrogen, have a relatively small C:N ratio, and contain more nitrogen than the decomposers need. These materials with small C:N ratios are excellent materials to add to soils to increase the available nitrogen for growing plants.

An addition of about 20 pounds of nitrogen to the soil as fertilizer per ton of straw is required to prevent a temporary nitrogen deficiency, as a result of the microbial decomposition of the straw. When a large amount of straw or other similar material is added to soils without supplemental nitrogen fertilizer, a month or so is required for the temporary nitrogen-deficient period to pass under conditions favorable for decomposition.

TABLE 12.4 The Carbon-Nitrogen Ratio of Some Organic Materials

Material	C/N Ratio
Soil humus	10
Sweet clover (young)	12
Barnyard manure (rotted)	20
Clover residues	23
Green rye	36
Cane trash	50
Corn stover	60
Straw	80
Timothy	80
Sawdust	400

Data are taken from several sources. The values are approximate's only, and the ratio in any particular material may vary considerably from the values given.

Denitrification

Denitrification is the reduction of nitrate or nitrite to molecular nitrogen or nitrogen oxides by microbial activity. Many of the denitrification products are gaseous and escape from the soil. Just as it is natural for nitrogen to be added to soils by fixation, it is natural for nitrogen to be lost from the soil by denitrification. In ecosystems where the total amount of nitrogen remains constant from year to year, the annual additions of nitrogen via fixation tend to balance the losses of nitrogen by denitrification. Thus, denitrification is one of the most significant processes in the nitrogen cy-

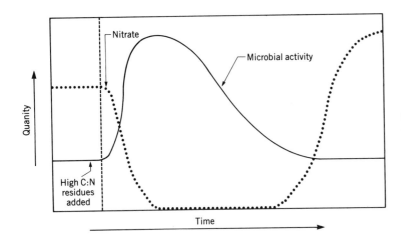

FIGURE 12.6 Addition of organic material with high C : N ratio results in depletion of soil nitrate. The activity of the decomposers is greatly stimulated and they use whatever nitrogen is present. In time, most of the added material is decomposed and the soil nitrate level increases.

cle, because denitrification accounts for large losses of nitrogen from soils.

Denitrification is carried out by certain facultative and anaerobic organisms. Nitrate-nitrogen in poorly drained soil is especially subject to denitrification. Anaerobic conditions may exist in well-drained soils after rains, when the interiors of soil aggregates are wet. Denitrification is shown as step 5 in Figure 12.2, and the reaction is as follows.

$$C_6H_{12}O_6 + 4NO_3^- = 6CO_2 + 6H_2O + 2N_2$$

Nitrite (NO_2^-), nitric oxide (NO), and nitrous oxide (N_2O) are also produced.

Normally, denitrification is detrimental to agriculture, because nitrogen is lost from soils. Denitrification, however, helps to prevent an excess of nitrate in the groundwater of irrigated valleys where high rates of nitrogen fertilizer have been used. In a large irrigated valley, denitrification occurs as nitrate in shallow groundwater moves slowly down the length of the valley. This denitrification contributes to a reduced nitrate content of the drainage water that exits the valley, and thus less threat of nitrate pollution. Where nitrate is carried downward below the biologically active zone where denitrifiers and an energy source exist, denitrification does not occur. Thus, nitrate in deep ground water is not denitrified.

Near Atlanta, Georgia, there is a large sewage disposal project that uses sprinklers to dispose of the effluent in a forest. The rate of effluent application is controlled to produce an anaerobic soil environment and denitrification of the nitrates in the effluent before the water leaves the disposal area. This use of denitrification is likely to increase as more and more communities become involved in disposing of sewage effluent without contaminating groundwater, and at the same time, contributing water to groundwater recharge.

Human Intrusion in the Nitrogen Cycle

The amount of nitrogen in a corn crop is large (see Table 12.2) compared to the nitrogen-supplying power of soils. The United States produces more than 50 percent of the world's corn, and this requires a large amount of nitrogen fertilizer. The balance sheet for the earth's nitrogen cycle in Figure 12.7 shows that the amount of industrially fixed nitrogen was equivalent to the terrestrial (natural) fixation and about twice as great as the amount of nitrogen fixed by legume crops. The biosphere now receives about 11 percent more nitrogen than is lost. It is obvious that humans have become an important factor in the nitrogen cycle through industrial nitrogen fixation. The long-term effects of a nitrogen buildup in

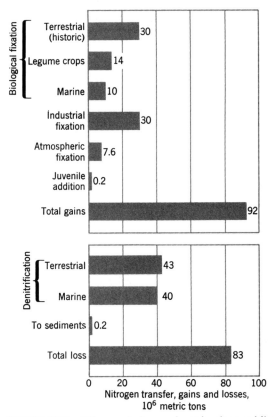

FIGURE 12.7 Nitrogen balance sheet for the world's biosphere. Nine million metric tons (92–83) more nitrogen is being added than is being removed by denitrification (and loss to sediments). (Data from C. C. Delwiche, *"The Nitrogen Cycle,"* 1970, Scientific American, Inc. All rights reserved.)

the biosphere are unknown. There is, however, increased potential for groundwater pollution and eutrophication of lakes. In agriculture, as long as the use of nitrogen fertilizer does not result in adding more nitrogen than crops and microorganisms can immobilize, there is little danger of groundwater pollution.

Summary Statement on Nitrogen Cycle

For the earth there is a quasi-equilibrium between fixation and denitrification that maintains a rather constant soil nitrogen content. It is difficult to increase the nitrogen content of soils because of the natural, large losses of nitrogen by denitrification (and also leaching and volatilization).

Within the soil there is an internal nitrogen cycle that shuttles nitrogen back and forth through mineralization, nitrification, and immobilization. The amount of nitrogen available for crops is the nitrogen in excess of the needs of the mineralizing microorganisms. The low nitrogen-supplying power of soils results in large additions of nitrogen to soils as fertilizers to meet the nitrogen needs of high-yielding nonleguminous crops. The large amount of industrially fixed nitrogen has resulted in a significant intrusion in the earth's nitrogen cycle by humans.

Plant Nitrogen Relations

Nitrogen is a part of chlorophyll, and nitrogen-deficient plants have light-green or yellow leaves. Nitrogen-deficient plants are stunted, producing low crop yields. The yellowish color of lawns in the growing season is usually an indication of nitrogen deficiency. Many plants develop distinct nitrogen-deficiency symptoms, including yellow midribs of the lower leaves of corn (maize) plants, as shown in Color Plate 1. Yellow and nitrogen-deficient corn leaves have at low nitrate content, according to tissue analysis (see Color plate 1).

An excess of nitrogen causes rapid vegetative growth and dark-green leaves. The vegetation is succulent and has reduced resistance to injury from insects, disease, and frost. For some crops, succulent tissue means greater damage from mechanical harvesting and reduced storability. Cereal grains develop tall, weak stalks that are easily blown over or broken during rainstorms, when the heads are heavy and filled with grain near harvest time.

PHOSPHORUS

The earth's crust contains about 0.1 percent phosphorus (P). On this basis, the phosphorus in a plow layer is equivalent to the phosphorus in

20,000 bushels of corn for an acre. This does not include phosphorus that could be absorbed by roots at depths below the plow layer. Phosphorus, however, commonly limits plant growth. The major problem in phosphorus uptake from soils by roots is the very low solubility of most phosphorus compounds, resulting in a low concentration of phosphate ions in the soil solution at any one time.

Soil Phosphorus Cycle

Most phosphorus occurs in the mineral *apatite* in igneous rocks and soil parent materials. Fluorapatite ($Ca_{10}(PO_4)_6F_2$) is the most common apatite mineral. Fluorapatite contains fluorine (F), which contributes to a very stable crystalline structure that has great resistance to dissolution or weathering. The structure is similar to that of teeth dentine. Fluoridation of water and tooth paste is designed to incorporate fluorine into dentine to increase resistance to tooth decay.

Apatite weathers slowly, producing the phosphate ion, $H_2PO_4^-$, which exists in the soil solution, as shown as step 1 of Figure 12.8. The $H_2PO_4^-$ is immobilized when roots and microorganisms absorb it and convert the phosphorus into organic compounds. This results in a significant amount of phosphorus in soils as organic phosphorus. Commonly, 20 to 30 percent of the phosphorus in plow layers of mineral soils is organic phosphorus. As with nitrogen, organic phosphorus is mineralized by microorganisms and is again released to the soil solution as $H_2PO_4^-$. The organic phosphorus cycle, a subcycle in the overall soil phosphorus cycle, is similar to the soil nitrogen cycle, where phosphorus is shuttled back and forth by mineralization and immobilization (steps 2 and 3 in Figure 12.8).

A major difference between the nitrogen and phosphorus cycles in soils is that the available form of nitrogen is mainly nitrate and is stable in the soil solution. Nitrate remains in solution and moves rapidly to roots by mass flow, unless denitrified or leached. The phosphate ion, by contrast, quickly reacts with other ions in the soil solution,

FIGURE 12.8 Major processes in the soil phosphorus cycle. The availability of phosphorus to plants is determined by the amount of phosphorus in the soil solution.

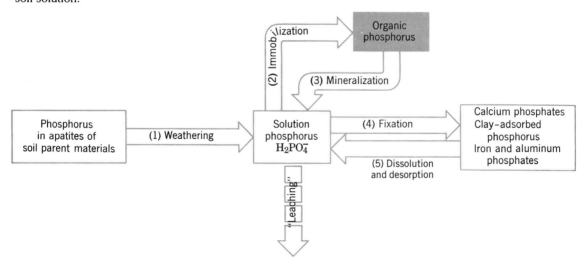

resulting in precipitation and adsorption to mineral colloids that convert the phosphorus to an unavailable or fixed form (step 4 in Figure 12.8). As a consequence, most of the phosphate ions from mineralization of organic phosphorus, or mineral weathering, may be converted to an unavailable form before plants have an opportunity to absorb the phosphorus and before loss by leaching can occur. The kinds of ions in the soil solution that render phosphate insoluble are related to soil pH.

Effect of pH on Phosphorus Availability

The ions in the soil solution are a function of pH. In calcareous parent materials and soils, the pH is commonly in the 7.5 to 8.3 range. In these soils, apatite is the dominant phosphorus mineral. Apatite has very low solubility, and even plants in virgin soils being cropped for the first time, have difficulty satisfying their phosphorus needs. The ions in the soil solution include Ca^{2+} from the hydrolysis of calcium carbonate. Any phosphate ions released, predominantly $HPO_4{}^{2-}$ above pH 7.2, tend to precipitate as insoluble calcium phosphate compounds that are slowly converted to apatite. Therefore, in calcareous soils most of the phosphorus tends to exist as apatite and tends to remain as apatite. Phosphorus availability to plants is low.

Over the course of soil evolution in humid regions, the calcium carbonate dissolves, the calcium is leached, and the soils become acid. In acid soils, there is much less Ca^{2+} and much more Al^{3+} and Fe^{3+} in solution than in calcareous soils. This results in precipitation of phosphate ions as insoluble aluminum and iron phosphates. Again, these phosphorus compounds have very low solubility, resulting in a low concentration of phosphate ions in solution.

The formation of these insoluble calcium, aluminum, and iron compounds is called phosphorus *fixation* (step 4 in Figure 12.8). In the case of phosphorus, the fixation, is usually the result of precipitation or adsorption. Phosphate ions are fixed by adsorption onto clays and other soil constituents. Fixed phosphorus dissolves or is released slowly into the soil solution (step 5 of Figure 12.8). The very low concentration of phosphorus in the soil solution results in (1) very little phosphorus being moved to roots by mass flow, and (2) minimal loss of phosphorus from soils by leaching. The optimum pH for phosphorus availability is near 6.5, where there is the least potential for phosphorus fixation. Even at this pH only a very small amount of phosphorus is found in the soil solution, the amount in solution being of the order of 0.04 pounds per acre-furrow-slice.

Newly formed aluminum and iron phosphates are amorphous (not crystalline) and have a large surface area per gram. With time, the amorphous forms become more crystalline, as shown in Figure 12.9. There is a large decrease in surface area with increasing crystallinity, and this is associated with decreased solubility or availability of the aluminum and iron phosphate minerals. In an experiment with Sudan grass grown in quartz sand, plants growing in soil that received crystalline forms of iron and aluminum phosphate grew somewhat like plants that had not received phosphate. By contrast, the plants growing in soil that received colloidal iron or aluminum phosphate grew much better and absorbed much more phosphorus, as shown in Table 12.5.

Changes in Soil Phosphorus Over Time

The forms of phosphorus in soils change during the long period of soil evolution. In young calcareous soil, the phosphorus is mainly apatite phosphorus. Plant growth and organic matter accumulation in the soil result in the conversion of a significant amount of apatite phosphorus into organic phosphorus. As soils become acid, phosphorus is fixed as iron and aluminum phosphate. In acid soils, phosphorus released by dissolution of apatite is gradually converted into iron and aluminum phosphates. When first formed, the iron and aluminum phosphates are amorphous and, over time, they gradually become crystalline.

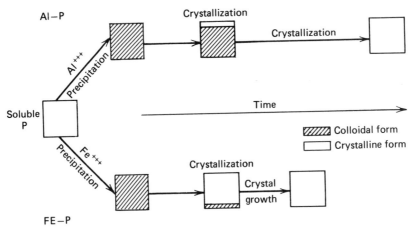

FIGURE 12.9 Transformation of soluble phosphorus into less available forms. Availability decreases with increases in crystallinity. (From A. S. R. Juo, and B. G. Ellis, 1968. "Chemical and Physical Properties of Iron and Aluminum Phosphates and Their Relation to Phosphorus Availability." *Soil Sci. Soc. Am. Proc.* **32**:216–221. Used by permission of the Soil Science Society of America.)

Eventually, in intensively weathered soils rich in iron and aluminum oxides, the crystalline forms of iron and aluminum phosphate become encapsulated (occluded) by iron and aluminum oxides. Occluded phosphorus is the least available form of soil phosphorus. Note in Figure 12.10 that most of the phosphorus is calcium phosphate in a minimally weathered North Dakota soil, little phosphorus is aluminum and iron phosphate, and there is no occluded phosphorus. By contrast, the intensively weathered soil from Hawaii, has no calcium phosphate, and a small amount of iron and aluminum phosphorus, but the phosphorus is mainly occluded. These soils have the least ability to supply plants with phosphorus. The moderately weathered Wisconsin soil is intermediate and is the best suited for supplying phosphorus to plants.

Plant Uptake of Soil Phosphorus

The dominant form of phosphorus available to plants exists in the soil solution mainly as $H_2PO_4^-$ below pH 7.2 and mainly as HPO_4^{2-} above pH 7.2. The potential for P fixation tends to maintain phosphorus in insoluble forms and, therefore, the phosphorus concentration in the soil solution is very low. Thus, mass flow cannot supply plants with sufficient phosphorus except in unusual cases. As a result, most of the $H_2PO_4^-$ at root surfaces has been moved there by diffusion. It has been reported that about 90 percent of the phosphorus in corn is supplied by diffusion and about

TABLE 12.5 Yield and Phosphorus Uptake by Sudan grass after Growing 30 Days in Quartz Sand Culture

Form of Phosphorus	Phosphorus Uptake[a]	Yield[a]
Colloidal Al-P	100	100
Colloidal Fe-P	107	96
Crystalline Al-P	22	47
Crystalline Fe-P	13	28
No P	13	27

Adapted from Juo and Ellis, 1968.
[a] Relative to colloidal aluminum phosphate as 100.

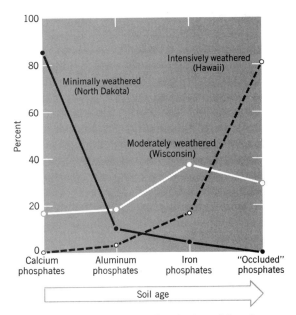

FIGURE 12.10 Percentage distribution of four forms of phosphorus as related to weathering or soil age. (Adapted from S. C. Chang and M. L Jackson. 1958. "Soil Phosphorus Fractions in Some Representative Soils." *Jour. Soil Sci.* **9**:109–119. Used by permission.)

10 percent is supplied by mass flow and root interception. Diffusion of phosphorus to roots is the major limiting step in phosphorus uptake from soils by roots.

Obviously, the greater the phosphate concentration of the soil solution, the greater is the availability of soil phosphorus. The problem in agriculture is to maintain an adequate phosphorus concentration in the soil solution, even though there is rapid fixation of phosphorus. An adequate supply of solution phosphorus can result from the use of phosphorus fertilizers. The first fixation compounds, whether in calcareous or acid soils, tend to be amorphous, and they are more available than the crystalline apatite and aluminum and iron phosphates, which form slowly over time. The insoluble crystalline forms require years to form, but in the meantime, the amorphous and

adsorbed phosphorus formed from the use of phosphorus fertilizers can provide adequate solution phosphorus for high crop yields on most soils.

Fixation of fertilizer phosphorus results in low uptake of the fertilizer phosphorus during the year of application. Commonly, only 10 to 20 percent of phosphorus in fertilizer is used by plants during the year of fertilizer application. Therefore, repeated use of phosphorus fertilizers results in an increase in soil phosphorus content, and in many instances, soils have become sufficiently high in phosphorus that further additions of phosphorus fertilizers do not increase plant growth.

Plant Phosphorus Relations

Phosphorus plays an indispensable role as a universal fuel for all biochemical activity in living cells. High-energy adenosine triphosphate (ATP) bonds release energy for work, when converted to adenosine diphosphate (ADP). Phosphorus is also an important element in bones and teeth. The relation of phosphorus in soils to plants and animal health, and the extensive occurrence of phosphorus deficiency in grazing animals, are well known.

Some of the phosphorus deficiency symptoms of plants are not specific. The growth of both shoots and roots is greatly reduced. One of the common diagnostic aids is foliage color. Phosphorus is mobile in plants and, when a deficiency of phosphorus occurs, phosphorus is moved from the older leaves to the younger leaves at the top of the plant. Deficiency symptoms first appear on the older, or bottommost, leaves. The deficiency symptom is commonly a purplish color as a result of increased anthocyanin development in phosphorus-deficient tissue, as shown for corn in Color Plate 2.

Low phosphorus in animal diets is a common problem. Some relationships between soil phosphorus levels, phosphorus levels in plant tissue, and phosphorus deficiency in dairy cattle are shown in Figure 12.11. As the amount of fertilizer

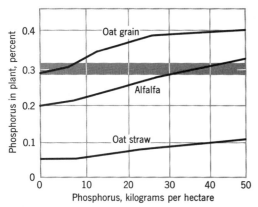

FIGURE 12.11 Effect of phosphorus fertilizer application on concentration of phosphorus in oats and alfalfa. About 0.3 percent of phosphorus (dashed line) is required by dairy cattle for normal growth. (From W. H. Allaway, 1975.)

phosphorus increased (or available soil phosphorus increased), the phosphorus content of oat grain and alfalfa hay was markedly increased. Because cattle require about 0.3 percent phosphorus in their diet for optimum growth, oat grain contained sufficient phosphorus without fertilization. Oat straw remained phosphorus deficient for cattle, even with phosphorus fertilization of the oat crop. Cattle diets consisting mainly of grasses usually require a phosphorus supplement to insure adequate dietary phosphorus. The use of phosphorus fertilizer changed alfalfa hay from an inadequate phosphorus source to an adequate source. It is interesting to note that the U.S. racehorse industry is centered on the high phosphorus soils of Kentucky and Florida.

Phosphorus deficiency is not a serious problem in human nutrition, because of the large amount of cereal grains and meat consumed, which are good phosphorus sources. For humans, the use of phosphorus fertilizer is more important for increasing the quantity of food than for increasing the phosphorus content of the food.

POTASSIUM

There is a wide range in the potassium content of soils and availability of potassium for plant growth. Some soils are very deficient in available potassium, whereas others are very sufficient. This is in stark contrast to the general needs of nitrogen and phosphorus for high-yielding crops. Basically, potassium in soils is found in minerals that weather and release potassium ions (K^+). These ions are adsorbed onto the cation exchange sites. The exchangeable potassium tends to maintain an equilibrium concentration with the soil solution from which roots absorb K^+. Organic soils tend to be deficient in available potassium because they contain few minerals with potassium.

Soil Potassium Cycle

The earth's crust has an average potassium content of 2.6 percent. Parent materials and youthful soils could easily contain in a plow layer 40,000 to 50,000 pounds per acre, or kilograms per hectare. The potassium content below the plow layer could be similar. About 95 to 99 percent of this potassium is in the lattice of the following minerals:

Feldspars
   Microcline $KAlSi_3O_8$
   Orthoclase $KAlSi_3O_8$
Micas
   Muscovite $H_2KАl_3(SiO_4)_3$
   Biotite $(H,K)_2(Mg,Fe)_2Al_2(SiO_4)_3$

Micas, especially biotite, weather faster and release their potassium much more readily than do feldspars. The feldspars tend to exist as larger particles than micas, with the feldspars largely in the sand and silt and the micas in silt and clay fractions.

Weathering of feldspar results in the dissolution of the feldspar crystal and release of K^+ to the soil solution. The potassium in micas is interlayer potassium. Weathering of micas results in the migration of K^+ out of the interlayer space along the edge of weathering mica particles; the weathered edge of mica is shown in Figure 12.12. The weathered edge of the mica particle in Figure 12.12 is vermiculite.

The K^+ appears in the soil solution or is adsorbed onto a cation exchange site (step 1 in Figure 12.13). An equilibrium tends to be established between the solution potassium and exchangeable potassium (steps 2 and 3). The amount of solution potassium is about 1 percent as large as the exchangeable potassium.

As weathering proceeds, in the absence of the removal of K^+ from the soil by plant uptake or leaching, there is a buildup of exchangeable potassium as a result of weathering. The exchangeable potassium maintains a quasi-equilibrium with the potassium entrapped or fixed, as shown in step 4 in Figure 12.13. Fixation is the reverse of the weathering of mica and occurs as the amount of exchangeable potassium increases and K^+ move back into voids where interlayer potassium weathered out at an earlier time. Release, the opposite of fixation and shown as step 5, is encouraged when plant uptake and leaching (steps 6 and 7 in Figure 12.13), result in reduced solution potassium. A reduced concentration of solution potassium is followed by reduced exchangeable potassium, which is then followed by the release of fixed potassium.

FIGURE 12.12 Weathered mica silt particle showing loss of potassium from the darkened area around the edges. The loss of potassium results in expansion of the 2:1 mica layers. The weathered zone is vermiculite. (Photograph courtesy M. M. Mortland.)

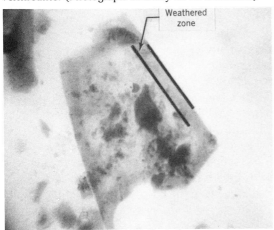

Plant removal and leaching tend to occur during the spring and summer, when the supply of available or exchangeable potassium is reduced. In winter, by contrast, plant uptake and leaching of potassium may be minimal and the release of potassium by weathering results in an increase in exchangeable potassium. In other words, in winter the direction of the net effect of the reactions is of the order 1, 2, and 4. In spring and summer, the direction is 6 (and 7), 3, and 5.

Potassium fixation is affected by the equilibrium conditions in the soil and soil drying and wetting. The application of potassium fertilizer greatly increases the amount of solution potassium. Drying the soil at this time results in increased concentration of K^+ in the soil solution at the edges of clay particles and the movement of potassium into the interlayer spaces, where it becomes entrapped or fixed. By contrast, soon after the harvest of a crop, the concentration of K^+ in the soil solution is low. Drying the soil at this time results in the migration of potassium from interlayer spaces and increases the amount of available or exchangeable potassium.

Summary Statement When plants are dormant and leaching is minimal, weathering of potassium minerals results in an increase in solution potassium, followed by an increase of exchangeable potassium, followed by an increase in the amount of fixed potassium. Normally, this occurs in fall and winter in minimally and moderately weathered soils. When plant uptake and leaching occur, normally in spring and summer, there is a decrease in solution potassium, followed by a decrease of exchangeable potassium, and then a decrease in the amount of fixed potassium.

Weathering and subsequent loss of potassium by leaching with time can deplete the soil of potassium minerals and create soils with few potassium minerals and very low available potassium. This is generally true for intensively weathered soils.

Potassium does not complex with organic com-

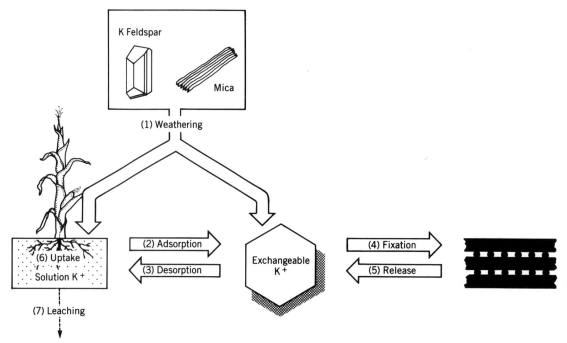

FIGURE 12.13 Major processes in the soil potassium cycle.

pounds; that is, it is not an integral part of any organic molecules in plants. Therefore, potassium availability is only minimally related to the soil's organic matter content in mineral soils. In fact, the most potassium-deficient soils are those composed mostly of organic matter, such as organic soils.

Plant Uptake of Soil Potassium

The soil's ability to maintain a desirable concentration of solution potassium is determined by the it's *potassium buffer capacity*. The potassium buffer capacity is the ability of the solid phase of the soil to buffer changes in the K^+ concentration of the soil solution. Soils with high potassium buffer capacity tend to be fine-textured and contain a large amount of fixed potassium or interlayer potassium (contain abundant mica and vermiculite)

that can be released. In these soils, the solution potassium is quickly replenished when removed by roots. It would seem that potassium fixation is completely beneficial because potassium is protected from leaching, yet has good potential for plant availability in the future. However, some soils have an unusually large potassium-fixing capacity (vermiculitic soils), which competes with plants for potassium in fertilizer that is applied to soils, and much of the applied potassium is "lost" to fixation.

The clays in intensively weathered soils are mainly kaolinite and oxidic clay. These soils contain little if any fixed potassium and have low potassium buffering capacity.

In many minimally and moderately weathered soils, the exchangeable potassium is about equal to the needs of the crop, and the amount in solution at any one time is equal to 1 to 3 percent of the exchangeable. As with the case of phosphorus, the amount of potassium in the soil water is small compared with plant needs. The result is that diffusion is very important for moving potassium to roots for uptake, because mass flow cannot provide an adequate amount. For corn, it has been reported that diffusion accounts for about 80 percent of the potassium uptake. Diffusion is the rate-limiting step for plant uptake of potassium from most soils.

Very high potassium fertilization may produce an unusually high concentration of solution potassium. Then, plants may absorb much more potassium than is needed for growth-producing *luxury uptake*, which is the absorption of nutrients by plants in excess of their need for growth. The luxury uptake of potassium reduces the uptake of other cations, especially magnesium, and, to a lesser extent, calcium (see Figure 12.14).

Plant Potassium Relations

Potassium enhances the synthesis and translocation of carbohydrates, thereby encouraging cell wall thickness and stalk strength. A deficiency is sometimes expressed by stalk breakage, or *lodging*. Potassium can be substituted for by sodium in some of its physiological roles in some plants.

The deficiency symptoms are the most severe on old growth, because potassium is mobile in plants. Potassium deficiency symptoms are quite characteristic on many plants as a yellowing, and eventually death, of leaf margins. For alfalfa, white spots first appear along leaf margins followed by a coalescing of the spots to give a marginal edge effect. The dying of the edges of older leaves is commonly called *necrosis*. Potassium deficiency symptoms of corn and alfalfa are shown in Color Plate 1.

CALCIUM AND MAGNESIUM

There are many similarities between the behavior of calcium and magnesium with those of potassium in soils. They are all released from miner-

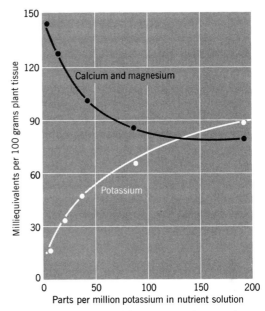

FIGURE 12.14 The calcium, magnesium, and potassium content of alfalfa plants are a function of the potassium concentration in the nutrient solution. (Adapted from Wallace, 1948.)

als by weathering and occur as exchangeable cations. They are the most abundant in young and minimally weathered soils and least abundant in intensively weathered and leached soils. All three elements are absorbed by roots as cations from the soil solution.

Some important calcium minerals include calcite, dolomite, gypsum, feldspar, apatite, and amphibole. Important magnesium minerals include: dolomite, biotite, serpentine, hornblende, and olivine. In contrast to potassium, there is no significant fixation of calcium or magnesium into unavailable forms. The cations set free by weathering are adsorbed onto the cation exchange sites. An equilibrium tends to be established between the exchangeable and solution forms. Compared to potassium, the amounts of exchangeable and solution calcium and magnesium are usually many times larger than the amounts of exchangeable and solution potassium. By contrast, plant needs for calcium and magnesium are much less than for potassium. Mass flow commonly moves more calcium and magnesium to root surfaces than plants need, whereas at the same time, plants need more potassium than is moved to root surfaces. Thus, calcium and magnesium limit plant growth only occasionally. In fact, calcium deficiencies in plants are rare, and magnesium deficiencies occur only occasionally. When deficiencies of calcium do occur, it is usually on intensively weathered soils.

Magnesium is less strongly adsorbed to cation exchange sites than is calcium. The cation exchange capacity of neutral soils may be 75 percent or more calcium saturated and only 15 per cent magnesium saturated. Magnesium deficiency is usually associated with acidic sandy soils. In Michigan, acidic sandy soils with less than 75 pounds of exchangeable magnesium per acre-furrow-slice are likely to be magnesium deficient for corn. By contrast, soils formed from serpentinite (magnesium rich) parent material contain abundant magnesium and very low exchangeable calcium. The extreme imbalance between calcium and magnesium causes severe calcium deficiency. Plant growth on such soils is meager and the soils are called *serpentine barrens*.

Plant Calcium and Magnesium Relations

Magnesium is a constituent of chlorophyll. As with several other nutrients, a magnesium deficiency results in a characteristic discoloration of leaves. Sometimes, a premature defoliation of the plant results. The chlorosis of tobacco, known as *sand drown,* is due to magnesium deficiency. Cotton plants suffering from a lack of magnesium produce purplish-red-colored leaves with green veins. Leaves of sorghum and corn become stripped; veins remain green with yellow intervein strips (as shown for corn in Figure 12.15). The magnesium is mobile in plants and is moved from the older, lower leaves to the younger, upper leaves. This causes the lower leaves to exhibit the deficiency symptoms. The magnesium-deficiency symptom on coffee leaves is shown in Color Plate 4.

Calcium is immobile in plants. When a deficiency occurs, there is malformation and disintegration of the terminal portion of the plant. The deficiency symptom has been established for many plants by the use of greenhouse methods. As indicated earlier, calcium-deficiency symptoms are seldom seen in fields. This is because other serious nutrition problems usually occur before calcium becomes sufficiently deficient.

Soil Magnesium and Grass Tetany

Grass tetany (grass staggers) is a major disease of cattle, which is characterized by muscle spasms, especially in the extremities. The disease is due to a magnesium deficiency. The incidence of grass tetany is more common when cattle consume a diet consisting mainly of grass rather than legumes. This is because the magnesium content of grasses is lower than that of legumes. The incidence of grass tetany is generally low when grasses have 0.2 percent or more magnesium. The incidence of grass tetany is related to levels of available soil magnesium, plant species, and temperature. Conditions that favor the incidence of grass tetany include: (1) very low exchangeable magnesium (2) reduced uptake of magnesium by plants caused by high levels of available potassium (see Figure 12.14), sometimes caused by the application of potassium fertilizers; and (3) cool, wet spring weather that encourages the growth of grasses, relative to legumes, in pastures where the animals graze.

SULFUR

Sulfur exists in some soil minerals, including gypsum ($CaSO_4 \cdot 2H_2O$). Mineral weathering releases the sulfur as sulfate (SO_4^{2-}) which is absorbed by roots and microorganisms. Sulfur is released as

FIGURE 12.15 Magnesium deficiency symptoms on corn. The veins are green and the intervein areas of the lower leaves are yellow.

sulfur dioxide (SO_2) into the atmosphere by the burning of fossil fuels and becomes an important constitutent of the precipitation. In many locations, more sulfur is added to soils (via precipitation) than plants need. Sulfur accumulates in soils as organic sulfur in plant residues and is then mineralized to SO_4^{2-}. The net effect is the accumulation of sulfur in soils as organic sulfur. Plants, depending on the SO_2 content of the air, may absorb sulfur through leaf stomates. In the past, sulfur has been an important element in

fertilizers used in the United States. For these reasons, sulfur is seldom limiting for plant growth even though plants require about as much sulfur as phosphorus. Sulfur deficiencies tend to occur where there is little input of sulfur by precipitation due to low precipitation and little input of sulfur into the atmosphere by burning of fossil fuels. The output of sulfur is so great near some mining and smeltering operations that nearby vegetation is killed.

The addition of sulfur to soils via precipitation contributes to the development of soil acidity. Sulfate is adsorbed onto the surfaces of iron and aluminum oxides and onto the edges of silicate clays, where aluminum is located. As soil pH decreases, the capacity for sulfur adsorption increases. This is one reason why acid rain has had only a minimal effect on many soils; the more acid the soil becomes, the more sulfur is adsorbed and fixed or inactivated. Sulfur is lost from soils as sulfate by leaching and as H_2S gas, which is produced by microbial reduction of sulfates in anaerobic soils.

Areas in the United States in which crops have responded to sulfur are shown in Figure 12.16. Most of the locations occur where low amounts of sulfur are added to soils by precipitation and/or soil parent materials have low sulfur content. The annual addition of sulfur to soils from precipitation was more than 200 pounds per acre near Chicago, Illinois, and only 6 pounds per acre in eastern Nebraska. Note in Figure 12.16 that no response to sulfur is shown for Illinois, whereas some response to sulfur has been obtained in eastern Nebraska. Generally, the industrial Northeast receives sufficient sulfur from precipitation to ensure sufficient amounts for crops. In many irrigated fields in semiarid and arid regions, enough sulfur is added by the irrigation water to ensure adequate sulfur for crops.

Plants that have the greatest need for sulfur include: cabbage, turnips, cauliflower, onions, radishes, and asparagus. Intermediate sulfur users are legumes, such as alfalfa and cotton and tobacco. Grasses have the lowest sulfur require-

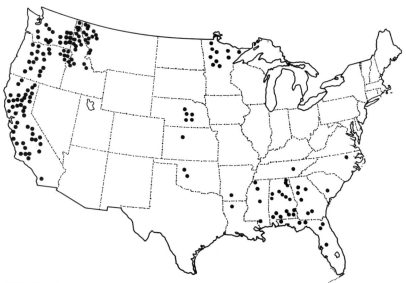

FIGURE 12.16 Locations where crops have responded to sulfur fertilizers. (Data from Jordan and Reisenauer, 1957.)

ment. Sulfur deficiency symptoms are similar to nitrogen in that sulfur-deficient plants are stunted and light-green to yellow in color.

SUMMARY

Macronutrients are used by plants in relatively large amounts and include nitrogen, phosphorus, potassium, calcium, magnesium, and sulfur.

Plant-deficiency symptoms for specific nutrients are related to the mobility of the nutrients within the plants.

Nitrogen availability in soils is controlled mainly by mineralization and immobilization. Nitrogen is added to soils by fixation. and an approximate amount is lost annually by denitrification. The dominant form of available nitrogen is nitrate, which is readily moved to roots by mass flow and lost from soils by leaching and denitrification.

Phosphorus originates from apatite weathering and exists in soils in many different insoluble minerals and in soil organic matter. Soil pH, by its control of the ions in solution, greatly affects phosphorus fixation and, therefore, phosphorus availability in soils. The low concentration of solution phosphorus limits phosphorus uptake by plants and results in very limited leaching of phosphorus.

Potassium is released by mineral weathering and the available supply is the exchangeable potassium. During periods of limited plant uptake and leaching, release of potassium by weathering results in increases in solution, exchangeable, fixed potassium in most minimally and moderately weathered soils. During periods of active plant uptake and leaching, there is a decrease in solution, exchangeable, and fixed potassium. The rate of potassium diffusion to roots is the limiting step in potassium uptake from soils.

Calcium and magnesium have many parallels with potassium in terms of the forms in soils. The amounts of exchangeable calcium and magnesium usually exceed the amount of exchangeable potassium, yet, plants use much more potassium

COLOR PLATE 1

Nitrogen deficiency on corn. Lower leaves turn yellow along midrib, starting at the leaf tip.

Left is nitrogen-deficient section of corn leaf and cut open section of stem that remains uncolored when treated with diaphenylamine, indicating low nitrate level in plant sap. On right the leaf is dark green and diaphenylamine produces a dark blue color indicative of high nitrate level in plant sap.

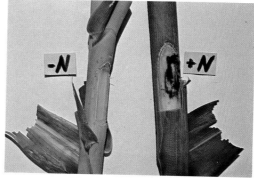

Potassium deficiency on alfalfa (and clovers) shows as a series of white dots near leaf margins; in advanced stages entire leaf margin turns white. (Courtesy American Potash Institute.)

Potassium deficiency on corn. Lower leaves have yellow margins.

COLOR PLATE 2

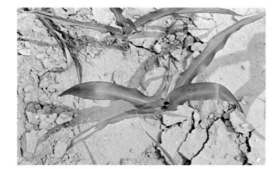

Phosphorus deficiency on corn is indicated by purplish discoloration.

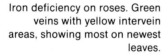

Iron deficiency on pin oak. The leaf veins remain green as the intervein areas lose their green color.

Iron deficiency on roses. Green veins with yellow intervein areas, showing most on newest leaves.

Manganese deficiency on kidney beans. Leaf veins remain green as the intervein areas lose their green color and turn yellow.

Zinc deficiency of navy beans grown on calcareous soil. The small unfertilized zinc-deficient plants stand in marked contrast to the taller plants that were fertilized with zinc. A case where zinc fertilization is necessary to produce a crop.

Boron deficiency on sugar beets causes heart rot. Most advanced symptom is on left.

COLOR PLATE 4

Magnesium deficiency on coffee. Veins remain green and intervein areas turn yellow.

Manganese-treated plants in rear showing no deficiency. Plants in foreground are unfertilized and difference in degree of manganese deficiency symptoms is indicative of varietal response to limited soil manganese.

Excess soluble salt symptoms on geranium. The leaf margins turn yellow and become necrotic.

than calcium or magnesium. Calcium is rarely deficient for plant growth, and such a deficiency occurs usually only in intensively weathered soils. Magnesium is sometimes deficient in plants growing on acid and sandy soils with low cation exchange capacity.

Sulfur is released from sulfur minerals. Much of the sulfur in soils has been added by precipitation. Near industrial areas, more sulfur is added to soils by precipitation than plants need. In some places, excessive additions of sulfur kill the vegetation.

REFERENCES

Allaway, W. H. 1975. "The Effect of Soils and Fertilizers on Human and Animal Health." *USDA Agr. Information Bul*. 378.

Barber, S. A. 1984. *Soil Nutrient Bioavailability*. Wiley, New York.

Black, C. A. 1968. *Soil-Plant Relations*. Wiley, New York.

Chang, S. C. and M. L. Jackson. 1958. "Soil Phosphorus Fractions in Some Representative Soils." *Jour. Soil Sci*. **9:**109–119.

Cook, R. L. and C. E. Millar. 1953. "Plant Nutrient Deficiencies." *Mich. Agr. Exp. Sta. Spec. Bull*. 353, East Lansing.

Delwiche, C. C. 1970. "The Nitrogen Cycle." *Scientific American,* September, pp. 137–146.

Foth, H. D. and B. G. Ellis. 1988. *Soil Fertility*. Wiley, New York.

Hanway, J. 1960. Growth and Nutrient Uptake by Corn. *Iowa State University Extension Pamphlet* 277.

Jordan, H. V. and H. M. Reisenauer. 1957. "Sulfur and Soil Fertilty." *USDA Yearbook Soil*, Washington, D.C.

Juo, A.S,R. and B. G. Ellis. 1968. "Chemical and Physical Properties of Iron and Aluminum Phosphates and Their Relation to Phosphorus Availability." *Soil Sci. Soc. Am. Proc*. **32:** 216–221.

Mengel, K. and E. A. Kirby. 1982. *Principles of Plant Nutrition*. 3rd. ed. International Potash Institute, Bern, Switzerland.

Mortland, M. M. 1958 "Kinetics of Potassium Release from Biotite." *Soil Sci. Soc. Am. Proc*. **22:**503–508.

Paul, E. A. 1984. "Dynamics of Organic Matter in Soils." *Plant and Soil*. **76:**275–285.

Stevenson, F. J. 1986. *Cycles of Soil*. Wiley, New York.

Wallace, A. et al. 1948. "Further Evidence Supporting Cation-Equivalent Constancy in Alfalfa." *J. A. Soc. Agron*. **40:**80–87.

Weber, C. R. 1966. "Nodulating and Nonnodulating Isolines." *Agron. Jour*. **58:**43–46.

CHAPTER 13

MICRONUTRIENTS AND TOXIC ELEMENTS

Micronutrients are required in very small amounts and function largely in plant-enzyme systems (see Table 13.1). The factors that determine the amounts of micronutrients available to plants are closely related to soil conditions and plant species. For example, a change in soil pH can change a deficient situation for plants into a toxic situation. Four of the seven micronutrients are used by plants as cations, and they will be considered first.

IRON AND MANGANESE

Iron and manganese are weathered from minerals and appear as divalent cations in solution; as such, they are available to plants. Generally, in acid soils, sufficient Fe^{2+} and Mn^{2+} exist in the soil solution to meet plant needs. In some very acid soils, manganese, and to a lesser extent iron, are toxic because of the high amounts in solution. Deficiencies are common in alkaline soils where oxidized forms of iron and manganese exist as insoluble oxides and hydroxides. As a consequence, deficiencies are common in arid regions where many soils are calcareous and alkaline. In humid regions, many calcareous soils exist on recent lake plains and on exposed subsoil or parent materials. At building sites, such as homesites, acid soil may be contaminated with enough mortar from brick wall construction, or the surface soils may be mixed with highly calcareous parent material during excavation and become alkaline. Shrubs and flowers grown on these locations commonly have deficiency symptoms for iron and/or manganese (see Figure 13.1).

Cereals and grasses, including sugarcane, tend to have a manganese deficiency when grown on alkaline soils. Plants particularly susceptible to iron deficiency include: roses, pin oaks, azaleas, rhododendrons, and many fruits and ornamentals.

Deficiency symptoms of iron and manganese are striking and, for some plants, are very similar. In some cases, it is impossible to know whether the symptom is due to an iron deficiency or a manganese deficiency unless trials are conducted to determine which element alleviates the symptoms. In other cases, it is known from experience which element is deficient when the deficiency symptoms occur. Typical iron-deficiency symptoms or *iron-chlorosis* appears on leaves as yellow interveinal tissue and dark-green-colored

TABLE 13.1 Essential Micronutrient Elements and Role in Plants

Element	Role in Plants
Boron (B)	Somewhat uncertain, but believed important in sugar translocation and carbohydrate metabolism.
Iron (Fe)	Chlorophyll synthesis and in enzymes for electron transfer.
Manganese (Mn)	Controls several oxidation-reduction systems, formation of O_2 in photosynthesis.
Copper (Cu)	Catalyst for respiration, enzyme constituent.
Zinc (Zn)	In enzyme systems that regulate various metabolic activities.
Molybdenum (Mo)	In nitrogenase needed for nitrogen fixation.
Chlorine (Cl)	Activates system for production of O_2 in photosynthesis.
Cobalt[a] (Co)	Essential for symbiotic nitrogen fixation by *Rhizobium*.

* Compiled from many sources.

[a] Not essential for all vascular plants

veins, as shown for pin oak and roses in Color Plate 2. The younger leaves tend to be most affected because iron is immobile in plants.

The absence of sufficient manganese stunts tomatoes, beans, oats, tobacco, and various other plants. Associated with this stunting is a chlorosis of the upper leaves. The interveinal tissue turns yellow, whereas the veins remain dark-green-colored, as shown for kidney beans in Color Plate 3. The gray-speck disease of oats is attributed to manganese deficiency and appears as gray-colored areas on stems. For many nutrients, varietal or cultivar differences result in differing plant responses to the same level of available nutrients. This is illustrated by varietal response to manganese by soybeans as shown in Color Plate 4.

Many chlorotic plants growing on alkaline soils have an iron deficiency and many have a manganese deficiency. Zinc may also be a contributing factor. Healthy green pin oak trees were found to need 55 centimeters of acid soil when growing in an area where many soils were alkaline and the trees were chlorotic (see Figure 13.2).

Plant "Strategies" for Iron Uptake

Iron is fourth in abundance in the earth's crust, following oxygen, silicon, and aluminum. Yet, iron deficiencies in plants are common because the amount of soluble iron in the soil solution is frequently too low to meet plant needs. Plants have developed special "strategies" for coping with this situation in alkaline soils where the oxidized or ferric form of iron has very low solubility.

Plants increase their ability to absorb iron from calcareous and highly aklaline soils in two ways. First, plant roots decrease the pH in the rhizosphere by the excretion of H^+ that solubilizes ferric iron. Second, for monocots (grasses), *siderophores* are excreted by the roots. Siderophores (iron bodies) are metabolites secreted by organisms that form a highly stable coordination compound (organic chelate) with iron. The siderophores solublize ferric iron, which is subsequently absorbed by the roots and the iron used by the plant. Microorganisms also excrete siderophores. These mechanisms enable certain plant species to satisfy their iron needs when growing on alkaline soils. Many plants, however, are not

FIGURE 13.1 Iron-deficient symptoms frequently occur on isolated pin oak trees growing in areas where soils are normally acid. Because of local contamination of the soil with alkaline materials from road and building construction, the soil is alkaline. Note die-back at the top of the tree. Chlorotic leaves are yellow with green-colored veins.

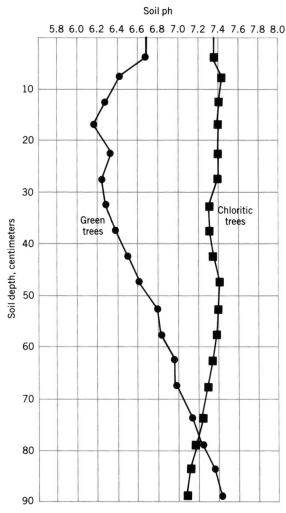

FIGURE 13.2 Soil reaction (pH) profiles associated with untreated green and chlorotic pin oak trees. Pin oak trees were healthy if the upper 55 centimeters of soil was acid. (Data from Messenger, 1983.)

efficient in utilizing ferric iron and develop deficiency symptoms or fail to grow.

COPPER AND ZINC

Copper and zinc are released from mineral weathering to the soil solution as Cu^{2+} and Zn^{2+}. These micronutrient cations can be adsorbed onto cation exchange sites. Little Cu^{2+} exists as exchangeable copper, however, but tends to be strongly adsorbed to the inorganic fraction or complexed with organic matter. As a result, copper is quite immobile in soils and the copper concentration of soil solutions tends to be very low. Strongly adsorbed and nonexchangeable copper equilibrate with Cu^{2+} in solution, and much of the copper in solution is complexed with soluble organic matter of low molecular weight. Even though there is a very low concentration of copper in solution, plants require very little copper, which is generally obtained in sufficient quantity by root interception and mass flow. Copper deficiencies in plants tend to occur on newly developed organic soils and leached sandy soils that are characterized by low amounts of total copper. Copper is not easily leached and, after about 20 to 40 pounds of copper per acre have been applied to deficient soils, little additional copper is needed.

Zinc, as Zn^{2+}, occurs as an exchangeable cation, is strongly absorbed onto several soil constitutents, and is complexed by organic matter. There is considerable similarity in the forms of zinc and copper in soils and plant uptake relationships. Zinc, however, appears to be complexed with organic matter to a lesser degree and has sharply reduced availability with increasing soil pH.

Plant Copper and Zinc Relations

Copper is quite immobile in plants and deficiency symptoms are highly variable among plants. In cereals the deficiency shows first in the leaf tips at tillering time. The tips become white and the leaves appear narrow and twisted. Growth of the internodes is depressed. In fruit trees, gum pockets under the bark and twig dieback occur.

Zinc deficiency was first recognized in plants growing on organic soils in Florida. Zinc deficiencies occur similarly to copper on organic soils and leached acid sandy soils from which zinc has been leached. Zinc is commonly deficient on alkaline and/or calcareous soils. An extreme case of zinc deficiency of navy beans on calcareous soil is shown in Color Plate 3. Application of phosphorus fertilizers at high rates has also been found to reduce zinc availability in soils, especially when levels of soil zinc are only marginally sufficient (see Figure 13.3).

Zinc deficiencies are widespread in the United States on corn, sorghum, citrus and other fruits, beans, vegetable crops, and ornamentals. Pecan rosette, the yellows of walnut trees, the mottle-leaf of citrus, the little-leaf of the stone fruits and grapes, the white-bud of corn, and the bronzing of tung tree leaves are all due to zinc deficiency. In tobacco plants, a zinc deficiency is characterized by a spotting of the lower leaves.

FIGURE 13.3 Navy bean plants growing on soils low in available zinc grew less well with increasing amount of applied phosphorus (zero to 720 pounds per acre) fertilizer owing to reduced availability of zinc to the plants.

BORON

The most abundant boron mineral in soils is tourmaline, a borosilicate. The boron is released by weathering and occurs in the soil solution mostly as undissociated boric acid, H_3BO_3. The boron in solution tends to equilibrate with adsorbed boron.

Boron is leached from acid sandy soils, and this results in low availability. Boron availability is reduced by increasing pH due to strong boron adsorption to mineral surfaces (fixation). Thus, boron availability is a maximum in soils with intermediate pH. Boron deficiency tends to occur in dry weather; which is likely because of reduced movement of boron to roots by mass flow in water.

Plants absorb boron as uncharged boric acid, and it appears to be a simple diffusion of the uncharged molecules into roots. Once the boron is stablized in the plant, it is quite immobile. Apical growing points and new plant leaves become chlorotic on boron-deficient plants, and growing points frequently die. Many physiological diseases of plants, such as the internal cork of apples, yellows of alfalfa, top rot of tobacco, and cracked stem of celery are caused by a boron deficiency. Boron deficiency of sugar beets causes heart rot of the beet, as shown in Color Plate 3.

CHLORINE

The chlorine requirement of plants is very small, even though chlorine may be one of the most abundant anions in plants. Chlorine is unique in that there is little probability that it will ever be deficient in plants growing on soils. Chlorine enters the atmosphere from ocean spray, is distributed around the world, and is added to soils via precipitation. The presence of chlorine in fertilizer, as potassium chloride, is an additional source of chlorine for many agricultural soils.

MOLYBDENUM

Weathering of minerals releases molybdenum that is adsorbed to various soil constituents. The adsorbed molybdenum maintains an equilibirum with molybdenum in solution. The solution form is mainly as the molybdate ion, MoO_4^{2-}. Molybdenum also accumulates in soil organic matter.

In acid soils, molybdenum is strongly adsorbed or fixed by iron oxides. In many instances, molybdenum deficiencies are corrected by liming, owing to increased molybdenum availability. The data in Figure 13.4 show that increasing soil pH by liming increased the molybdenum content of cauliflower leaves. Some calcareous soils in western United States developed from high molybdenum content parent materials and have very high levels of available molybdenum. The plants, however, are not injured by the high molybdenum level.

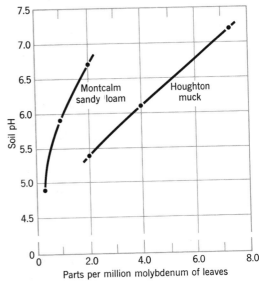

FIGURE 13.4 Increased soil pH due to liming increased the availability of soil molybdenum and resulted in an increased concentration of molybdenum in cauliflower leaves. (Data from Turner and McCall, 1957.)

Plant and Animal Molybdenum Relations

Molybdenum is needed for nitrogen fixation in legumes. When molybdenum is deficient, legumes show symptoms similar to those of nitrogen deficiency. In cauliflower, molybdenum deficiency results in cupping of leaf edges. It is caused by a reduced rate of cell expansion near the leaf margin, compared with that in the center of the leaf. Leaves also tend to be long and slender, giving rise to the symptom called *whiptail*. Interveinal chlorosis, stunting of plants, and general paleness are also exhibited, depending on the kind of plant.

Plants and animals need very small amounts of molybdenum. Plants deficient in the molybde-

num needed for animal consumption have been found growing on certain acid soils. The major nutritional problem is molybdenum toxicity, *molybdenosis*, which develops in grazing animals when forage has over 10 to 20 ppm molybdenum.

The molybdenosis problem arises because forage plants have a wide range of tolerance for molybdenum, whereas animals do not. Legumes accumulate more molybdenum than do common grasses, and they take up more molybdenum in wet than in dry soils. The data in Figure 13.5 show that large increases in molybdenum (and manganese) can occur in plants with large increases in molybdenum (and manganese) in the growing medium. By contrast, note the virtual lack of an increase in iron content of the plant with an in-

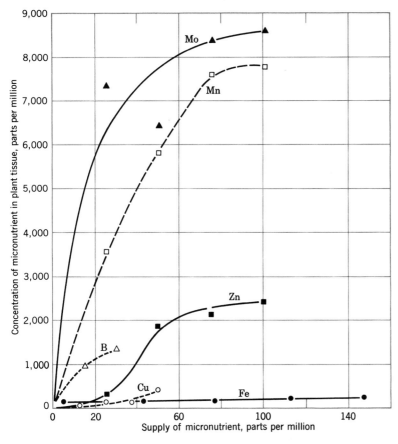

FIGURE 13.5 The iron (and copper) content of tomato leaflets was little affected by a large increase in the supply of nutrient in the growing medium. The molybdenum and manganese content of leaflets was greatly increased. (From Beeson, 1955.)

creasing amount of available iron. For soils with very high available molybdenum, there is no effective method for reducing molybdenum uptake from soils by plants.

Most problem areas of molybdenosis are related to the geologic origin and wetness of soils. Common sources of molybdenum in the western United States are granite, shale, and fine-grained sandstone. Soils producing forages with high molybdenum level are generally confined to valleys of small mountains streams (western United States). Only a very small part of any valley actually produces high molybdenum forages. These soils are wet, or poorly drained, alkaline, and high in organic matter. Molybdenosis in Nevada and California is associated with soils formed from the high-molybdenum-content granite of the Sierra Nevada Mountains. A typical location where high molybdenum content forage is produced is shown in Figure 13.6.

COBALT

Cobalt is required by microorganisms that symbiotically fix nitrogen. This is the only known need for cobalt by plants. Microorganisms in the stomachs of ruminants incorporate cobalt into vitamin B_{12}, and this vitamin provides these animals with cobalt. Single-stomached animals, including humans, must obtain their cobalt from foods containing vitamin B_{12}, such as milk or meat products. Only persons who eat strictly vegetarian foods are likely to have a cobalt-deficient diet. Cattle and sheep that are not fed legumes usually need supplementary cobalt.

Cobalt deficiency of animals was observed by the earliest settlers in parts of New England. In New Hampshire, it was called *Chocorua's Curse* and *Albany Ail*, from local place names. In southern Massachusetts, it was known as *neck's ail* because the disease was most common on necks of land that projected into bays. It is now known that the cobalt deficiency in New England had its origin in the geologic history of the soils. The low-cobalt soils formed in glacial deposits derived from the granite of the White Mountains, which contained very little cobalt. The soils are also sandy and tend to lose cobalt by leaching. The location of low-cobalt soils in New England is shown in Figure 13.7.

Cobalt deficiency among grazing animals is a problem on the wet, sandy soils of the lower coastal plain in the southeastern United States.

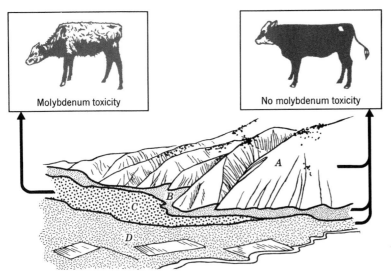

FIGURE 13.6 Plants containing toxic levels of molydbenum are found only on wet soils fromed from high-molybdenum parent materials, area C, in this mountain valley in Nevada. Areas A and B have well-drained soils in which the molybdenum is not readily available to plants. Area D is wet, but has soils formed low-molybdenum parent materials. (Data from Allaway, 1975.)

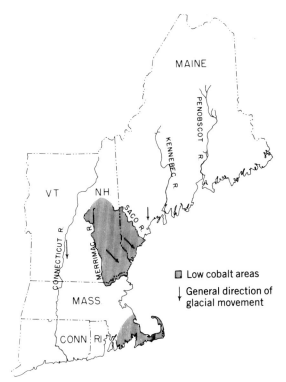

FIGURE 13.7 Low cobalt areas (shaded) in New England. The arrows indicate the direction of glacial ice movement from the White Mountains containing low-cobalt rocks. (Data from Kubota, 1965.)

The low cobalt soils are very sandy, poorly drained, and have humic or organic pans (Bh horizons). The original sandy parent materials had little cobalt, and the cobalt was leached from the upper soil horizons during soil development. These soils have the lowest cobalt content of any soils in United States. Consequently, little cobalt is available to forage plants growing on these soils.

SELENIUM

Selenium (Se) is not required by plants but is needed in small amounts by warm-blooded animals and, probably, humans. Large areas of the United States have soils so low in selenium that forages for livestock are deficient in selenium. Acid soils formed from low-selenium parent materials produce forages that are selenium deficient for good animal nutrition. In the western states, a selenium toxicity is a problem associated with many soils.

In 1934, mysterious livestock maladies on certain farms and ranches of the Plains and Rocky Mountain states were discovered to be caused by plants with high selenium content, which were toxic to grazing animals. Affected animals had sore feet, excessive shedding, and some died. During the next 20 years, scientists found that the high levels of selenium occurred only in soils derived from certain geologic formations with a high selenium content. Another important discovery was that a group of plants, *selenium accumulators*, had an extraordinary ability to extract selenium from the soil. Selenium accumulators contained about 50 ppm of selenium compared with less than 5 ppm in nonaccumulator plants. One selenium accumulator is a palatable legume, *Astragalus bisulcatus*. This plant was found to have a selenium content that was about 140 times more than that of adjacent plants. The adequacy of selenium in forages for farm animals in the United States is shown in Figure 13.8.

POTENTIALLY TOXIC ELEMENTS FROM POLLUTION

Potential poisoning of humans by arsenic, cadimum, lead, and mercury are real concerns today. Arsenic has accumulated in soils from sprays used to control insects and weeds and to defoliate crops before harvesting. Although arsenic accumulates in soils and has injured crops, it has not created a hazard for humans or animals.

Cadmium poisoning has occurred in Japan from the dumping of mine waste into rivers, where fish ingested the cadmium from the mine waste. Cadmium also appears in some sewage sludges. Using soils for the disposal of sewage represents a

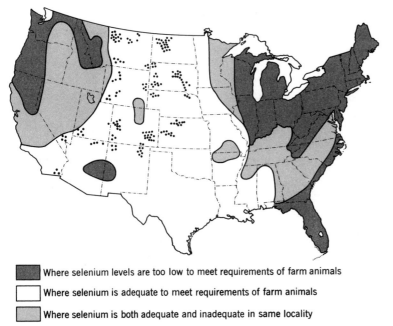

☐ Where selenium levels are too low to meet requirements of farm animals

☐ Where selenium is adequate to meet requirements of farm animals

☐ Where selenium is both adequate and inadequate in same locality

· Where selenium toxicity may be a problem

FIGURE 13.8 Areas where forages and feed crops contain various levels of selenium. (Data from Allaway, 1975.)

potential danger. Mine spoils may have toxic levels of cadmium, but natural agricultural soils do not contain harmful levels of cadmium.

Lead is discharged into the air from automobile exhausts and other sources, and it eventually reaches the soil. In soil, lead is converted to forms unavailable to plants. Any lead that is absorbed tends to remain in plant roots and is not transported to the shoots. Soils must become very polluted with lead before significant amounts move into the tops of plants.

Mercury is discharged into the air and water from use in pesticides and industrial activities. Under conditions of poor aeration, inorganic mercury is converted to methyl mercury, which is very toxic. Plants do not take up mercury readily from soils; however, soils should not be used to dispose of mercury because of the highly toxic nature of methyl mercury.

The soil is being used increasingly for sewage disposal. The application of high levels of heavy metals, and their potential uptake by plants, rep-

resents an important limitation in the use of soils for sewage and industrial waste disposal. Sewage known to be lacking in heavy metals can be added to soils without any danger of heavy metal poisoning of humans or other animals.

RADIOACTIVE ELEMENTS

Rocks and soils naturally contain radioactive elements. Currently, there is concern about radon. Many radioactive elements have very short half-lives, and contamination of the soil with these elements is of little concern. Radioactive cesium (Cs) is produced by atomic bombs and has a long half-life. Radioactive cesium appears to be fixed in vermiculitic minerals much like potassium. This limits its availability to plants. The soil tends to slow down the movement of radioactive cesium from the soil into plants and, later, into animals.

The comtamination of soils is via the atmosphere, and radioactive elements that fall on vegetation are absorbed by the leaves. Testing of nuclear weapons prior to 1962 resulted in significant contamination of tundra vegetation in Alaska. Increased radioactivity was found in the caribou that grazed on the tundra. People whose major food was caribou were found to have about 100 times more radioactivity in their bodies than people in the continental United States. More recently, a similar contamination occured in northern Europe as a result of the Chernobyl nuclear reactor explosion in the Soviet Union.

Nuclear bomb tests were conducted on Bikini Atoll of the Marshall Islands between 1946 and 1958. The soils were heavily contaminated with radioactive cesium-137. Several approaches have been suggested to decontaminate the soil. First, mix a large quantity of potassium with soil and irrigate it with seawater to increase the likelihood of potassium uptake by plants relative to cesium. A second approach is to remove the top 16 inches of topsoil and replace it with uncontaminated soil. The Bikinians favor the second approach.

SUMMARY

Micronutrients are needed by plants in very small amounts and function largely in plant-enzyme systems.

The reduced (divalent) forms of iron and manganese are available to plants. Plant deficiencies are associated with the oxidized forms, especially in alkaline soils. Toxic concentrations may occur in very acid soils. Plants have developed special ways to obtain iron from alkaline soils, including the excretion of hydrogen ions and siderophores.

Copper and zinc are released to the soil solution by mineral weathering. Uptake is mainly as divalent cations. Copper deficiencies tend to occur on acid sandy soils and organic soils. Plant deficiencies of zinc are common on alkaline mineral soils and organic soils.

Boron is released by the weathering of tourmaline, and plant uptake is mostly as undissociated boric acid. Boron is leached from acid mineral soils, resulting in plant deficiency. In alkaline soils, boron is fixed.

Chlorine is picked up by winds from seawater and distributed via precipitation throughout the world. Chlorine deficiency in plants growing on soils is highly unlikely.

Molybdenum availability is greatly affected by soil pH. In acid soils molybdenum is fixed by iron, and plant deficiencies occur. If some of these soils are limed and the pH is increased, molybdenum may become toxic for plants. Molybdenum toxicity for grazing animals is related to soil pH, soil organic matter content, and soil wetness.

Cobalt is needed by microrganisms that fix nitrogen.

Selenium is not needed by plants but is needed for animals. The selenium content of forages is highly variable, depending on soil, producing forages for farm animals that range from deficient to toxic.

Potentially toxic elements sometimes added to soils include arsenic, cadimum, lead, and mercury.

Plants and soils receive readioactive elements via the precipitation or fall-out from the atmosphere.

REFERENCES

Allaway, W. H. 1975. "The Effects of Soils and Fertilizers onHuman and Animal Health." *USDA Agr. Inform. Bull.* 378, Washington, D.C. .

Anonymous. Fallout in the Food Chain. *Time,* September 13, 1963.

Beeson, K. C. 1955. "The Effects of Fertilizers on the Nutritional Quality of Crops," in *Nutrition of Plants, Animals, and Man Symposium Proc.* Michigan State University, East Lansing.

Bienfait, H. F. 1988. "Mechanisms in Fe-Deficient Reactions of Higher Plants." *J. Plant Nut.* **11**(6–11): 605–629.

Chaney, R. L. 1988. "Recent Progress and Needed Research in Plant Iron Nutrition." *J. Plant Nut.* **11**(6–11):1589–1603.

Elliott, S. 1988. "Bikini Still Atomic No-man's Land." *Lansing State Journal,* May 22, 1988.

Foth, H. D. and B. G. Ellis. 1988. Soil Fertility. Wiley, New York.

Kubota, J. 1965. "Soils and Animal Nutrition." *Soil Con.* November, pp. 77–78.

Kubota, J. 1969. "Trace Element Studies Link Animal Ailments with Soils." *Soil Conservation.* Vol. 35, November, pp. 86–88.

Lewin, R. 1984. "How Microorganisms Transport Iron." *Science.* 225:401–402.

Messenger, A. S. 1983. "Soil pH and the Foliar Macro-nutrient/Micronutrient Balance of Green and Inter-veinally Chlorotic Pin Oaks." *J. Environ. Hort.* 1(4): 99–104.

Turner, F. and W. W. McCall. 1957. "Studies on Crop Response to Molybdenum and Lime in Michigan." *Mich. Agr. Exp. Sta. Quart. Bul.* **40:**268–281.

CHAPTER 14

FERTILIZERS

Manures and other materials have been applied to soils for thousands of years to increase plant growth. About 150 years ago, some of the essential plant nutrients were discovered and, subsequently, fertilizers containing essential plant nutrients were manufactured and applied to soils to increase fertility. By 1855, researchers at the Rothamsted Agricultural Experiment Station, near London, England, declared that soil fertility could be maintained permanently with mineral fertilizers. Today, fertilizer use has been estimated to account for 20 to 25 percent, and 30 to 40 percent, of the total crop production in the world and United States, respectively. The close parallel between the yields of corn (maize) and the application of nitrogen fertilizer in Illinois is shown in Figure 14.1.

FERTILIZER TERMINOLOGY

A *fertilizer* is any organic or inorganic material of natural or synthetic origin (other than liming materials) that is added to soils to supply one or more essential nutrient elements. Most of the fertilizers considered in this chapter are inorganic materials that are the product of sophisticated

manufacturing methods and have a uniform composition.

Grade and Ratio

The *fertilizer grade* expresses the nutrient content of fertilizers. The grade is the percentage composition, expressed in the order: $N-P_2O_5-K_2O$. The grade, 27-3-9, contains 27 percent nitrogen, phosphorus equivalent to 3 percent P_2O_5, and potassium equivalent to 9 percent K_2O. Most of the nitrogen and potassium in fertilizers is water soluble. In the United States the phosphorus in the fertilizer must be sufficiently soluble to dissolve in neutral ammonium citrate. Other countries have different laws regarding the solubility or availability of the nutrients in fertilizers. Comparisons between fertilizers can be made only on the basis of their nutrient content, not according to the kinds of compounds in the fertilizer.

The Soil Science Society of America favors a fertilizer grade based on the percentages of nitrogen, phosphorus, and potassium. In some cases, the fertilizer composition is expressed on this basis.

The *fertilizer ratio* is the relative proportions of primary nutrients in a fertilizer grade divided by

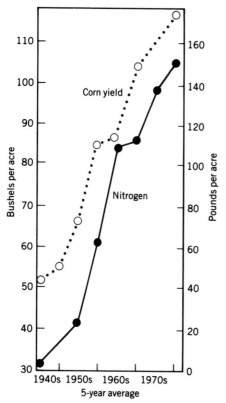

FIGURE 14.1 Corn yields and nitrogen fertilizer use in Illinois; 5-year averages. (Data from Welch, 1986.)

the highest denominator for that grade. A 27-3-9 fertilizer has a ratio of 9-1-3. This fertilizer is high in nitrogen, low in phosphorus, and moderate in potassium, relatively speaking. Such a fertilizer provides good maintenance for lawns in which the soil phosphorus level from past fertilizer use is high and the major need is for nitrogen.

General Nature of Fertilizer Laws

In general, the laws controlling the sales of fertilizers in various states are similar. They all require periodic registration of the brands or grades and accurate labeling. Most states require that the following information be made available to purchasers: (1) name or brand; (2) grade, which be-

comes the guaranteed analysis; and (3) name and address of manufacturer.

Fertilizers that are offered for sale may be sampled by inspectors at any time, and analyzed to determine whether the fertilizers meet the guaranteed analysis. The inspection and analysis may be under the control of the Department of Agriculture, the Agricultural Experiment Station, or a state chemist. In this way purchasers are assured that manufacturers comply with the fertilizer laws.

Types of Fertilizers

Fertilizers that contain only one compound and supply one or two plant nutrients are called *fertilizer materials or carriers*. Fertilizer carriers or materials that are mixed together and processed to produce fertilizers containing at least two or more nutrient elements are *mixed* fertilizers. Fertilizer may be in the form of a solid, liquid, or gas.

FERTILIZER MATERIALS

The first fertilizer material was manufactured when ground bones were treated with sulfuric acid to increase the solubility (availability) of the phosphorus. This occurred in England in about 1830, and was the beginning of the modern fertilizer industry. This was followed by the development of potassium materials from dried sea deposits late in the nineteenth century, and the manufacture of nitrogen materials from the fixation of atmospheric N_2 in the early twentieth century.

Nitrogen Materials

The source of essentially all industrial nitrogen, including fertilizer nitrogen, results from the fixation of atmospheric N_2, according to the following generalized reaction:

$$N_2 + 3H_2 \xrightarrow[\text{pressure, and catalysts}]{\text{proper temperature,}} 2NH_3$$

The hydrogen is usually obtained from natural gas (CH_4) and the overall reaction, using $8N_2$ and $2O_2$ to represent air, is

$$7CH_4 + 10H_2O + 8N_2 + 2O_2 = 16NH_3 + 7CO_2$$

The NH_3 is *ammonia,* commonly called anhydrous ammonia. Much of the ammonia is produced in the southern states near petroleum and natural gas fields and is transported by pipeline to various parts of the United States. Being 82 percent nitrogen, ammonia is the most concentrated nitrogen material and the least expensive on a per pound of nitrogen basis. Low cost has contributed to its widespread use.

At normal temperature and pressure, ammonia is a gas, but it is stored and transported under sufficient pressure to keep it a liquid. Locally, ammonia is handled by tank trucks and trailers for direct application to fields, as shown in Figure 14.2. When the ammonia is released from injectors at a depth of about 15 centimeters and reacts with the water of moist soil, NH_4^+ is formed, which is adsorbed onto cation exchange sites. However, application of ammonia at shallow depths in dry soil can result in considerable loss of ammonia by direct escape from the soil as a gas.

About 98 percent of the nitrogen fertilizer produced in the world is ammonia or one of its derivatives. The manufacture of ammonia and five of its derivatives is shown diagrammatically in Figure 14.3. Oxidation of NH_3 to nitric acid and neutralization of the HNO_3 with NH_3 produces ammonium nitrate (NH_4NO_3). The CO_2 from NH_3 manufacture is used to produce urea [$CO(NH_2)_2$]:

$$CO_2 + 2NH_3 = CO(NH_2)_2 + H_2O$$

Urea [$CO(NH_2)_2$] hydrolyzes in soil in the presence of urease to form ammonia. The application of urea, in effect, is the same as applying ammonia. The placement of urea on the soil surface (as for lawn fertilization) results in the significant loss of NH_3 to the atmosphere by volatilization unless rain or irrigation leach the urea into the soil. Urea and ammonium nitrate should not be mixed or stored together, because the mix adsorbs much water from the air and forms an unfavorable physical condition.

Reaction of NH_3 with (1) sulfuric acid produces ammonium sulfate, (2) phosphoric acid produces ammonium phosphate, and (3) water and ammonium nitrate produce nitrogen solutions (see Figure 14.3). All of these nitrogen materials are water soluble.

In general, water-soluble nitrogen materials are about equally effective when they are used on the basis of an equivalent amount of nitrogen and are placed within the soil. For ammonium nitrate (NH_4NO_3), dissolution into the soil solution produces both NO_3^- and NH_4^+ for plant uptake. Under conditions of well-aerated soil and a pH 6 or above, NH_4^+ is quickly nitrified to NO_3^-. The result is that it often makes little difference, whether the nitrate or ammonium form of nitrogen is applied. A few exceptions include the use of ammonium for flooded rice production to avoid denitrification and the strong acid-forming effect of ammonium sulfate. For this reason, $(NH_4)_2SO_4$ is preferred if an increase in soil acidity is beneficial.

The application of a soluble nitrogen fertilizer typically results in large, rapid increases in the nitrate concentration of the soil solution. Under these conditions, plant uptake of nitrogen is rapid and the fertilizer nitrogen is quickly depleted. An application of nitrogen fertilizer on a lawn in May

FIGURE 14.2 Injection of anhydrous ammonia into soil. The ammonia is handled as a liquid under pressure.

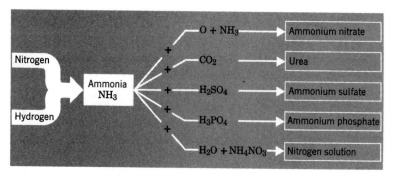

FIGURE 14.3 A diagrammatic representation of the manufacture of ammonia and some of its fertilizer derivatives.

may result in little, if any, fertilizer nitrogen left in June. Consequently, there is a need for a less soluble nitrogen fertilizer. Urea-formaldehyde and sulfur-coated urea are insoluble or slow-release fertilizers designed to supply nitrogen slowly and reduce the frequency of nitrogen application. These fertilizers are more expensive and their use is restricted to lawns, golf courses, and horticulture.

A brief summary of some properties of nitrogen materials is given in Table 14.1.

Phosphorus Materials

The starting point for the manufacture of phosphorus fertilizers is *phosphate rock*, which is a calcium fluorophosphate of sedimentary or igneous origin. Huge deposits of phosphate rock, also called rock phosphate, are located east of Tampa, Florida, where about 70 percent of the phosphate ore in the United States is mined. The sedimentary deposits were formed 10 to 15 million years ago and are buried under 5 to 10 meters of sand. Mining consists of removing the sandy overbur-

TABLE 14.1 The Principal Fertilizers Supplying Nitrogen

Nitrogen Carrier	Nitrogen %	Remarks
Anhydrous ammonia	82	Used directly, or for ammoniation, nitrogen solutions, etc.
Ammonium nitrate	33	Conditioned to resist adsorption of water
Ammonium sulfate	20	Also produced as a by-product of the coking of coal
Ammonia liquor (aqua ammonia)	20	Formed when ammonia is absorbed in water
Nitrogen solutions	Variable	Many kinds formed from solutions of aqua ammonia, ammonium nitrate, urea, etc.
Urea	45	$CO(NH_2)_2$ is hydrolyzed to ammonium in soils
Urea-formaldehyde	35–40	Contains insoluble, slowly available nitrogen
Sulfur-coated urea	39	Slow release, variable N and S content

den and using a hydraulic gun to break up the ore to form a slurry, as shown in Figure 14.4. The slurry is pumped to a recovery plant, where rock phosphate pebbles are separated from sand and clay by a washing and screening process. The rock phosphate pebbles are ground to a powder, which is sometimes used directly as a fertilizer.

The primary phosphorus mineral in phosphate rock is *fluoroapatite*, $Ca_{10}F_2(PO_4)_6$. Apatite is also the source of phosphorus in most soil parent materials. Apatite is quite insoluble, and when phosphate rock is applied to soil, very little phosphorus dissolves. Its use is restricted to areas close to the mines, where it is inexpensive, or to its application in very acid soil, where soil acids promote dissolution. Rock phosphate is a preferred source of phosphorus by some organic farmers.

As with ground bones, acidulation of rock phosphate increases solubility and plant availability of the phosphorus. The acidulation of rock phosphate with sulfuric acid produces ordinary *superphosphate,* as is shown by the following equation:

$$Ca_{10}F_2(PO_4)_6 + 7H_2SO_4 =$$
(phosphate rock) (sulfuric acid)

$$3Ca(H_2PO_4)_2 + 7CaSO_4 + 2HF$$
(monocalcium phosphate) (calcium sulfate) (hydrogen fluoride)

The ordinary superphosphate produced in the preceding reaction consists of about one half monocalcium phosphate and one half calcium sulfate. Hydrogen fluoride is one of the most hazardous pollutants if released into the atmosphere. The gases from manufacture of the phosphorus fertilizer are passed through a scrubber and the fluorine is recovered. Some of the fluorine is used to fluorinate water. Ordinary superphosphate is about 9 percent phosphorus, which equals 20 percent P_2O_5 equivalent and has a grade of 0-20-0. The phosphorus in monocalcium phosphate is water soluble. Very little 0-20-0 is used today, because other phosphorus fertilizers are less expensive on an equivalent phosphorus basis.

FIGURE 14.4 Mining rock phosphate. The overburden is removed and a hydraulic gun is used to break up the ore layer and form a slurry that is pumped to a recovery plant. (Photograph courtesy International Minerals and Chemical Corporation.)

Under proper conditions, the reaction of rock phosphate with excess sulfuric acid will produce phosphoric acid (H_3PO_4). Treatment of phosphate rock with phosphoric acid produces superphosphate with a higher content of phosphorus. The reaction is:

$$Ca_{10}(PO_4)_6F_2 + 14H_3PO_4 = 10Ca(H_2PO_4)_2 + HF$$

The same water-soluble monocalcium phosphate is produced, but without the $CaSO_4$. The grade is 0-45-0 and has been called concentrated, or triple superphosphate. Early production of superphosphate was mainly with sulfuric acid, and its use added a significant amount of sulfur to soils. Now, the extensive use of 0-45-0 has reduced sulfur additions to soils.

Ammonium phosphates are produced by neutralizing phosphoric acid with ammonia. Two popular kinds are produced, depending on the extent of neutralization—monoammonium phosphate (MAP), 11-48-0, and diammonium phosphate (DAP), 18-46-0. The actual grade is variable, depending on the proportions of the reactants. Ammonium polyphosphate (APP), 10-34-0, is manufactured by reacting NH_3 and H_3PO_4 in a cross-pipe reactor. These phosphates contain both nitrogen and phosphorus in water-soluble form. Routes for the manufacture of the major phosphorus materials are shown in Figure 14.5.

In the washing process to remove rock phosphate pebbles, the very small rock phosphate particles (many colloidal-sized) are put into waste ponds. Here, the particles settle out and the water evaporates. These low-soluble and low-value materials are processed and marketed under various names, including colloidal phosphate. Bonemeal is also a phosphorus source, but it is too expensive to use on farm fields. Both of these materials are natural products and are preferred by some gardeners. A summary of major materials containing phosphorus, including their phosphorus content and some remarks, is given in Table 14.2.

Potassium Materials

Wood ashes contain several percent potassium and are an important source of this element. Potassium mines were opened in Germany in the late nineteenth century and at Carlsbad, New Mexico, in the early twentieth century. Mining of the hugh potassium deposits in Saskatchewan, Canada, began in 1959 (see Figure 14.6.). These potassium deposits, which were formed when ancient seas evaporated, are called *evaporite* deposits. The salts in the ocean water precipitated as the water evaporated and the salts in the water became more concentrated. Later, these salt de-

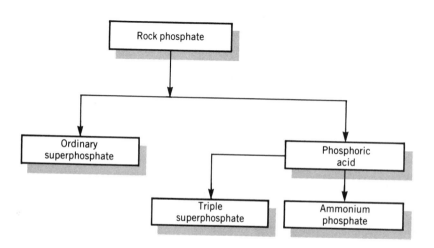

FIGURE 14.5 Routes for the manufacture of phosphorus materials.

TABLE 14.2 The Principal Phosphatic Materials

Material	Percentage Available		Remarks
	P_2O_5	P	
Rock phosphate	25–35[a]	11.0–15.4[a]	Effectiveness depends on degree of fineness, soil conditions, and crop grown
Superphosphate (ordinary)	20	8.7	Made by treating ground phosphate rock with H_2SO_4
Triple superphosphate	46	20	Made by treating ground phosphate rock with H_3PO_4
Monoammonium phosphate	48	21	Made by neutralizing H_3PO_4 with ammonia
Diammonium phosphate	46	20	Made by neutralizing H_3PO_4 with ammonia
Ammonium polyphosphate	34	15	Made by reacting NH_3 and H_3PO_4 in a pipe reactor
Basic slag	5–20	2.2–8.8	By-product obtained in the manufacture of steel
Colloidal phosphate	18–23[a]	7.9–10.1[a]	A finely divided, relatively low-grade rock phosphate or phosphatic clay
Bonemeal	17–30[a]	7.5–13.2[a]	Includes raw as well as steamed bone meals

[a] Total phosphoric acid instead of amount that is available.

posits were buried under various kinds of overburdens and rocks.

The minerals in the deposits are representative of the salts in seawater, mainly salts of sodium and potassium. *Sylvite* is KCl and *sylvinite* is a mixture of KCl and NaCl. *Langbeinite* is a mixture of potassium and magnesium sulfates. Processing of most of the ores consists of separating KCl from the other compounds in the ore; more than 95 percent of the potassium in fertilizers is KCl.

Some of the Canadian deposits are relatively pure KCl and require little, if any, processing.

The most popular method of processing potassium ores is flotation. The ore is ground, suspended in water, and treated with a flotation agent that adheres to the KCl crystals. As air is passed through the suspension, KCl crystals float to the top and are skimmed off. The material is dried and screened to obtain the proper particle size. The material, KCl, is marketed as 0-0-60.

FIGURE 14.6 A continuous mining machine at work about 1 kilometer underground in potash mine at Esterhazy, Saskatchewan, Canada. (Photograph courtesy American Potash Institute and International Minerals and Chemical Corporation.)

Some KCl is obtained from brine lakes at Searles Lake in California, and at the Great Salt Lake in Utah. Minor potassium materials include K_2SO_4 and KNO_3. All of the potassium materials contain water-soluble potassium.

Micronutrient Materials

Although the bulk of fertilizer consists of the materials containing nitrogen, phosphorus, and potassium, micronutrients are sometimes added to fertilizer to supply one or more of the micronutrients. In these instances, the percentage of the element in the fertilizer becomes a part of the guaranteed analysis. Some common micronutrient fertilizers are listed in Table 14.3.

MIXED FERTILIZERS

Large quantities of materials containing only one or two nutrients are applied directly to the land. This is particularly true for NH_3, which is handled as a liquid under pressure and applied with special applicators. Because of the unique timing of nitrogen fertilizer applications, much of the nitrogen is applied alone. The mixing together of different nutrient-containing materials produces *mixed* fertilizers that contain two or more plant nutrients. Mixed fertilizer is commonly applied in relatively small amounts with or near seeds for the purpose of accelerating early growth of crop plants.

The formula is a recipe for making the fertilizer. The mixing together of 364 pounds of NH_4NO_3 (33-0-0), 1,200 pounds of ordinary superphosphate (0-20-0), and 400 pounds of KCl (0-0-60), plus 36 pounds of conditioner, would produce 2,000 pounds of fertilizer with a grade of 6-12-12. The major systems for mixing fertilizers are granulation, bulk blending, and fluid.

Granular Fertilizers

In the manufacture of granular fertilizers, dry materials are mixed together with a little water, or dry materials are mixed with acids and other fertilizer solutions to form a thick slurry. The mixing occurs in a heated, revolving drum where particles of different size are formed as the product dries, producing granules. Materials like diatomaceous earth (a powdery siliceous material formed from the remains of diatoms) or kaolinitic clay are added to produce a coating on the granules that inhibits adsorption of water from the air and prevents caking (agglomeration or lumping). Caking is due to soft granules that coalesce at contact points, causing a lumpy or massive condition. Oil is added to reduce dust. A desirable granular fertilizer contains hard granules of uniform size and composition and has good storage and handling qualities. Such materials are required by today's modern fertilizer applicators. Granular fertilizers are sold in bags or as bulk materials for direct spreading onto fields, and are also used to make bulk blended fertilizers.

TABLE 14.3 Some Common Micronutrient Carriers

Carrier	Nutrient Composition
Borax ($Na_2B_4O_7 \cdot 10H_2O$)	11% B
Copper sulfate ($CuSO_4 \cdot 5H_2O$)	25% Cu
Ferrous-sulfate ($FeSO_4 \cdot 7H_2O$)	20% Fe
Manganous oxide (MnO)	48% Mn
Manganese sulfate (variable hydration)	23–25% Mn
Zinc sulfate ($ZnSO_4 \cdot 7H_2O$)	35% Zn
Sodium molybdate ($Na_2MoO_4 \cdot 2H_2O$	40% Mo

Bulk Blended Fertilizers

Granular fertilizers can be mixed together without additional processing to produce bulk blends. Such fertilizers, when handled in bulk form, are considerably cheaper than regular granular mixed fertilizers. Other advantages include the ease with which a wide variety of grades can be produced quickly and with negligible need for storage of finished products. In many bulk-mixing plants, the formula is calculated by using a computer program, and the bulk blend is loaded into a truck and immediately applied to fields. Currently, 54 percent of the mixed fertilizers in the United States are bulk blends.

Each of the materials in a bulk blend can separate from each other, or segregate, because no chemical reaction between the materials has occurred. Thus, a major precaution in bulk bending is that materials be of similar particle size; otherwise, segregation will occur during handling and application. The result is nonuniform distribution of the fertilizer nutrients in the field. In addition, urea should never be added to a blend containing ammonium nitrate, because the blend adsorbs water from the atmosphere and cakes.

Fluid Fertilizers

Fluid fertilizers are easily made and handled. Micronutrients and pesticides are readily incorporated. Fluid fertilizers can be applied to irrigation water and can be used for direct application to plant foliage. They are either solutions or suspensions.

The principal materials used for making solutions include urea, urea-ammonium nitrate solution, ammonium polyphosphate solution (10-34-0), and finely ground KCl. Liquid fertilizer grades are usually lower than dry fertilizers because of the limited solubilities of the materials in water. In general, they are more expensive than dry fertilizers. Additionally, low temperature can cause crystallization, or salting-out of the salts. Most popular liquid fertilizers have a salting-out temperature of freezing or below. A salting-out problem occurs only if liquids are stored in cold climates during the winter months. The application of liquid fertilizer is shown in Figure 14.7.

Suspension fertilizers are manufactured from liquid fertilizers similar to those that are used to make solution fertilizers. Suspensions, however, have more material added to the water than can dissolve, so it becomes a thin slurry. One to 2 percent clay (attapulgite, an alumino-Mg-silicate clay) is added to help hold particles in suspension. This allows for preparation of higher grades, as compared with liquid fertilizers. Conversely, salt crystals in suspension may grow over time and particles may settle out. This limits the length of the storage period.

The same amount of nutrients in fluid or dry form generally produces similar results when time of application and soil placement are similar. Although fluids are very easy to handle, they need

FIGURE 14.7 Applying liquid fertilizer broadcast. (Photograph courtesy Ag-Chem Equipment Company.)

specialized application equipment, especially for suspensions. The solution fertilizers are quite popular and often are surface applied with herbicides mixed into the solution. Suspension fertilizers have attained only modest acceptance.

NATURAL FERTILIZER MATERIALS

The emphasis in this section is on those generally available materials that have had minimal processing and that are sources of nutrients. They are of both organic and inorganic origin.

Animal manures have been used almost from the beginning of agriculture and are a good source of plant nutrients. A ton of wet (fresh) animal manure contains approximately 10 pounds of nitrogen, phosphorus equivalent to 5 pounds of P_2O_5, and potassium equivalent to 10 pounds of K_2O. Animal manures are an excellent source of nutrients, because of the availability of nutrients in the liquid portion and the decomposability of the solid fraction. Other organic ma-

TABLE 14.4 Nutrient Composition of Some Natural Fertilizer Materials

Material	Composition
	N, Percent
Cottonseed meal	6
Blood meal	15
Fish scrap	4–6
Soybean meal	7
Animal tankage	7–8
	P_2O_5, percent
Rock phosphate	25–30
Waste pond tailings	10–15
Bonemeal (steamed)	23–30
Animal tankage	10
	K_2O, percent
Greensand (glauconite)	7
Granite fines	3
Corncob ash	40
Wood ashes	5–7

Complied from several sources, including Taber, Davison, and Telford, 1976.

terials with a C:N ratio of 20 or less are also valuable as nitrogen sources. If material such as sawdust is used, it can be valuable as a mulch, but additional nitrogen is needed to prevent competition for nitrogen between plants and microbes. Materials high in protein are high in nitrogen. Some nitrogen materials are listed in Table 14.4.

Ground rock phosphate and waste pond or colloidal phosphate are sources of phosphorus. The phosphorus is in calcium phosphate compounds, which are very slowly soluble in alkaline soils; the same is true for bonemeal. These compounds are more satisfactory when used on acid soils. Animal tankage (refuse from animal meat processing) is also a source of phosphorus. The approximate amount of phosphorus in some materials is given in Table 14.4.

Potassium is not an integral part of plant tissues, and sources of potassium are inorganic or from the ashes resulting from the burning of organic materials. The potassium in ashes is in the form of K_2CO_3. Repeated use of wood ashes may tend to cause dispersion of soil structure; the alkalinity effect may produce an undesirable increase in soil pH, because many plants grown on alkaline soil are prone to iron deficiency.

Finely-ground granite is available, however; it contains highly insoluble potassium minerals such as feldspar and mica. The potassium in greensand (glauconite) is interlayer potassium in a micaceous mineral and similar to the potassium in the mica of granite fines. These sources of potassium have low availability. The composition of some potassium materials is given in Table 14.4.

SUMMARY

Fertilizers contain mainly soluble forms of plant nutrients. The grade indicates the nutrient composition. A grade of 27-3-9 contains on a weight basis 27 percent nitrogen, phosphorus equivalent to 3 percent P_2O_5, and potassium equivalent to 9 percent K_2O.

Most of the nitrogen in fertilizers comes from fixation of atmospheric nitrogen. Reacting nitrogen with hydrogen produces ammonia, NH_3. The reaction of ammonia with other chemicals produces a wide array of nitrogen fertilizers.

The phosphorus in fertilizers comes from phosphate rock. The rock is acidulated with sulfuric or phosphoric acids to produce superphosphates that contain citrate soluble phosphorus. Phosphoric acid is also used in the manufacture of ammonium polyphosphates.

The potassium in fertilizers comes mainly from evaporite deposits. The potassium chloride in the deposits is recovered and marketed as 0-0-60.

Many micronutrient carriers are available to supply micronutrients.

Mixed fertilizers are marketed mainly as granular, bulk blended, or fluid (liquid and suspension) fertilizers.

Various naturally occurring organic and inorganic materials are used as sources of nitrogen, phosphorus, and potassium. The materials have lower nutrient content and generally fair to low availability.

REFERENCES

Engelstad, O. P., ed. 1985. *Fertilizer Technology and Use*, 3rd ed. Soil Sci. Soc. Am., Madison, Wis.

Foth, H. D. and B. G. Ellis. 1988. *Soil Fertility*. Wiley, New York.

International Fertilizer Development Center. 1979. *Fertilizer Manual*. TVA. Muscle Shoals, Ala.

Taber, H. G., A. D. Davison, and H. S. Telford. 1976. "Organic Gardening." *Washington State University Ext. EB 648*. Pullman, Wash.

Welch. L. F. 1986. "Nitrogen: Ag's Old Standby." *Solutions: J. Fluid Fert. and Ag. Chem*. July/August, pp. 18–21.

CHAPTER 15

SOIL FERTILITY EVALUATION AND FERTILIZER USE

Soil fertility is *the status of a soil with respect to its ability to supply elements essential for plant growth without a toxic concentration of any element.* Soil fertility is determined or evaluated with various techniques, and this information is the basis for making fertilizer recommendations. The emphasis in this chapter is the evaluation of soil fertility, making fertilizer recommendations, and the use of fertilizers.

SOIL FERTILITY EVALUATION

Evaluations of soil fertility are based on visual observations and tests of both plants and soils. The major techniques used to evaluate soil fertility are plant deficiency symptoms, analysis of the nutrient content of plant tissue, and soil tests for the amount of available nutrients.

Plant Deficiency Symptoms
Color Plates 1 to 4 in Chapter 12 contain deficiency symptoms for several nutrients on several different plants. By the time plant nutrient deficiency symptoms become obvious, it may be too late in the season to apply fertilizer. Such informa-

tion, however, is useful in determining which nutrients are deficient. Lawns commonly turn light-green or yellow when nitrogen is deficient, and this indication of nitrogen deficiency can be very useful for home owners. Many trees, shrubs, and flowers develop iron deficiency symptoms when growing on alkaline soils and these symptoms are very important to many gardeners. Deficiency symptoms can be useful to researchers in determining the location of soil fertility experiments on a deficient nutrient. Care must be exercised in using deficiency symptoms to evaluate soil fertility, because several different deficiencies can produce similar symptoms.

Plant Tissue Tests
Tissue or plant analyses are frequently used as the basis of fertilizer recommendations for crops that require one or more years to mature. This includes many tree crops, sugarcane, and pineapple. The nutritional status of the plant can be assessed and the appropriate fertilizer can be applied long before harvest, which allows plenty of time for the plants to benefit from the fertilization. As a result, plant analyses are popular with horticulturists and foresters who grow tree crops. In

addition, it is difficult to evaluate soil fertility for tree crops by using soil samples, because of the widespread and deeply penetrating root systems of trees and the difficulty of getting a representative soil sample.

The use of diphenylamine as an indicator of the concentration of nitrate in the sap of corn is shown in Color Plate 1. Although the test is qualitative, it is very useful. Other similar easy-to-use tests are available to verify plant nutrient deficiency symptoms. Simplicity of use and immediate results are beneficial to consultants and others who diagnose plant growth problems in the field.

Exacting quantitative chemical tests of plant tissue are used to determine the degree of deficiency or sufficiency of nutrients in plants. An enormous amount of research data has been summarized into tables and figures that relate nutrient composition of particular plants to growth or yield. For example, corn plants have sufficient nitrogen for maximum yield when the nitrogen content of the leaf opposite and below the uppermost ear is 3 percent.

The relationship between the percent of potassium in first-year needles of Red Pine and growth, or plant height, is shown in Figure 15.1. Note that needles with less than 0.35 percent potassium had deficiency symptoms. Growth was maximum, with about 0.5 percent or more potassium in needles. To obtain the information needed to make a specific potassium fertilizer recommendation based on tissue analysis, the percentage of potassium in needles must be related to the plant's response to potassium fertilizer. For example, suppose in an experiment in a Red Pine plantation where the potassium content of the needles was 0.25 percent potassium, that 50 pounds of K_2O per acre resulted in growth represented by 60 percent of the relative maximum yield, and the application of 100 pounds of K_2O produced growth equal to 100 percent of relative yield (maximum yield). A forester having a particular goal for growth, could then fertilize accordingly, based on the experimental data.

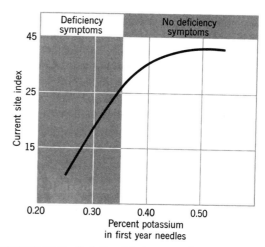

FIGURE 15.1 Relationship between the potassium content of first-year needles of Red Pine, *Pinus resinosa,* and site index or tree growth. (Data of Stone et al., 1958.)

Plant analyses have been used to determine locations where the soil has sufficient available nutrients to produce maximum yields in orchards without the use of additional fertilizer. For long-maturing crops, such as sugarcane and pineapples, periodic plant analyses are used to produce a log of the nutritional status of the crop, and this serves as a guide for proper timing of fertilizer applications.

Soil Tests

One of the weakest steps in making good fertilizer recommendations by using soil tests is the obtaining of a representative soil sample. A soil sample of 1 pint (about 1 pound or 500 grams) when used to represent 10 to 15 acres, means that 1 pint of soil is used to characterize 10,000 to 30,000 tons of soil.

Fairly uniform areas of fields should be sampled separately on the basis of uniformity in soil characteristics and past soil management. Sampling of unusual areas, including the location of manure or lime piles, should be avoided. Collect about 20 individual samples of the plow layer and mix them together in a clean plastic pail (see Figure 15.2). From this large soil sample, about 1 pint is sent to the testing lab. Because different labs use different kinds of tests and procedures, persons interested in soil tests should consult a soil testing lab for proper sampling procedures and care of samples after they have been collected. Most states have soil testing laboratories associated with both the Agricultural Experiment Station and agribusiness firms. Many fertilizer distributors provide soil testing services for their customers.

Shallower sampling is advised in no-till systems where nutrients accumulate in the upper few inches of soil as a result of the surface application of fertilizer (see Figure 15.3). In some instances, samples of the subsoil are taken.

Soil tests have been developed that measure the fraction of nutrients (available fraction) in the plow layer, which is closely related to plant growth. Soil tests for potassium, calcium, and magnesium measure the exchangeable amounts. As with the procedures for developing fertilizer recommendations based on plant tissue tests, soil tests must be related to plant growth or crop yield. For example, the relationship between phosphorus soil tests and the relative yield of corn in north-central Iowa is shown in Figure 15.4. Note that a soil test of 61 or more pp2m (pounds of phosphorus per acre-furrow-slice) resulted in a

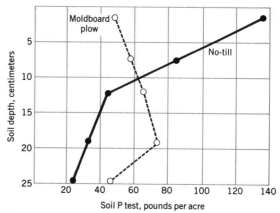

FIGURE 15.3 Distribution of soil phosphorus after three years of tillage and fertilization. (From Potash and Phosphate Institute, 1988.)

relative yield of nearly 100 percent. This means that a soil test of 61 or more pp2m is high and indicates a situation where the application of phosphorus fertilizer is not expected to increase yield. The various soil test levels are categorized as high, medium, and low. For example, low phosphorus tests range from 16 to 30 pounds of phosphorus pp2m, and medium from 30 to 61 pp2m. These results are based on the use of a soil extractant composed of dilute HCl and NH_4F, called Bray's P1. This test is used in almost all labs in the midwestern region in the United States.

FIGURE 15.4 Relationship between phosphorus soil test and the growth of corn in north-central Iowa. (Data courtesy of R. Voss, Iowa State University.)

FIGURE 15.2 Guides for taking soil samples. Avoid unsual areas and take about 20 random samples per area.

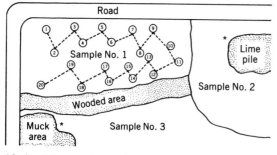

*Omit or take separate samples

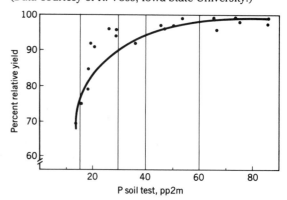

Some regions use other extractants because the soils contain different forms of phosphorus, and the results from Bray's P1 test do not relate well to crop yields.

The next question involving the use of soil tests to make fertilizer recommendations is: If the soil test for phosphorus is low, how much phosphorus fertilizer will be required to obtain a particular crop-yield goal? In general, the most profitable rate of fertilizer application is the amount needed to obtain about 90 to 95 percent of the relative maximum yield. Data from experiments relating crop response to varying amounts of fertilizer for various soil test levels are placed in tables, which are used to make fertilizer recommendations. The phosphorus fertilizer recommendations for corn in north-central Iowa are given in Table 15.1. The data in this table also take into consideration the level of subsoil phosphorus. A soil phosphorus test of 35 pp2m is medium, and the phosphorus recommendation for corn would be 70 pounds of P_2O_5 per acre, if subsoil phosphorus is very low or low, and 60 pounds of P_2O_5 per acre if the subsoil test is medium or high in phosphorus. A series of tables are constructed to make recommendations for various crops in different climate and soil regions. The procedures for developing fertilizer recom-

mendations for potassium and some of the micronutrients are the same as those for phosphorus. Nitrogen fertilizer recommendations are usually based on consideration of the nitrogen mineralizing power of the soil, the kind of previous and current crop, the application of manure, and the return of crop residues to the soil.

An example of soil test results and fertilizer recommendations is given in Figure 15.5. The soil testing laboratory report in this figure contains three sections: First, there is an identification section, giving sample numbers and record of the previous crop. The soil analysis section gives the soil test results for soil pH (and lime index), phosphorus, potassium, calcium, magnesium, cation exchange capacity, and percentage of organic matter. The fertilizer recommendations section contains the yield goals and recommendations for two or three years for various crops.

Fertilizer recommendations for home gardens and lawns are much like those made for commercial producers. For home gardens and lawns, however, there is usually little, if any, profit motive involved, so a less rigorous procedure will satisfy the needs of many persons. For example, in the absence of a soil test, a reasonable fertilizer application for a lawn or garden is approximately 15 to 20 pounds of 10-10-10 per 1,000 square feet. This is equivalent to 1.5 to 2 pounds of nitrogen, P_2O_5, and K_2O per 1,000 square feet. Many lawns and gardens have been heavily fertilized for many years. Frequently, soil tests are high for certain nutrients, which indicates that further fertilization is not needed. In these instances, the soil test indicates which nutrients are sufficient.

Computerized Fertilizer Recommendations

The data in fertilizer recommendation tables are put into equations and fertilizer recommendations are generated by using a computer. Information regarding crop, soil, yield goals, soil pH, and other critical factors are entered, and the computer generates the recommendations and pro-

TABLE 15.1 Phosphorus Fertilizer Recommendations for Corn or Sorghum Grain in Iowa

Soil Test Class	Basic Recommended Amounts for Soil Test Class and Subsoil P Levels (P_2O_5, lb/acre)			
	Subsoil P			
	VL	L	M	H
Very low	120	110	100	80
Low	100	90	80	70
Medium	70	70	60	60
High[a]	45	45	45	45
Very high	0	0	0	0

Data from R. Voss, 1982.
[a] Maintenance recommendation.

SOIL TESTING LABORATORY
AGRONOMY DEPARTMENT, PURDUE UNIVERSITY
WEST LAFAYETTE, IN 47907 (317) 494–8080

Sent to:

    Phil Kolp
    South St.
    Strawtown IN

County:

Extra copies sent to:

County Ext. office: No
ASCS: No
Other: _____

Date received: Tue 1-Jul-89
Date sent: Tue 1-Jul-89

IDENTIFICATION

Laboratory sample number	80000	80001	80002
Sample ID	1	2	3
Previous crop	Soybeans	Soybeans	Grass maint

SOIL ANALYSIS

Soil pH	4.6		6.7		6.9	
Lime index	6.2		7.1		7.0	
Phosphorus (Bray P1)	P LB/A	12	Low	140	V. High	190	V. High
Potassium	K LB/A	150	Low	360	V. High	650	V. High
Calcium	Ca LB/A	1600	Adeq.	2400	Adeq.	8700	Adeq.
Magnesium	Mg LB/A	90	Inadeq.	260	Adeq.	690	Adeq.
Cation Exchange Capacity (CEC)	14		8		25	
Organic matter (%)	1.2		1.6		5.3	

FERTILIZER RECOMMENDATIONS

PROPOSED CROP		Corn		Wheat		Grass LSeed	
Yield goal	Bu or T/A	180	Footnotes*	70	Footnotes*	6	Footnotes*
Nitrogen	N LB/A	210	1, 3	90		0	6
Phosphorus	P2O5 LB/A	120	20, 21	20		40	
Potassium	K2O LB/A	160	31	0		120	
Magnesium	Mg LB/A	0	40	0		0	
Limestone	Tons/A	7	50, 51	0		0	

ALTERNATIVE CROP		Soybeans		Soybeans		Grass seed	
Yield goal	Bu or T/A	50		50		6	
Nitrogen	N LB/A	0		0		50	6
Phosphorus	P2O5 LB/A	70	21	0		20	
Potassium	K2O LB/A	90	31	0		0	
Magnesium	Mg LB/A	0	40	0		0	
Limestone	Tons/A	7	50, 51	0		0	

NEXT CROP		Soybeans		Corn		Legume main	
Yield goal	Bu or T/A	50		200		6	
Nitrogen	N LB/A	0		230	1, 3	0	7
Phosphorus	P2O5 LB/A	70	21	0		40	
Potassium	K2O LB/A	90	31	0		120	
Magnesium	Mg LB/A	0	40	0		0	
Limestone	Tons/A	7	50, 51, 52	0		0	

* See explanation of footnotes on reverse side.

FIGURE 15.5 Sample of a computer-generated laboratory report showing soil test results and fertilizer recommendations. (Courtesy Dr. J. Alhrichs, Purdue University.)

vides a printout of soil test results and the recommendations (see Figure 15.5). Some specific illustrations of the equations that are used for computing fertilizer recommendations follow.

The K_2O recommendation for potatoes grown in Michigan is based on the following equation:

$$Ks = 175 + (0.5 \times YG) - (1.0 \times ST)$$

where Ks is pounds of K_2O recommended per acre for loamy sand and sandy loam soils, YG is yield goal in hundred weights per acre, and ST is the soil test value in pounds of K pp2m(pounds per acre-furrow-slice). This recommendation considers: (1) differences in the soil's ability to supply potassium by relating the recommendation to soil texture, (2) yield goal sought by the grower, and (3) the soil test level of the field. For a yield goal of 500 cwt/acre and a soil test of 200 pounds of potassium pp2m, the K_2O recommended per acre is 225 pounds $(175 + 250 - 200)$.

For micronutrients, availability is closely related to soil pH so that soil tests for nutrients, plus soil pH, are used as the basis for recommendations. An example of a manganese recommendation in pounds of manganese per acre for responsive crops is:

$$\text{pound of manganese per acre} = 136 + (6.2 \times pH) - (0.35 \times ST)$$

Here, the amount of manganese recommended is directly related to soil pH and inversely related to the manganese soil test (ST). In addition, differential crop response to manganese is considered; responsive crops include oats, beans, soybeans, and sugar beets.

No commonly accepted soil test exists for making nitrogen fertilizer recommendations. Rather, a nitrogen balance system is commonly used, in which estimates are made of the amount of nitrogen needed by the crop and the amount of nitrogen likely to be available. The amount of nitrogen needed minus the amount of nitrogen available equals the nitrogen fertilizer recommendation. The recommendations vary greatly between crops

and yield expectations. For example, the nitrogen recommendation for corn grain in Michigan is calculated with the following equation:

$$\text{pounds of fertilizer nitrogen per acre} = [-27 + (1.36 \times YG)] - [40 + (0.6 \times PS)] - (4 \times TM)$$

In this equation, YG is the yield goal in bushels per acre and PS refers to percent stand for legumes, as alfalfa and clover, that precede the corn crop. Five or six alfalfa plants per square foot is a 100 percent stand. A nitrogen credit of 100 pounds is made when an alfalfa crop with 100 percent stand precedes the corn crop $[40 + (0.6 \times 100) = 100$ pound credit]. The TM refers to tons of manure, crediting each ton with 4 pounds of nitrogen fertilizer equivalent. Thus, corn with an expected yield of 150 bushels per acre and following an 80 percent stand of alfalfa with 10 tons of manure applied per acre, will have a nitrogen recommendation of 49 pounds per acre $(-27 + 204 - 88 - 40)$.

Nitrogen does not accumulate and buildup in soils where leaching occurs as do phosphorus and potassium. For this reason, the nitrogen need for a lawn remains nearly the same year after year. Many home owners can maintain a good lawn fertilization program by applying only nitrogen fertilizer. A moderate recommendation is 1 pound of nitrogen per 1,000 square feet in the spring and again in late summer and/or fall. A serious gardener will be able to assess the nitrogen need of a lawn by careful observation of the color and the rate of growth when water does not seriously limit growth.

APPLICATION AND USE OF FERTILIZERS

After a fertilizer recommendation has been made, the major factors to consider in the use of fertilizer are time of application and placement. High and very high soil tests mean that fertilizer is not expected to increase plant growth or yields significantly. Obviously, under such conditions, a consideration of time of application and placement of fertilizer is meaningless. In other words, the higher the soil fertility level, the less important are time of application and placement. In many farm fields, however, time of application and placement are important considerations.

Time of Application

The closer the time of fertilizer application is to the time of maximum plant need, the more efficient plants will be in using the fertilizer nutrients. This is especially true for nitrogen, which is lost by denitrification whenever the soil is anaerobic. Nitrate is mobile and lost by leaching during periods of surplus water, and NH_3 is subject to volatilization when it exists on the soil surface. This is shown in Figure 15.6, where nitrogen is applied as a side dress, placement of the nitrogen fertilizer in

FIGURE 15.6 Effect of fall, spring, and sidedress nitrogen rates on corn yields at DeKalb, Illinois. Average for 1966–1969. (Reproduced from American Society of Agronomy, Vol. 63, No. 1, p. 119–123, copyright © 1971.)

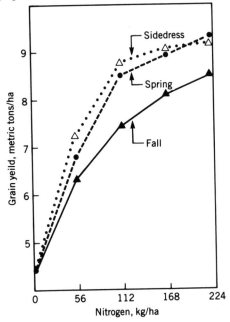

the soil along the rows when the corn is a foot or two high, and as a preplanting broadcast application onto the soil surface in early spring were much more effective than application the previous fall. Side dress application resulted in providing nitrogen near the time of maximum plant need and with the least amount of time available for nitrogen loss.

The time of application of phosphorus and potassium fertilizer is less critical than for nitrogen. However, fixation of phosphorus, and to a lesser extent potassium fixation, reduce their availability and reduce the plant's ability to recover the applied fertilizer phosphorus and potassium. The longer the period of time between application and plant use, the greater is the fixation and inefficiency of the nutrient use. In general, the levels of phosphorus and potassium in many soils in United States have increased to the point where time of application has little effect, and much of the phosphorus and potassium fertilizer is bulk-blended and applied broadcast (spread uniformly on the soil surface) at a time when it is most convenient.

Methods of Fertilizer Placement

Side dressing of nitrogen fertilizer for corn has already been mentioned as being an application of fertilizer along the rows after the corn is a foot or two high. A *starter fertilizer* application is one that contains a relatively small amount of nutrient elements and is applied with or near the seed in a band for the purpose of accelerating early growth of crop plants. Starter fertilizer is commonly applied an inch or two below and to the side of seeds, so that the nutrients will be near young roots and in moist soil. The result can be a faster start for the plants and a greater ability to compete with weeds. The earlier start, however, does not necessarily mean higher yields at harvest time. Another benefit of band placement is less contact of the soil with the phosphorus, thus reducing phosphorus fixation.

The amount of fertilizer that can be safely

placed in a band near seeds is related to the salt effect of the fertilizer. As shown in Figure 15.7, fertilizer in contact with seed may reduce or prevent germination. If, however, the seed and fertilizer are separated by a distance of only 2 to 3 centimeters (an inch), germination is barely affected.

Placement of nitrogen fertilizers is of little concern because nitrate has great solubility and mobility. Essentially, all of the nitrate is in solution and moves readily to roots by mass flow. The uptake of fertilizer phosphorus and potassium, compared with nitrogen, is more affected by low concentrations of phosphorus and potassium in the soil solution, and their uptake is more dependent on rate of diffusion. The diffusion rates of phosphorus and potassium are so low that they move only 0.02 and 0.5 centimeters or less, respectively, to roots in loam soils during the entire growing season.

The major effect of the phosphorus and potassium fertilizer placement results from the net effect of the amount of roots in the fertilized soil

FIGURE 15.7 Effects of fertilizer placement (shown by arrows) on the germination of wheat. On left it is with seed and, on the right, the fertilizer and seeds are separated.

volume versus the concentration of phosphorus and potassium in the soil solution. Consider that a given amount of phosphorus fertilizer is placed in a small volume of soil, and it produces a very high concentration of phosphate ions in solution. Phosphorus uptake could be rapid, even with a small amount of root length or surface area-density for absorption. This situation is representative of band placement. As the soil volume increases, with a constant amount of fertilizer, more soil and phosphorus contact results in greater phosphorus adsorption (fixation), reducing the phosphorus concentration in the soil solution. Conversely, a greater soil volume means more roots and more root surface for phosphorus absorption. Theoretically, there is a most desirable volume for each soil into which the fertilizer should be mixed or placed to achieve maximum uptake of phosphorus, and potassium, and crop yield. Experimental data in Figure 15.8 show that when 50 pounds of P_2O_5 per 2 million pounds of soil were placed with varying amounts of the soil volume, the maximum predicted phosphorus uptake occurred with soil volumes ranging from 2 to 20 percent of the soil volume, depending on soil. The greater the amount of available phosphorus already in the soil, the larger the volume of fertilized soil for maximum phosphorus uptake. Localized placement of phosphorus in a band along the crop rows is especially effective in intensively weathered soils. These soils are rich in iron and aluminum oxides, have high phosphorus fixation capacity and, generally, have a very low phosphate concentration in the soil solution.

In summary, localized placement of fertilizer in a band puts fertilizer in contact with a small soil volume in which there are few roots; however, there is a high concentration of fertilizer nutrients in solution. By comparison, when fertilizer is broadcast uniformly over the entire soil surface, the fertilizer is eventually mixed with a very large soil volume. This results in many roots in the soil volume but a much lower concentration of fertilizer nutrients in solution.

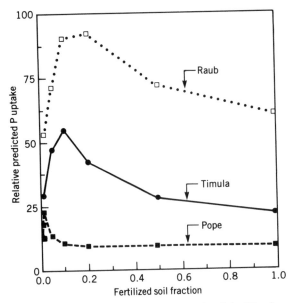

FIGURE 15.8 The effect of fraction of soil fertilized on predicted uptake of phosphorus by corn plants from the application of 50 pounds of P_2O_5 per acre. Uptake values are relative. The Raub soil has the highest soil test for phosphorus, and the Pope soil has the lowest. (Data from Kovar and Barber, 1987.)

Broadcast fertilizer application is the surface application in a more or less uniform rate (see Figure 15.9). If the broadcast application is made after the crop has been planted, it is called *top dressing*. Broadcasting or top dressing is the typi-cal method used to fertilize a lawn, forage field, or forest. The fertilizer is placed on the soil surface. It may be ineffectively used if the soil surface remains dry. Rain and irrigation water move nitrogen rapidly into the soil, but they move potassium to a lesser extent, and phosphorus very little. In general, the mobility of nitrogen as nitrate in soils results in little if any placement effect for nitrate fertilizers. Urea, and other NH_3-producing fertilizers, including animal manures, are subject to significant nitrogen loss by volatilization when broadcast onto the soil surface, and the losses cannot be easily avoided.

Fertilizer application by airplane or helicopter is common for flooded rice fields, forests, and rugged pasturelands. Fertilizer is added to irrigation water and is applied to leaves as a foliar spray. Foliar sprays are important for supplying plants with certain micronutrients, such as iron, that are rapidly precipitated when a soluble form is added to the soil as fertilizer.

Capsules can be inserted into tree trunks to supply certain micronutrients. Fertilizer "spikes" can be inserted in the soil of container grown plants to increase the supply of nutrients in the soil solution.

Salinity and Acidity Effects

Fertilizers are soluble salts and their effect is to lower the water potential of the soil. The rate of

FIGURE 15.9 Broadcast application of dry fertilizer. (Photograph courtesy Ag-Chem Company.)

water uptake by seeds and roots is reduced. Fertilizers placed directly with seeds can delay and/or reduce germination, as shown in Figure 15.7. Biological oxidation of ammonia or ammonium, NH_3 or NH_4^+, produces H^+ and contributes to soil acidity. Continuous corn grown on acid soils that is fertilized mainly with NH_3 can eventually result in very acid soils and accompanying acid soil infertility. Toxicities of manganese and/or aluminum and a deficiency of molybdenum are likely.

Each pound of nitrogen from ammonium sulfate produces acidity equal to that neutralized by about 5.5 pounds of $CaCO_3$. Ammonium nitrate, anhydrous ammonia (NH_3), and urea produce acidity equivalent to about 1.8 pounds of $CaCO_3$ for each pound of nitrogen. Where increased soil acidity is desirable, $(NH_4)_2SO_4$ is recommended as the nitrogen fertilizer; it is frequently used for blueberry production. The acidity effects of phosphorus and potassium fertilizers are generally unimportant, except for that of the ammonium nitrogen in ammonium phosphates. The wheat shown in Figure 15.10 failed to grow because of acidity produced by heavy and continued application of ammonmium chloride over several years.

FIGURE 15.10 Very little wheat was produced on the plot in the foreground with a pH of 3.7, because of long and continued application of ammonium chloride. Note excellent wheat growth on the plot in the rear with a favorable soil pH.

Ammonium chloride is not a common fertilizer; however, in an experiment it produced similar results on soil pH and crop yields as did ammonium sulfate.

ANIMAL MANURES

After centuries during which animal manure was highly prized for its fertilizing value, manure on many American farms has become a liability. Specialized animal farms with many animals produce more manure than can be used effectively on the cropland. The emphasis has changed from the application of manure onto soils to increase soil fertility to the use of the soil for animal waste disposal, as shown in Figure 15.11. The increasing shortage of fuel wood has increased the use of manure as a source of energy in developing countries.

Manure Composition and Nutrient Value

Many factors affect the quantity of manure produced and its composition, including: (1) the kind and age of the animal, (2) the kind and amount of feed consumed, (3) the condition of the animal, and (4) the milk produced or work performed by the animal.

Farm manures consist of solids and liquids. The phosphorus and dry matter are concentrated in the solid portion (feces). On average, the liquid portion of cattle manure contains about two thirds of the potassium and one half of the nitrogen as shown in Figure 15.12. Because the composition of manure is so variable, data on manure composition presented in Table 15.2 can only be approximate. Earlier, it was shown that a ton of average farm manure was considered the equivalent of 4 pounds of nitrogen in fertilizer for corn grain production. The phosphorus and potassium in a ton of manure are equated to about 2 pounds of P_2O_5 and 8 pounds of K_2O from fertilizer. The nutrient value of the manure from 500 beef cattle is

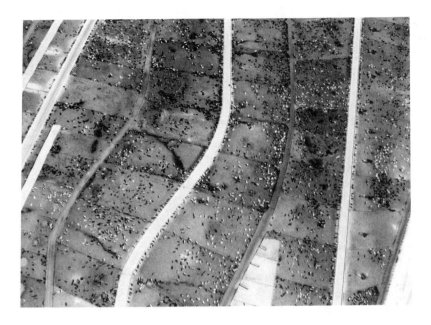

FIGURE 15.11   Aerial view of large feedlot at Coalinga, California. The waste-disposal problem for each 10,000 cattle is equal to that of a city of 164,000. (Photograph courtesy Environmental Protection Agency.)

$19,600 per year. This is based on the assumptions that the cattle weigh 750 pounds, and the value of each pound of nutrients is worth 18 cents for nitrogen, 21 cents for P_2O_5, and 9 cents for K_2O.

Nitrogen Volatilization Loss from Manure

Large amounts of NH_3 are produced in manure, especially by the decomposition of urea and uric acids. The pungent odor in a closed barn is usually due to NH_3. Manure contains many microbial decomposers, and there is a continuous formation and loss of NH_3 as long as favorable conditions exist. After manure is spread in the field, drying and/or freezing can accentuate NH_3 loss. As a result, only a few practical things can be done to conserve the nitrogen in manure during storage and application. During storage, one or a few large piles is better than many small piles of manure. The sooner the manure is incorporated into the soil after spreading, the more nitrogen is conserved. An ideal handling of manure occurs

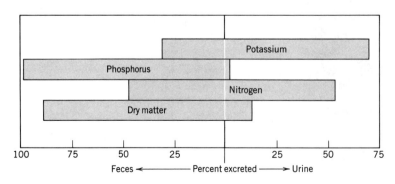

FIGURE 15.12   Distribution of nitrogen, phosphorus, potassium, and dry matter in cattle manure. (From Azevedo and Stout, 1974.)

TABLE 15.2   Quantity and Composition of Fresh Manure Excreted by Various Kinds of Farm Animals

Animal	Excrement	Pounds per Ton[a]	Water, %	Nitrogen, lb	P_2O_5, lb	K_2O, lb	Tons Excreted[b] per Year
Horse	Liquid	400	—	5.4	Trace	6.0	—
	Solid	1,600	—	8.8	4.8	6.4	—
	Total	2,000	78	14.2	4.8	12.4	9.0
Cow	Liquid	600	—	4.8	Trace	8.1	—
	Solid	1,400	—	4.9	2.8	1.4	—
	Total	2,000	86	9.7	2.8	9.5	13.5
Swine	Liquid	800	—	4.0	0.8	3.6	—
	Solid	1,200	—	3.6	6.0	4.8	—
	Total	2,000	87	7.6	6.8	8.4	15.3
Sheep	Liquid	660	—	9.9	0.3	13.8	—
	Solid	1,340	—	10.7	6.7	6.0	—
	Total	2,000	68	20.6	7.0	19.8	6.3
Poultry	Total	2,000	55	20.0	16.0	8.0	4.3

Compiled from Van Slyke, 1932.
[a] Multiply by 0.5 to convert to kilograms per metric ton.
[b] Clear manure without bedding; tons excreted by 1,000 lb of live weight of various animals.

when manure in pen or stall barns is made anaerobic by cattle tramping, the manure is occasionally removed from these barns and then plowed under immediately after application.

Many farmers are moving to a liquid manure-handling system. The manure is accumulated in pits and the liquid manure is injected into the soil.

Manure as a Source of Energy

Biogas production using manure and other organic wastes is based on anaerobic decomposition of the wastes. During anaerobic decomposition, the major gas produced is methane (CH_4). The remaining manure sludge and liquid effluent contain the original plant-essential elements of the digested materials. Because much carbon is lost as CH_4, the sludge and effluent are enriched with nitrogen and phosphorus, relative to the original materials.

Basically, a slurry of manure and other organic wastes is fed into a hopper that is connected to a microbial digestion chamber. The gas produced is piped off to light lamps, cook food, or run small engines. The sludge and effluent are excellent for use as fertilizer. A biogas generator, with a metal floating digestion chamber, is shown in Figure 15.13. The dairy farm with the largest herd of registered Holstein cattle in the United States has a biogas generation plant that produces all the electricity needed to operate the farm.

Several million biogas generators are used in China, where the primary interest is in the nutrient value of the digested sludge and the effluent. Because most harmful disease organisms are killed in digestion, biogas generation is also a good means for disposing of human wastes and for keeping the environment more sanitary. The gas is combustible and so safety precautions are needed to prevent an explosion resulting from exposure of the gas to an open flame.

During the settlement of the Great Plains in the United states, bison manure (chips) was used for fuel. Today, many people live in areas that are not forested and fuel wood is scarce. In India, about 80 percent of the cattle manure is formed in cakes, dried, and used as fuel (see Figure 15.14).

FIGURE 15.13 Biogas generator with intake hopper (at left), floating metal gas holder and anaerobic digestion chamber in the center, and sludge discharge at the right.

LAND APPLICATION OF SEWAGE SLUDGE

Land application of wastes to soils has been practiced since ancient times and still represents an important part of many soil management programs. The federal mandate on water pollution control and waste disposal, however, marks the beginning of an era when land disposal of sewage wastes is being intensively studied and accepted by the public. Sewage sludge is the solid residue remaining after sewage treatment. The increasing cost of disposal by sanitary landfills has increased interest in land disposal. Seventy to 80 percent of the municipal sewage treatment plants in Michigan dispose of sludge by land application. About 2 percent of the land in the United States would be required to dispose of all of the sludge produced.

FIGURE 15.14 Cow dung shaped into cakes, plastered on a wall, and allowed to dry. The dried chips are used for cooking food where fuel wood is in short supply.

Sludge as a Nutrient Source

Sewage sludge is used as a source of nutrients for crop production. The median contents, on a dry-weight basis, of sewage sludge produced in the United States is about 3 percent nitrogen, 2 percent phosphorus, and 0.3 percent potassium. A 10-ton application per acre would contain 600 pounds of nitrogen, which is mainly organic nitrogen, and about 20 to 25 percent is estimated to be mineralized the first year. A 10-ton application would supply about 120 to 150 pounds of nitrogen for plant uptake. The regular nitrogen recommendation for crops could then be reduced by this amount. About 3 percent of the nitrogen remaining in the sludge will be mineralized in the second, third, and fourth years after application. Where sludge is applied repeatedly, allowance must be made for the release of nitrogen from residual-sludge. Excessive applications of sludge result in reduced yields, so there is a limit on how much sludge can be applied to land and still produce acceptable yields. Excessive application of sludge also poses a hazard for pollution of groundwater with nitrate.

The phosphorus in sludge is equated to that in fertilizers. The phosphorus content of sludge is typically high enough when using sludge to satisfy all or meet most of the nitrogen needs of crops, that more phosphorus will be applied in the sludge than is needed. This is not a serious problem, because soils have great capacity to fix phosphorus. On coarse-textured soils very high levels of soil phosphorus could eventually pose a hazard for leaching of some phosphorus to the groundwater.

The potassium content of the 10 tons of sludge would be equated to about 60 pounds of fertilizer potassium.

Solid dry-sludge application to the land as a broadcast application may encounter an odor problem, which is why sludge is sometimes handled as a slurry and injected directly into the soil. This has the added advantage of greatly reducing nitrogen loss by volatilization.

Heavy Metal Contamination

Certain industrial processes, such as metal plating, contribute heavy metals to sludges. Among these are cadimum (Cd), lead (Pb), zinc (Zn), nickel (Ni), and copper (Cu). The greatest detrimental impact of applying sludge to agricultural land is likely to be associated with the cadmium content of the sludge. Many crops may contain elevated concentrations of cadmium in their vegetative tissues without showing symptoms of toxicity. However, cadmium increases in the grain have been much lower than in the vegetative tissues. Potential hazards of heavy metals in the environment are discussed in Chapter 13.

The heavy metal content of the sludge limits the total amount of sludge that may be applied to land. Metal loading values have been developed, which are the total amounts of the heavy metals that can be applied before further sludge application is disallowed. The reaction and fixation of heavy metals into insoluble forms are related to soil texture. Therefore, loading values for heavy metal application on agricultural land have been tied to the cation exchange capacity, which is an indication of the organic matter and clay content in soils. Loading values are given in Table 15.3.

An average loading value for cadmium is 10 pounds per acre. For a sludge containing 14 ppm of cadmium, there is 0.028 pounds of cadmium per ton. The amount of sludge that would need to be applied to attain the cadmium loading value of 10 pounds per acre is 357 tons.

Heavy metals are less soluble in alkaline soils, and maintaining high soil pH is one way to reduce the likelihood of heavy metal contamination of food crops. Soil pH should be maintained at 6.5 or higher. The lime requirement of soils receiving sludge is that amount of lime needed to increase soil pH to 6.5. The great importance of soil pH in deactivating heavy metals has overshadowed the

TABLE 15.3 Total Amount of Sludge Metals Allowed on Agricultural Land

Metal	Soil Cation Exchange Capacity, cmol (+) /kg ^{a}		
	<5	5-15	>15
	Maximum amount of metal (lb/acre)		
Lead (Pb)	500	1,000	2,000
Zinc (Zn)	250	500	1,000
Copper (Cu)	125	250	500
Nickel (Ni)	50	100	200
Cadmium (Cd)	5	10	20

Data from Jacobs, 1977.
^{a} Determined by the pH 7 ammonium acetate procedure.

importance of cation exchange capacity as the basis for determining heavy metal loading values.

FERTILIZER USE AND ENVIRONMENTAL QUALITY

Environmental concerns regarding fertilizer, manure, and sludge use center around the pollution of surface waters with phosphate and pollution of groundwater with nitrate-N.

Phosphate Pollution

Phosphate from fertilizer and organic wastes reacts with soil and becomes adsorbed (fixed) onto soil particles. The adsorbed $H_2PO_4^-$ is carried along and deposited in surface waters as soil sediment because of erosion. This adsorbed phosphate in the sediment establishes an equilibrium concentration of phosphate with the surrounding water, just as it does in the soil with the soil solution. The result is an increased level of available phosphorus for plants growing in the water because of the desorption of the $H_2PO_4^-$. In-

creased growth of algae and other aquatic plants, and their subsequent microbial decomposition, depletes the water of oxygen and kills fish living in the water. Sometimes, excess weed growth hampers swimming and boating. The real problem is caused by erosion of soil rather than the use of phosphorus fertilizer. In fact, much of the pollution occurs at construction sites where unprotected soil is exposed to serious erosion. Where phosphorus fertilizer increases the vigor of a vegetative cover, erosion and pollution of surface waters are reduced.

In some states, one half or more of the soil samples test high or very high for phosphorus. More and more farmers are receiving fertilizer recommendations for only small maintenance levels of phosphorus, which should help reduce the danger of phosphorus pollution of waters.

Nitrate Pollution

Groundwater naturally contains some nitrate (NO_3^-). Since ancient times, the disposal of human and animal wastes from villages created varying degrees of nitrate pollution. Today,

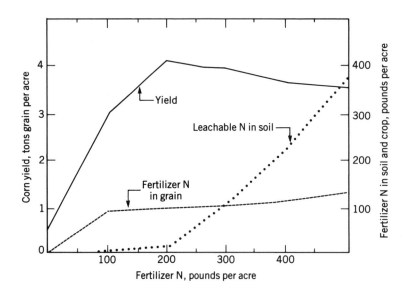

FIGURE 15.15 Relationship between the amount of applied nitrogen fertilizer, corn yield, and the amount of nitrogen recovered in grain and remaining in soil as leachable nitrogen. (Data of Broadbent and Rauschkolb, 1977.)

however, the problem is more serious because of the extensive and high rates of nitrogen fertilizer used for crop production, and the concentration of animals on large farms and ranches. Some of the initial concerns of nitrate pollution in the United States were the result of the pollution of water wells on farms. These wells became polluted with nitrate produced in barnyards from manure.

The extent to which nitrate moves to the water table as water passes through soil is a function of the amount of nitrate in the soil and the amount of nitrate immobilized by the crop and microorganisms. Thus, the use of nitrogen fertilizer could have no effect or it could increase or decrease the amount of nitrate leaching through the soil to the water table. Nitrate pollution (an increase in the nitrate level of groundwater above the normal level) is a problem when more nitrogen fertilizer is added than crops and microorganisms can immobilize or absorb during growth. The data in Figure 15.15 show that when more than 220 pounds of nitrogen per acre were applied, the crop yield and uptake of nitrogen did not continue to increase significantly, and additional fertilizer nitrogen appeared as nitrate that could be leached from the soil.

The nitrate pollution problem can be solved by making a realistic assessment of yield goals and the application of nitrogen fertilizer and organic wastes accordingly. Efforts are being made by many Agricultural Experiment Station and Agricultural Extension personnel in various states to encourage formers to avoid excessive application rates of both nitrogen and phosphorus fertilizers. Recommendations of nitrogen for soybeans are being reduced or eliminated.

Nitrate Toxicity

Nitrate is absorbed by roots, translocated to the shoots, and converted into proteins. Usually, the nitrate content of food and feed crops is not toxic to animals. Whenever nitrate moves into the plant more rapidly than it is converted into protein,

nitrate accumulates in the plant. Sometimes as much as 5 percent or more of a plant's dry weight may be nitrate-nitrogen. Under these conditions, a high concentration of *nitrite* may be produced in the digestive tract of ruminants (cud-chewing animals). Nitrites are very toxic and interfere with the transport of oxygen by the bloodstream.

High nitrate levels in plants usually result from high soil nitrate plus some environmental factor that results in nitrate accumulation in plants. During a drought, plants absorb nitrates, but the lack of water interferes with the normal utilization of the nitrate for growth. Then, nitrate accumulates in plants. Other factors that may contribute to nitrate accumulation are cloudy weather and reduced photosynthesis.

Various plants show different tendencies to accumulate nitrate. Annual grasses, cereals cut at the hay stage, some of the leafy vegetables, and some annual weeds are most likely to contain high nitrate levels. Pigweed is a nitrate accumulator. High concentrations of nitrate are less likely to accumulate in legumes and perennial grasses, but even these species are not completely free from the problem. Within any one plant, the nitrate level in the seeds or grain is almost always lower than in the leaves. High levels of nitrate have not been found in the grains used as food and feed crops.

Monogastric (single-stomached) animals are not likely to develop a nitrite problem from the ingestion of nitrates. No authenticated cases of human poisoning have been traced to high levels of nitrate in U.S. food crops. Cases of nitrite poisoning have been reported in European infants as a result of consumption of high nitrate-content vegetable baby foods that were improperly stored.

SUSTAINABLE AGRICULTURE

Western agriculture, characterized by high technology, has been considered to be efficient. The quote "one farmer producing enough food to feed 70 people" fails to convey the fact that four tractors, thousands of dollars of capital investment, large trucks, miles of concete highway, and people in processing and marketing activities are required. Now, there is growing concern about future energy supplies and costs. Additionally, today's agriculture experts are finding that the solution to one problem frequently creates other, more difficult problems. Many farmers are seeking alternative farming methods, because costs seem to be overtaking the benefits from the adoption of newer technology.

Another aspect of the interest in sustainable agriculture is the concern over food contamination from harmful chemicals. For many years, organic gardeners have been devising methods to grow crops and animals without chemical fertilizers, pesticides, and animal growth stimulators. For some, farming and food production have a "holistic" component that is reflected in a certain personal philosophy and way of life.

Slash-and-burn farmers in the humid tropics have a self-sustaining system that is ecologically sound and permanent, provided the population does not overwhelm the system and make fallow periods too short to regenerate the soil. Many subsistence farmers throughout the world have minimal reliance on outside production inputs. Agricultural development in Europe and the United States was based largely on self-sustaining systems. During the nineteenth century, crop rotations, including legumes to fix nitrogen, and animals to provide both power and cycling of nutrients in manure, resulted in fairly satisfactory control of pests and fairly high yields. Farm manures were highly valued for their nutrient content.

Crop production research in this area has produced promising results. In some cases, the results of reduced crop yields have been canceled out by lower input costs. This problem, however, is not only a technological one and it is not an isolated one within society. Rather, it is an expression of societal concerns that invades all areas of life. Technology has brought apparent abundance to millions, yet millions are as improvished as

ever. Significant movement toward and acceptance of a restructing of agriculture to a sustainable mode will require a change in society's choice indicators. That is, total costs of production must include offsite costs and environmental damage, and a desire for economic profit in the short term will need to be tempered by a greater concern for the quality of life over the long term.

SUMMARY

Soil fertility is the soil's ability to provide essential elements for plant growth without a toxic concentration of any element. Soil fertility is assessed by use of deficiency symptoms, plant tissue analysis, and soil testing.

Plant tissue analysis and soil tests are the basis for making fertilizer recommendations.

Fertilizer nutrients are absorbed most efficiently when they are applied near the time of maximum plant uptake, owing to the potential loss of nitrogen by leaching and volatilization and the fixation of phosphorus and potassium. These factors are also important considerations for placement of fertilizers.

In some cases, the acidity and salinity effects of fertilizers are important considerations in the selection and application of fertilizers.

Manure contains nutrients and is a substitute for chemical fertilizers. Manure is also an energy source for biogas generation and as a substitute for fuel wood.

Much of the sewage sludge produced in the United States is applied to land. This sludge contains nutrients that can substitute for fertilizers. Some sludges contain heavy metals, and the amount of sludge that can be applied is limited to the heavy metal content of the sludge. Soil pH should be maintained at 6.5 or higher to inactivate heavy metals.

Modern agriculture is energy intensive and has been becoming increasingly so. Sustainable agriculture concerns stem from the uncertainty of the future supply and cost of energy.

REFERENCES

Anonymous. 1986. *Utilization, Treatment, and Disposal of Waste on Land. Soil Science Soc. Am.*, Madison, Wis.

Anonymous. 1988. *Fertilizer Management for Today's Tillage Systems.* Potash and Phosphate Institute, Atlanta, Ga.

Azevedo, J. and P. R. Stout. 1974. "Farm Animal Manures: An Overview of Their Role in the Agricultural Environment." *Ca. Agr. Exp. Sta. Manual* 44, Davis, Cal.

Bezdicek, D. F. and J. F. Power (eds.) 1984. "Organic Farming: Current Technology and its Role in a Sustainable Agriculture." *ASA Spec. Pub. No 46,* Amer. Soc. Agronomy, Madison, Wis.

Broadbent, F. E. and R. S. Rauschkolb. 1977. "Nitrogen Fertilization and Water Pollution." *California Agriculture, 31* (5):24–25.

Cressman, D. R. 1989. "The Promise of Low-input Agriculture." *J. Soil and Water Con.* **44**:98.

Edens, T. C., C. Fridgen, and S. L. Battenfield. 1985. *Sustainable Agriculture and Integrated Farming Systems.* Michigan State University Press, East Lansing.

Engelstad, O. P. (ed.) 1985. *Fertilizer Technology and Use.* 3rd ed. Soil Sci. Soc. Am., Madison, Wis.

FAO. 1978. China: Azolla Propagation and Small-scale Biogas Technology. *FAO Soils Bull,* 41, Rome.

Foth, H. D. and B. G. Ellis. 1988. *Soil Fertility.* Wiley, New York.

Illinois Agronomy Department. 1984. *Illinois Agronomy Handbook 1985–1986.* University of Illinois, Urbana.

Jacobs, L. W. 1977. "Utilizing Municipal Sewage Wastewaters and Sludges." *North. Cent. Reg. Ext. Pub. 52.* Michigan State University, East Lansing.

Kovar, J. L. and S. A. Barber. 1987. "Placing Phosphorus and Potassium for Greatest Recovery." *Jour. Fert. Issues.* **4**(1):1–6.

Neely, D. 1973. "Pin Oak Chlorosis-Trunk Implantations Correct Iron Deficiency." *Jour. For.* **71**:340–342.

Page, A. L., T. J. Logan, and J. A. Ryan. 1989. *Land Application of Sludge.* Lewis Publishers, Chelsea, Mich.

Stoeckeler, J. H. and H. F. Arneman. 1960. Fertilizers in Forestry. *Adv. Agron.* **12**:127–165.

Stone, E. L., G. Taylor, and J. DeMent. 1958. "Soil and Species Adaption: Red Pine Plantations in New York." *First North Am. Forest Soils Conf. Proc.* Michigan State University, East Lansing.

Van Slyke, L. L. 1932. *Fertilizers and Crop Production.* Orange Judd, New York.

Voss, R. 1982. "General Guide for Fertilizer Recommendations in Iowa." *Agronomy 65.* Iowa State University, Ames.

Warncke, D. D., D. R. Christensen, and M. L. Vitosh. 1985. "Fertilizer Recommendations: Vegetable and Field Crops in Michigan." *Mich. State Univ. Ext. Bull.* E-550. East Lansing, Mich.

Welch, L. F., D. L. Mulvaney, M. G. Oldham, L. V. Boone, and J. W. Pendleton. 1971. "Corn Yields with Fall, Spring, and Sidedress Nitrogen." *Agron. Jour.* **63**:119–123.

Welch. L. F. 1986. "Nitrogen: Ag's Old Standby." *Solutions: J. Fluid Fert. and Ag. Chem.* July/August, pp. 18–21.

Wolcott, A. R., H. D. Foth, J. F. Davis, and J. C. Shickluna. 1965. "Nitrogen Carriers: Soil Effects." *Soil Sci. Soc. Am. Proc.* **29**:405–410.

CHAPTER 16

SOIL GENESIS

Soil genesis deals with the factors and processes of soil formation. The formation of soil is the result of the interaction of five soil-forming factors: parent material, climate, organisms, topographic position or slope, and time. The dominant processes in soil genesis are: (1) mineral weathering, (2) humification of organic matter, (3) leaching and removal of soluble materials, and (4) translocation of colloids (mainly silicate clays, humus, and iron and aluminum oxides). The net effect of these processes is the development of soil horizons, that is, the genesis of soil.

ROLE OF TIME IN SOIL GENESIS

Soils are products of evolution, and soil properties are a function of time or soil age. The age of a soil is expressed by its degree of development and not the absolute number of years. In a sense, soils have a life cycle that is represented by various stages of development. Reference has been made to minimally, moderately, and intensively weathered soils. To study time as a soil-forming factor,

the other soil-forming factors must remain constant, or nearly so.

A Case Study of Soil Genesis

At the end of the Wisconsin glaciation, about 10,000 years ago in the upper Midwest of the United States, the Great Lakes were larger than they are today. The lowering of lake levels occurred in stages to produce a series of exposed beach surfaces of different ages. In the case of the study area, there are three different soils on lake beaches aged 2,250 years, 3,000 years, and 8,000 years. All slopes were nearly level. Similar calcareous, quartzitic sand parent material occurred at each location, as shown in Figure 16.1.

For a short period of time after the glacial ice melted, a tundra climate existed. Soon after, the climate became favorable for trees. The climate went through a warming trend between 4,000 and 6,000 years ago; today, the climate is humid-temperate. In general, the vegetation was a mixed coniferous-deciduous forest.

Two factors that greatly affected soil genesis in

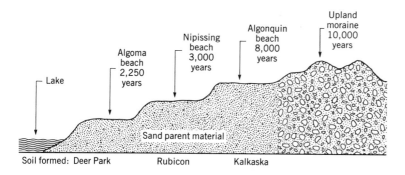

FIGURE 16.1 Relationship between lake beach locations and ages for soil genesis in northern Michigan. Ages of deposits given as years before the present (BP). (Data from Franzmeier and Whiteside, 1963.)

this study were (1) the quartzitic sand-parent material, and (2) a humid climate that produces much leaching. Very little clay existed in the parent material and very little clay was formed via weathering. Clay translocation was insignificant, and Bt horizons could not develop. Under the conditions of a high-leaching regime, colloidal humus and oxides of iron and aluminum (sesquioxides) were translocated downward and accumulated in the subsoil.

The changes that occurred during soil genesis are summarized in Table 16.1. During the first 2,250 years, humification and accumulation of organic matter near the soil surface resulted in the formation of an A horizon. The lime present in the upper part of the parent material was leached out and the A horizon became acid. The acid A horizon overlies a slightly discolored parent material layer. This young soil, which has only A and C horizons, is the Deer Park soil.

During the next 750 years, leaching continued to remove lime, and acidity developed in the soil to a depth of 1 meter or more. Sufficient sesquioxides accumulated in the subsoil to form a Bs horizon (s from sesquioxide). Simultaneously, an E horizon evolved. This 3,000-year-old soil on the Nipissing beach has A, E, Bs, and C horizons and is called the Rubicon soil.

During the next 5,000 years, there was considerable movement and accumulation of humus in the subsoil, and the Bs horizon became a Bhs

horizon (h from humus). This 8,000-year-old soil has A, E, Bhs, and C horizons and is called the Kalkaska soil. A soil similar to the Kalkaska is shown in Figure 16.2 (also see book cover).

Note from Table 16.1, that as colloidal and amorphous sesquioxides and humus (noncrystalline materials with high specific surface) accumulated in the subsoil, the tree species changed. The change was toward species more demanding of water and nutrients, and this parallels the soil's ability to retain more water and nutrients as amorphous colloidal materials accumulated in the B horizon. Therefore, as the soils evolved, soil properties changed and this caused a change in the composition of the forest species on each terrace. The most productive soil in the sequence for forestry is the Kalkaska soil.

Time and Soil Development Sequences

It should be recognized that the rate of soil genesis in a particular situation is highly dependent on the nature of the soil-forming factors. In some situations, soil genesis is rapid, whereas in others, soil genesis is very slow.

Some soil features or properties develop quickly, but the development of other properties requires much more time.

A horizons can develop in a few decades. Consequently, soils with only A and C horizons can

TABLE 16.1 Major Processes and Changes During Soil Evolution in
Northern Michigan From Sand Parent Material

Beach	Age, years	Soil and Major Processes	Dominant Species
Algoma	2,250	Deer Park, A-C soil. Humification of organic matter and leaching of lime resulting in formation of an acid A horizon, beginning of movement of iron and aluminum oxides to subsoil.	Red oak, red maple, and white pine
Nipissing	3,000	Rubicon, A-E-Bs-C soil. Translocation of iron and aluminum oxides to form Bs and E horizons. Continued leaching of lime to produce an acid solum.	Aspen, birch, red oak, red maple, and balsam fir
Algonquin	8,000	Kalkaska, A-E-Bhs-C soil. Translocation of humus (with some iron and aluminum oxides) to form Bhs horizon.	Birch, red and sugar maple, and balsam fir

Adapted from Franzmeier and Whiteside, 1963.

develop in 100 years or less. The formation of a Bw horizon involves changes in color and/or structure and requires about 100 to 1,000 years. Soils with an A-Bw-C horizon sequence, therefore, require commonly several hundred years for development. The development of the Bt horizon is dependent on some clay formation within the soil and significant translocation of clay. These processes usually require a minimum of several thousand years. Soils with an A-E-Bt-C horizon sequence need about 5,000 to 10,000 years to form under favorable conditions. The time to form intensively weathered soils containing a high content of kaolinite and oxides of iron and aluminum is approximately 100,000 years. Some of these soils in the tropics are more than a million years old.

ROLE OF PARENT MATERIAL IN SOIL GENESIS

Parent material has a great influence on the properties of young soils, including color, texture, structure, mineralogy, and pH. Over time the effects of the parent material decrease, but some effects of the parent material still persist in old soils.

A

E

Bhs

C

FIGURE 16.2 Soil with A, E, Bhs, and C horizons. (The metal frame is used to collect a vertical slab of soil to mount on a board and treat with plastic, to be used for display purposes.)

Consolidated Rock as a Source of Parent Material

Consolidated rocks are not parent material but serve as a source of soil parent material. Soil formation may begin immediately after the deposition of volcanic ash but must await the physical disintegration of hard rocks when granite, basalt, or sandstone are exposed to weathering. During the early stages of rock weathering and soil formation, the formation of parent material and soil may occur simultaneously as two overlapping processes. Soils formed in parent materials derived from the weathering of underlying rocks are common in the U.S. Appalachian Highlands and Interior Highlands in the eastern states and in the Cordilleran Region in the western states (see Figure 16.3). Within these regions, however, many soils have formed in the sediments that occur at the base of steep slopes or in small depressions, basins, or areas beside streams and rivers.

Soil Formation from Limestone Weathering

Parent material formation from most hard rocks is by the physical disintegration and chemical decomposition of the mineral particles in the rock. A parent material derived from sandstone is, essentially, composed of the sand particles that were cemented together in the sandstone. In the case of limestone, the soil develops in the insoluble impurities that remain after the calcium and magnesium carbonate have been dissolved and leached from the weathering environment. Clay is a common impurity in limestone, giving rise to the fine texture of soils derived from limestone weathering. Chert is a compact, siliceous material that occurs in some limestones. The chert resists dissolution, and soils formed from the weathering of cherty limestone are stony.

Many centimeters of limestone are required to form a centimeter of soil, because the impurities (residues) typically comprise only a small percentage of the limestone and some of the residues are carried away. It has been estimated that a foot of soil accumulation in the Blue Grass region of Kentucky required limestone of the order of 100 feet and more than half a million years. Most soils of the Kentucky Blue Grass region of Kentucky and the Nashville Basin of Tennessee have developed from limestone and were prized by settlers for their suitability for agriculture. A soil devel-

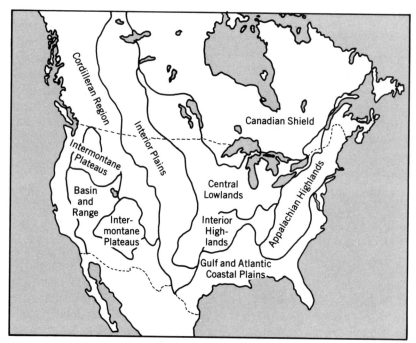

FIGURE 16.3 Physiographic regions of North America.

oped from limestone weathering is shown in Figure 16-4.

Sediments as a Source of Parent Material

Most soils have developed from sediments that were transported by water, wind, ice, or gravity. *Colluvial* sediments occur at the base of steep slopes where gravity is the dominant force, causing movement and sedimentation. Colluvial sediments are common and are an important parent material in mountainous areas.

Alluvial sediments are ubiquitious because of the widepread existence of streams and rivers (see Figure 16.5). The most extensive area of alluvial sediments in the United States occurs along the Mississippi River. Most of the agricultural soils

FIGURE 16.4 Parent material derived from limestone is the residual accumulation of impurities in the limestone after water has leached away the carbonate materials. Because clay is a major impurity, parent materials and soils derived from limestone weathering tend to be fine textured.

in California occur in valleys where alluvium is the dominant parent material. Colluvial and alluvial sediments occur as inclusions in areas dominated by hard rocks or other kinds of sediments. Regions of the United States in which sediments are extensive and important sources of parent material are shown in Figure 16.3. The regions include the Gulf and Atlantic Coastal Plains, the Central Lowlands, the Interior Plains, and the Basin and Range. Sediments were the most common source of parent material for the soils of the United States.

Gulf and Atlantic Coastal Plains The sediments of the Gulf and Atlantic Coastal Plains are of marine origin. They were derived from the weathered products of the adjacent highlands that were deposited when the area was submerged below the sea. Where calcareous clays rich in smectite were deposited, soils developed with high clay content and great expansion and contraction during alternating wet and dry seasons. Soils have developed in this kind of parent material in Texas, Mississippi, and Alabama. Properties of the soils are the result of the dominant effect of the parent materials.

The majority of the other soils on the coastal plains developed in sediments consisting mainly of quartz sand and kaolinite clay. The quartz and kaolinite represent the "end products" of highly weathered sediments and soils. The sediments accumulated over a long period of time and some of the material in the sediments participated in several weathering cycles before coming to rest in the sea. The coastal plains slope downward toward the sea, resulting in increasing age of sediments and a period of soil formation from the seacoast inward. In the southeastern United States this has produced three major surfaces— the lower, middle, and upper coastal plains. The soils generally have sandy A horizons and low natural fertility. Clay translocation has produced Bt horizons in most of the soils. Many of the subsoils are deficient in calcium for good root growth.

Central Lowlands Most of the Central Lowlands (see Figure 16.3) was glaciated, with the Ohio and Missouri rivers forming the approximate southern boundary of the glaciation. Four major advances of the ice occurred during a period of about 1.5 million years. Each advance of the ice

FIGURE 16.5 Alluvial sediments, deposited along streams and rivers, are common parent materials for soil formation throughout the world.

was followed by a long interglacial period. The glacial and interglacial periods and approximate times of their occurrence are listed in Table 16.2.

During deglaciation, rivers and streams of water flowed from the melting ice and carried and deposited material called *outwash* near the ice front. Outwash consists of coarse-textured sediments. Clay, and to a lesser extent silt, were carried by water into glacial lakes, where the clay settled to form calcareous, fine-textured *lacustrine* (lake) sediments. The melting of ice produced moraine sediments by the direct deposition of material in the ice onto the ground surface. The moraine sediments are unsorted and are called *till*. Sometimes the ice front readvanced over previously deposited sediments. As a result of these varied activities, glacial parent materials are very heterogenous, variable in composition, and commonly stratified, as shown in Figure 16.6.

The nature of the sediments was affected by the kinds of rocks the glacier moved over and the types of materials that were incorporated into the ice. As a result, some glacial materials contain some boulders and stones, but most are composed of highly variable amounts of sand, silt, and clay. Another important component was calcium carbonate whose content was related to the extent to which the glacier moved over limestone and other calcareous rocks. In the northern part of Wisconsin, Michigan, and the New England states, most of the glacial sediments have a high sand content.

During deglaciation, there was rapid ice melting in summer and large rivers carried away the melt water. As a result, stream valleys with very wide floodplains were produced on which sediments accumulated. During the winter, there was little meltwater, and the sediments on the floodplains were exposed to the winds. The floodplain materials dried, and in the absence of a vegetative cover, the silt-sized particles were preferentially picked up by westerly winds and deposited along the eastern sides of the major rivers in the deglaciated area. These sediments contain mainly silt, with smaller amounts of clay, and are called *loess*. The loess is thickest near the river and decreases in thickness with distance away from the river. In some cases, the loess was more than 100 feet thick. Loess about 50 feet thick along the Mississippi River in western Tennessee is shown in Figure 16.7.

Loess is the dominant parent material in the soils of Iowa and Illinois. The high silt content of the loess parent material is reflected in the abundance of silt loam A horizons and the high water-holding capacity and fertility of many of the Corn Belt soils. Many soils beyond the so-called loess area have been influenced by a thin layer of loess. In some of these soils, the upper soil horizons have developed in loess and the lower horizons have developed in till or limestone.

Much loess exists outside of the glaciated part of the Central Lowlands and is found in the Interior Plains, especially in Nebraska and Kansas. This is an area of limited precipitation and steppe (short grass) vegetation. The loess is believed to have resulted from transport of silt from a source area located to the west and under conditions of limited protection of land by a vegetative cover. Extensive areas of nonglacial loess also occur in China and Argentina. Thick loess overlies hills of basalt in the Palouse area of eastern Washington and western Idaho. The loess originated from the deserts of central Washington. About 10 percent

TABLE 16.2 Glacial and Interglacial Ages in North America

Period	Years Before Present[a]
Wisconsin glacial	125 000
Sangamon interglacial	
Illinoian glacial	360 000
Yarmouth interglacial	
Kansan glacial	780 000
Aftonian interglacial	
Nebraskan glacial	1 150 000

Data from Ericson, Ewing, and Wollin, 1964.
[a] Approximate time of furthest advance.

Till deposited by ice

Sand deposited by water

FIGURE 16.6 Sand deposited by water and later overridden by glacial ice, which melted and left a layer of till; an example of stratified parent material.

of the world's land has been influenced by loess deposition.

Interior Plains The Interior Plains region is comparable to the western part of the Great Plains. After the Rocky Mountains were formed, erosion and subsequent sedimentation produced sediments along the eastern side of the mountains in a manner analogous to the formation of the Coastal Plains sediments east and south of

FIGURE 16.7 A loess deposit about 50 feet (15 ms) thick on the uplands along the eastern side of the Mississippi River floodplain in western Tennessee.

the Appalachian Mountains. These sediments, however, contain more weatherable minerals, and the resulting soils are much more fertile than those formed from the coastal plains sands. There are considerable areas of sand and, in some places, erosion has exposed older sediments underneath. Glacial-derived parent materials cover the Interior Plains region north of the Missouri River.

Basin and Range Region The Basin and Range region consists of a series of low mountains separated by broad basins. The area was subjected to block-faulting with the ranges representing the uplifted areas and the down-faulted blocks representing the basins. Weathering and erosion have produced sediments of coalescing alluvial fans, which fill most of the basins. Lacustrine sediments are of minor extent.

Volcanic Ash Sediments Volcanic ash distribution parallels that of volcanic mountains, which were the source of the ash. The 1980 eruption of Mount St. Helens brought a new awareness of the existence and importance of volcanic ash in the northwestern part of United States. Though much of the material ejected in 1980 was crystalline in nature, ash is commonly amorphous and weathers to form amorphous allophane and iron and aluminum oxides. Allophane has a high pH-dependent cation exchange capacity. Soils with amorphous minerals lack the distinct sand and silt particles found in most mineral soils. The major agricultural area in the United States where soils have developed from ash occurs on Hawaii, the large island of the Hawaiian chain, where sugarcane is the dominant crop. Although ash is a minor parent material near Honolulu, Hawaii, on the island of Oahu, the famous landmark, Diamond Head, is an old volcanic cone composed of consolidated ash, as shown in Figure 16.8.

Ash is widely distributed by winds and is a minor but frequent addition to many soils a great distance from the ash source. Ash is a significant component of the loess in the Palouse in the northwestern United States and on the Pampa in Argentina. In some instances, an occasional addition of ash significantly increases the fertility of highly weathered soils as occurs on the Serengeti Plain in Kenya, Africa.

Effect of Parent Material Properties on Soil Genesis

Horizon development, especially many B horizons, depends on the translocation of fine-sized particles in water. A parent material composed of 100 percent quartz sand would not weather to produce mineral colloidal particles. In such parent material, little else can happen than some organic matter accumulation due to plant growth. Early in this chapter, it was noted that a sand-textured parent material in a humid environment led to the development of Bhs horizons. Within the parent material there was a small amount of primary minerals that weathered to release some iron and aluminum. There was insufficient clay to form Bt horizons, however. When soils develop in a parent material that is essentially all clay, little water will migrate downward to transport particles. Parent materials of intermediate texture lead to the development of a wide variety of soils.

The texture and structure of a soil will affect soil genesis by affecting water infiltration and erosion, permeability and translocation of materials in water, protection and accumulation of organic matter, and solum (A plus B horizon) thickness. The thicker solum of the Spinks soil, as compared with the St. Clair soil, is shown in Figure 16.9. The thicker solum is due to the lower clay content, greater water permeability, and greater downward migration of clay before lodgement during the formation of the Bt horizon in the Spinks soil. Greater permeability of sandy material also contributes to the more rapid removal of lime (cal-

FIGURE 16.8 Diamond Head is a famous landmark near Honolulu, Hawaii, and is an old volcanic cone composed of consolidated layers of ash (tuff). Note that the cone is highest on the right because the prevailing winds were from the left during ash fall.

cium carbonate), resulting in more rapid development of acidity and a greater leaching depth of the lime. Note that the texture of the A and B horizons in Figure16.9 is related to the clay content of the C horizon (parent material).

Stratified Parent Materials

A stratified parent material is shown in Figure 16.6. A layer of loess over till was shown to be a condition found in the glaciated part of the Central Lowlands. Frequently, the A or the A and B

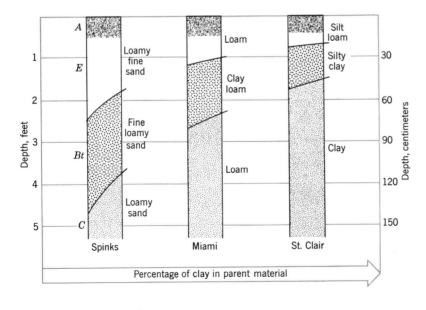

FIGURE 16.9 Relationship between texture of the parent material and the thickness and texture of horizons of three forest soils in the north-central part of the United States.

horizons develop in the loess and the lower part of the soil develops from till. Soil permeability to water (hydraulic conductivity) changes at the boundary of materials of different texture, which affects the downward movement of water. Alluvial parent materials typically are stratified. Where streams dissect the landscape, various materials may be exposed for soil formation, giving rise to a series of soils whose differences are due mainly to parent material differences. Such a sequence of soils is a *lithosequence,* as shown in Figure 16.10.

In soils with horizons developed from different parent materials, numbers are used to indicate this feature. For example, a soil with A and E horizons developed in loess, and Bt and C horizons developed in till has a horizon sequence of A-E-2Bt-2C. The 2 indicates that the Bt and C horizons developed in a different or second parent material.

Parent Material of Organic Soils

In locations where considerable quantities of plant material grow and where decay is limited because of water saturation and anoxia, low temperature, or acidity, an abundance of organic matter may accumulate. Sometimes, this results in a mat of organic matter on the surface of mineral soil. In permanently ponded areas throughout the world, thick deposits of peat and muck have formed. The Florida Everglades is one of the best-known areas. Organic deposits are the parent material for organic soils.

ROLE OF CLIMATE IN SOIL GENESIS

Climate greatly affects the rate of soil genesis. In areas that are permanently dry and/or frozen, soil does not form. The two components of climate to be considered are precipitation and temperature.

Precipitation Effects

Water is necessary for mineral weathering and plant growth. Water in excess of field capacity (surplus water) participates in the downward translocation of colloidal particles and soluble salts. The limited supply of water in deserts re-

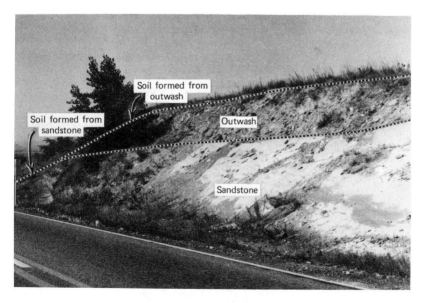

FIGURE 16.10 A roadcut showing outwash overlying sandstone. Strata of different materials can produce a lithosequence of soils on a hillside where the materials are exposed, as shown in the left side of the photograph.

sults in soils that tend to be alkaline, relatively unweathered, low in clay and organic matter content, and cation exchange capacity. Generally, soils of the arid and subhumid regions tend to be fertile, except for the limited ability of the soil microbes to mineralize soil organic matter and to supply available nitrogen. Where there is sufficient water for only limited leaching, the carbonates tend to be moved downward only a short distance, where they accumulate and form a zone or horizon of calcium carbonate accumulation.

FIGURE 16.11 Grassland soil developed under subhumid climate. Limited leaching is expressed by the presence of a k layer (accumulation of carbonates) in the upper part of the C horizon between depths of 70 and 110 centimeters. (Photograph courtesy USDA.)

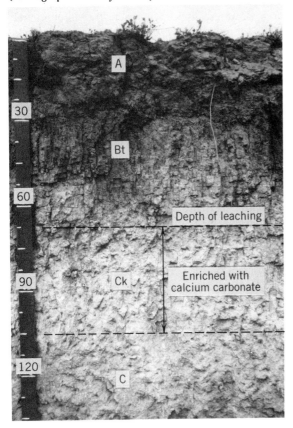

The symbol k is used to indicate this calcium carbonate accumulation feature. This feature is shown in Figure 16.11 for a soil that has a Ck horizon. Soils in deserts commonly have k horizons.

Increases in precipitation have been positively related with greater: (1) leaching of lime and greater depth to a k layer, (2) development of soil acidity, (3) weathering and clay content, and (4) plant growth and organic matter content. A study of surface soils developed in loess along a traverse from Colorado to Ohio showed that as precipitation increased, the depth to the k layer increased until the annual precipitation was about 26 inches (65 cm). This occurred near the Nebraska–Iowa border. At this location, pH of A horizons was about 7. Eastward along the traverse, sufficient leaching occurred to move calcium carbonate to the water table, and k layers were generally absent. In addition, the surface soils became more acid and those in Ohio had a pH of about 5.

Increasing precipitation from eastern Colorado to Indiana is associated with a change in native vegetation. The annual precipitation increases from about 16 to 36 inches (40 to 90 cm). The vegetation changes from widely spaced, bunch and short grasses in Colorado, to short and tall grasses in Nebraska, and to tall grasses in Illinois and Indiana. These changes are associated with an increase in organic matter from 180 to 360 metric tons per hectare (80 to 160 tons per acre) to a depth of 100 centimeters (see Figure 16.12). Similar changes in organic matter content occur on the grasslands in the Pampa of Argentina. Along these traverses with increasing precipitation, increasing soil organic matter contents of well-drained upland soils are associated with greater annual production of the grasses. East of the grassland area the vegetation changes to forest, and the soil organic matter content is considerably less. The organic matter content of tall grass soils in Illinois decreases from 360 metric tons per hectare to 150 metric tons per hectare for forest soils in Ohio (see Figure 16.12).

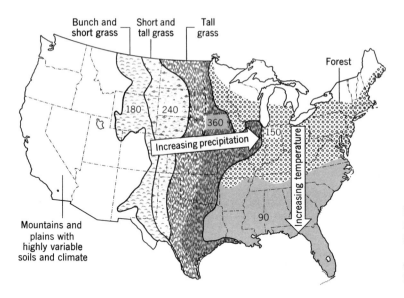

FIGURE 16.12 Generalized map showing organic matter content of soils, in metric tons per hectare 100 centimeters, as related to climate and vegetation. (Adapted from Schreiner and Brown, 1938.)

Increasing precipitation in the tropics also results in increasing amounts of organic matter in tropical soils. The relationship between grassland and forest soils in the temperate region, however, is reversed in the tropics. The forested, humid, tropical soils have a greater organic matter content than the soils of the adjacent drier grassland-shrub savannas. It has been suggested that high acidity and infertility of savanna soils are partially responsible for the lower production of plant material on the tropical savannas than that in temperate grasslands.

Temperature Effects

Every 10°C increase in temperature increases the rate of chemical reactions approximately two times. Increased weathering and clay formation occur with an average increase in soil temperature.

The relationship between average temperature and plant growth and the accumulation of organic matter is complex. The organic matter content of soil is the net result of plant growth or the addition of organic matter, the rate of organic matter mineralization, and the soil's capacity to protect organic matter from mineralization. Many

soils of tundra regions are rich in organic matter even though plant growth rates are very slow. Even slower are the rates of microbial decomposition of organic matter due to both low temperature and soil wetness. The grassland soils of the Great Plains have a gradual decrease in organic matter with increasing annual temperature. (see Figure 9.4).

The forested soils of the southeastern United States are shown to contain only 60 percent as much organic matter as do the forested soils further north (see Figure 16.12). This difference is partially due to the generally more sandy nature of the A and E horizons of the southern soils on the coastal plains as compared with the finer-texture of the forest soils further north.

Based on the observations of the relation between temperature and total soil organic matter content in the eastern United States, one would expect tropical soils to be low in organic matter. Tropical rain forests produce more organic matter and have greater organic matter decomposition than temperate region forests, because of year-round favorable temperature and precipitation. Temperate region forests have a long dormant winter period, resulting in less plant growth and decomposition of organic matter than that in tropical rain forests. Therefore, the content of soil organic matter tends to be similar in the temperate forests and tropical rain forests. Although researchers have found slightly more organic matter in the tropical counterparts, the differences are of little or no importance. The greater dominance of amorphous clays in tropical soils may also be a factor in the protection of organic matter.

Climate Change and Soil Properties

Some soils have been influenced by more than one climate because of climate shifts. A study of buried soils, *paleosols,* reveals the nature of the climate that existed many years ago when ancient soils formed. In the southwestern United States the uplift of the Rocky Mountains cut off moisture-laden winds and created a drier climate east of the mountains. Today, some of the soils on the southwestern deserts have a strongly developed Bt horizon, which indicates that a humid climate existed there when the Bt horizon was formed.

ROLE OF ORGANISMS IN SOIL GENESIS

Plants affect soil genesis by the production of organic matter, nutrient cycling, and the movement of water through the hydrologic cycle. Microorganisms play an important role in organic matter mineralization and the formation of humus. Soil animals are consumers and decomposers of organic matter; however, the most obvious role of animals appears to be that of earth movers.

A most obvious effect of organisms on soil genesis is that caused by whether the natural vegetation is trees or grass. The following discussion will emphasize the differential effects of trees and grass on soil genesis.

Trees Versus Grass and Organic Matter Content

Figure 16.12 shows that the forested soils of Ohio and Wisconsin contain significantly less organic matter than grassland soils located to the west. A study of forest and grassland soils developed from similar parent material in southern Wisconsin revealed that grassland soils contain about twice as much total organic matter as forest soils. In addition there is a marked difference in the distribution of organic matter within the soil profile (see Figure 16.13).

The explanation for the differences in the amount and distribution of organic matter is related to the differences in the growth habits of the two kinds of plants. The roots of grasses are short-lived and each year contribute to the soil large amounts of organic matter that become humified. Furthermore, there is a gradual decrease in root density with increasing soil depth, which parallels the gradually decreasing organic matter content. In the forest, by contrast, roots are long-lived and the annual addition of plant residues is largely as leaves and dead wood that fall directly onto the soil surface. Some of the plant materials decompose on the soil surface, and small animals transport and mix some of the organic matter with a relatively thin layer of topsoil. In the hardwood forest of southern Wisconsin earthworms were active and 81 metric tons of organic matter was in the top 6-inch layer per hectare and only 25 metric tons in the next deeper 6-inch layer. Thus, in the forest soil 47 percent of the soil organic matter was in the top 6 inches as compared to 35 percent for the grassland soil. The dark-colored surface layer enriched with organic matter, the A horizon, is thicker in the grassland soil than in the forest soil.

Another interesting feature shown in Figure 16.13 is the similar amount of total organic matter in each ecosystem. In the forest, however, most of the organic matter is found in the standing trees. When settlers cleared and burned the trees, more than one half of the organic matter in the forest ecosystem was destroyed. By contrast the conversion of the Prairie to use for cropping resulted in little loss of organic matter. In addition, erosion of the topsoil was more detrimental to forest soils than to grassland soils, because of the relatively greater concentration of organic matter near the surface of the soil.

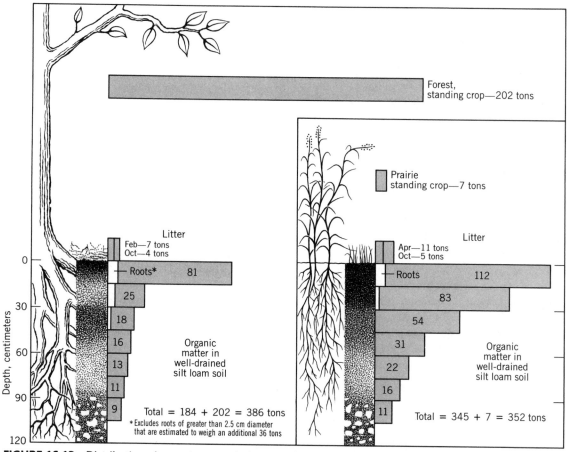

FIGURE 16.13 Distribution of organic matter in forest and grassland ecosystems in south-central Wisconsin in metric tons per hectare. (Adapted from Nielson and Hole, 1963, and used by courtesy of F. D. Hole, Soil Survey Division of the Wisconsin Geological and Natural History Survey, University of Wisconsin.)

Vegetation Effects on Leaching and Eluviation

Wide differences in the uptake of ions and, consequently, in chemical composition of plants have been well documented. These differences play a role in soil development. Species that normally absorb large amounts of the cations calcium, magnesium, potassium, and sodium will delay the development of soil acidity because they recycle more of these cations to the soil surface through the addition of organic matter. The effect will be to delay their removal from the cation exchange sites and to delay the development of acidity and the accumulation of exchangeable aluminum and hydrogen. The data in Table 16.3 support the fact that hardwoods maintain a higher pH than do spruce trees when grown on parent material with the same mineralogical composition.

Under the same climatic conditions, where for-

TABLE 16.3 Effect of Tree Species on Soil pH

Forest Type	Horizons	pH
Spruce	O	3.45
	E	4.60
	Bs1	4.75
	Bs2	4.95
	C	5.05
Hardwood	O	5.56
	A	5.05
	Bw1	5.14
	Bw2	5.24
	C	5.32

Adapted from R. Muckenhirn, 1949.

ests and grasslands soils exist side by side and have comparable parent material and slope, forest soils show a greater leaching of exchangeable calcium, magnesium, potassium, and sodium and greater eluviation of clay from the A to B horizon. Forest soils are more acid than grassland soils, due to greater leaching of basic cations; they also have less clay in the A horizon and more clay in the B horizon. A comparative study of soils on 2 to 8 percent slopes in Iowa found that the B/A clay ratio (percent clay in A horizon divided by percent clay in B horizon) of the forest soil was 2.0 compared with 1.04 for the grassland soil. Claypans develop in forest soils in less time than is required in grassland soils. The discussion of the effect of trees versus grass on soil genesis shows that the same fundamental processes occur in both kinds of soils but differ in degree of expression. Soils developed under trees and grass become strikingly similar when both have developed impermeable clay-pan subsoils.

Role of Animals in Soil Genesis

The role of animals in incorporating organic matter into the soil surface has been mentioned. Gophers and prairie dogs are well known for their burrowing and nest-building activities. These animals transport and deposit material on the sur-

face of the soil. In some tropical soils that are more than a million years old, and that have had significant termite activity, the soil, in a sense, has been *turned over* many times. This is one probable cause for the uniformity in properties in many intensively weathered soils in the tropics. For additional consideration of soil animals, see Chapter 8.

ROLE OF TOPOGRAPHY IN SOIL GENESIS

Topography is used here to refer to the configuration of the surface of a local or relatively small area of land. Differences in topography can cause wide variations in soils within the confines of a single field. Topography determines the local distribution or disposal of the precipitation and determines the extent to which water tables influence soil genesis. Water-permeable soils on broad, level areas receive and infiltrate almost all of the precipitation. Soils on sloping areas infiltrate less than the normal precipitation; hence, there is runoff. Depressions and low areas receive additional water, making more water available for soil genesis than the normal precipitation. The effects of water relations on soil genesis are more pronounced in humid than in arid regions.

Effect of Slope

Both the length and steepness of slope affect soil genesis. As the steepness of slope increases, there is greater water runoff and soil erosion. The net effect is a retardation of soil genesis. Generally, an increase in slope gradient is associated with less plant growth and organic matter content, less weathering and clay formation, and less leaching and eluviation. Consequently, soils have thinner sola and are less well developed on steeper slopes. Even on the native prairies of the central United States, some grassland soils on steep slopes have thin A horizons. The major cause of different soil colors in many fields is soil erosion. On eroding slopes the removal of soil produces light-colored soils and the sedimentation and deposition of the eroded soil; where changes in slope occur, dark-colored soils are produced (see Figure 16.14).

Many soils on sloping land are in equilibrium in terms of erosion rate and rate of horizon formation. As erosion reduces the thickness of the A horizon, the upper part of the B horizon is incorporated into the lower part of the A horizon. The upper part of the C horizon is slowly incorporated into the lower part of the B horizon. Under these conditions, the soil essentially maintains its character over a long period of time as the landscape evolves.

Effects of Water Tables and Drainage

For this discussion, it is assumed that soils are permeable to water, so that landscape position plays an important role in affecting the depth to the water table and soil drainage. Drainage is a measure of the tendency of water to leave the soil. Well-drained soils occur on slopes where the water table is far below the soil surface. The well-drained soils shown in Figure 16.15 are represented by the Gritney and Norfolk soils. Poorly drained soils tend to occur in the low parts of the landscape where water tables exist close enough to the soil surface to cause various degrees of anoxia and reduction. This landscape position is represented by the Bibb soils in Figure 16.15. The Rains soils are poorly drained because of a nearly level or slightly depressional position on a broad upland area. The Goldsboro soils (Figure 16.15) are intermediate, being moderately-well drained or somewhat poorly drained.

Poorly drained soils have an A horizon that is usually dark-colored because of the high organic matter content; the subsoil tends to be gray-colored. Well-drained soils have lighter colored A horizons and have uniformly bright-colored subsoils. Intermediate drainage conditions produce soils with intermediate properties. Whenever the soil is water saturated, downward translocation of clay and removal of soluble materials are inhibited.

As a result of differences in topography, soils developed from similar parent materials may vary greatly within a small area. A sequence of soils along a transect that differs because of topogra-

FIGURE 16.14 Cultivated land showing light-colored soil on actively eroding slopes and dark-colored soils where eroded material has been deposited and is accumulating.

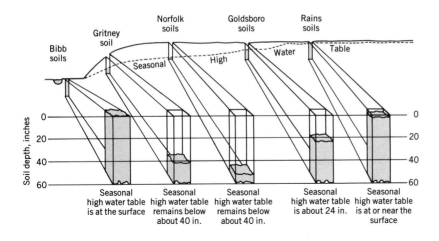

FIGURE 16.15 Representative landscape showing the relative location of some important soils and depth to the seasonal high water table in Wilson County, North Carolina. (From Wilson County, North Carolina, soil survey report, USDA.)

phy is a soil catena, or *toposequence.* Locally, variations in soils are due primarily to variations of topography and parent material.

Topography, Parent Material, and Time Interactions

On slopes, water runoff and soil erosion occur while simultaneously run-on water and soil deposition occur on areas at a lower elevation. Thus, in a landscape, young soils tend to occur where both sedimentation and erosion are active. The oldest soils tend to occur on broad upland areas where neither erosion nor sedimentation occurs.

About 500,000 or more years ago in the north-central United States, the Kansan glacier melted, leaving a relatively flat land surface composed of till and other glacial materials. During a long interglacial period, an old, well-developed soil with high clay content formed in the surface of the till, as shown in Figure 16.16a. Loess was later deposited on the soil developed from till, which was buried and became a paleosol (see Figure 16.16b). Erosion then dissected the till plain and removed the loess from the sloping areas, which reexposed the paleosol on the upper slopes. Erosion and dissection also exposed unweathered till on the lower slopes in the valleys (see Figure 16.16c). Today, the result is: (1) soils of moderate age formed in loess on broad, nearly level up-

lands; (2) young soils on upper slopes that developed in the paleosol; and (3) young (less developed) soils that developed from unweathered till on the lower valley slopes. In stream valleys, soils are very young and have developed from recent alluvium and colluvium. Within the landscape there are four major kinds of parent material (each of different age) and about six different topographic positions, as well as some microclimatic differences. In such a landscape in southern Iowa,these soil-forming factors result in formation of 11 major kinds of soils, as shown in Fig. 16.17.

Uniqueness of Soils Developed in Alluvial Parent Material

Weathering typically is more intense near the soil surface than in underlying layers, causing the mineral material to be less weathered with increasing soil depth. For alluvial soils, the surface horizon is the least weathered because it forms from the most recently deposited alluvium.

Alluvium accumulates along the Nile River at the rate of about 1 millimeter annually, resulting in a meter of alluvium every 1,000 years. It is common for alluvial sediments to be more than several hundred meters thick. After subsoil material is buried by more recent alluvium, the subsoil receives very little additional organic matter from root growth and the organic matter present con-

(a) Soil Developed from Kansan till

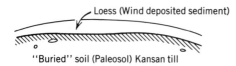

(b) Loess Deposited on Landscape of the Buried
Soil Developed from Kansan till

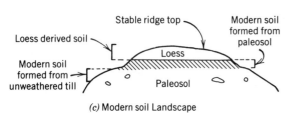

(c) Modern soil Landscape

FIGURE 16.16 Steps in the formation of a soil landscape in southern Iowa. (Adapted from Oschwald et al., 1965.)

FIGURE 16.18 Earth-moving activities of humans to construct terraces on slopes for paddy rice production have resulted in great modification of the natural soils.

tinues to mineralize. This results in a gradual decrease in soil organic matter with increasing depth. Alluvial soils tend to be young soils, and the horizons are caused mainly by differences in the properties of the layers of alluvium.

FIGURE 16.17 Relationship of slope, vegetation, and parent material to soils of the Adair-Grundy-Haig soil association in south-central Iowa. (From Oschwald et al., 1965.)

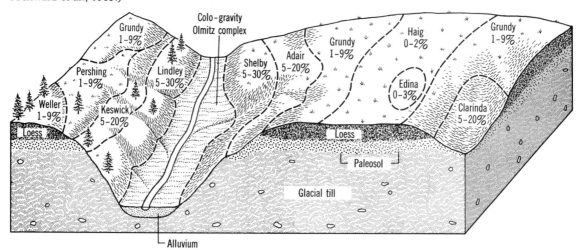

On floodplains, sand and the coarser particles are deposited near the river and may form a levee. Further from the river, the silt and clay tend to be deposited from slower-moving and, in some cases, very stagnant water. The result is a general increase in the clay content of soils with increasing distance from the river.

When rivers cause rapid down cutting, alluvium at the outer edges of the floodplain may become elevated relative to the river, so that they no longer flood. This results in the formation of terraces, and in the absence of flooding and deposition, the soils develop normally with time.

HUMAN BEINGS AS A SOIL-FORMING FACTOR

One of the most obvious and early effects of human activity on soils has occurred at campsites and villages. Organic residues have been added to the soils from food preparation and preservation, together with other materials such as bones and shells. Human excrement and other materials have similarly contributed to the formation of thick, dark-colored layers greatly enriched with phosphorus.

The use of land for agriculture, forestry, grazing, and urbanization has produced extensive changes in soils. Many of these changes have been discussed, including soil erosion, drainage, salinity development, depletion and addition of organic matter and nutrients, compaction, and flooding. Sometimes, solid waste disposal produces new soils that are little more than an accumulation of trash. Millions of acres of land have soils with properties that are due more to human activities than to soil-forming factors. In China, land has essentially been created to grow food crops. The enormous movement of soil and the shaping of the landscape for production of paddy rice is illustrated in Figure 16.18.

SUMMARY

Soils are the products of evolution. Over time, the soil properties are less affected by inheritance from the parent material, and soil properties become more closely related to the changes that occurred during soil genesis. The age of a soil is expressed by the degree of development and not the number of years. In one sense, soils have a life cycle that is represented by soils expressing different degrees of development.

Some soils form in parent material derived from the direct weathering of rocks. Most soils have formed in sediments.

Soil genesis, in general, is accelerated by increases in precipitation and temperature.

Soil genesis is accelerated by tree rather than grassland vegetation. Similar processes occur under both types of vegetation, and old soils are similar when developed under trees or grass. Grassland soils have much greater content of organic matter within the soil than forest soils, and the organic matter content decreases less rapidly with increasing soil depth.

Topography affects soil genesis mainly by dispersal of precipitation, erosion, and accompanying deposition of eroded material, and soil drainage. Youthful soils tend to occur where soil erosion and soil deposition are active.

Human beings have affected many soils at habitation sites and through agriculture, waste disposal, and earth moving.

REFERENCES

Aandahl, A. R. 1948. "The Characterization of Slope Positions and Their Influence on the Total Nitrogen Content of a Few Virgin Soils of Western Iowa." *Soil Sci. Soc. Am. Proc.* **13**:449–454.

Bidwell, O. W. and F. D. Hole. 1965. "Man as a Factor in Soil Formation." *Soil Sci.* **99**:65–72.

Blizi, A. F. and E. J. Ciolkosz. 1977. "Time as a Factor in the Genesis of Four Soils Developed in Recent Alluvium in Pennsylvania." *Soil Sci. Soc. Am. J.* **41**:122–127.

Boul, S. W., F. D. Hole, and R. J. McCracken. 1980. *Soil Genesis and Classification,* 2nd ed., Iowa State Press, Ames.

Ericson, D. B., M. Ewing, and G. Wollin. 1964. "The

Pleistocence Epoch in Deep-Sea Sediments." *Science.* **146:**723–732.

Fanning, D. S. and M.C.B. Fanning. 1989. *Soil Morphology, Genesis, and Classification.* Wiley, New York.

Franzmeier, D. P. and E. P. Whiteside. 1963. A Chronosequence of Podzols in Northern Michigan: I, II, and III. *Mich. Agr. Exp. Sta. Quart. Bull.* **46:**2–57. East Lansing.

Jenny, Hans. 1980. *The Soil Resource.* Springer-Verlag, New York.

Muckenhirn, R. J., E. P. Whiteside, E. H. Templin, R. F. Chandler, and L. T. Alexander. 1949. "Soil Classification and the Genetic Factors of Soil Formation." *Soil Sci.* **67:**93–105.

Nielsen, G. A. and F. D. Hole. 1963. "A Study of the Natural Processes of Incorporation of Organic Matter into Soil in the University of Wisconsin Arboretum." *Wis. Acad. Sci.* **52:**213–227.

Oschwald, W. R. et al. 1965. "Principal Soils of Iowa." *Agronomy Special Report* 42. Iowa State University, Ames.

Ruhe, R. V., R. B. Daniels, and J. G. Cady. 1967. "Landscape Evolution and Soil Formation in Southwestern Iowa." *USDA Tech. Bull.* 1349, Washington, D.C.

Sanchez, P. A., M. P. Gichuru, and L. B. Katz. 1982. "Organic Matter in Major Soils of the Tropical and Temperate Regions." *12th Int. Congress Soil Sci. Symposia Papers Vol 1.* pp. 101–104, New Delhi.

Schreiner, O. and B. E. Brown. 1938. "Soil Nitrogen." *Soils and Men,* USDA Yearbook of Agriculture, Washington, D.C.

Twenhofel, W. H. 1939. "The Cost of Soil in Rock and Time." *Am. J. Sci.* **237:**771–780.

Vance, G. F., D. L. Mokma, and S. A. Boyd. 1986. "Phenolic Compounds in Soils of Hydrosequences and Developmental Sequences of Spodosols." *Soil Sci. Soc. Am. J.* **50:**992–996.

CHAPTER 17

SOIL TAXONOMY

Classification schemes of natural objects seek to organize knowledge, so that the properties and relationships of these objects may be remembered and understood for some specific purpose. Many soil classification schemes have been developed, and most have been on a national basis. In 1975, the United States Department of Agriculture, with the help of soil scientists from many nations, published *Soil Taxonomy*. Many of its terms and ideas also appear in the legend of the world soil map published by the Food and Agriculture Organization (FAO) of the United Nations. Knowledge of either system makes use of the other system possible with minimal adjustments.

DIAGNOSTIC SURFACE HORIZONS

Diagnostic horizons have unique properties and are used with other properties to classify soils. Diagnostic surface horizons (epipedons) form at the soil surface and diagnostic subsurface horizons form below the soil surface. There are seven

diagnostic epipedons that differ in thickness, color, structure, content of organic matter and phosphorus, mineralogy, and the degree of leaching.

Mollic Horizon

The word *mollic* is derived from *mollify,* which means to soften. The *mollic* horizon is a surface horizon that is soft rather than hard and massive when dry. Mollic epipedon formation is favored where numerous grass roots permeate the soil and a moderate to strong grade of structure is created. Except for special cases, mollic horizons are dark-colored, contain at least 1 percent organic matter, and are at least 18 centimeters thick. They are only minimally or moderately weathered and leached. The cation exchange capacity is 50 percent or more saturated with calcium, magnesium, potassium, and sodium when the cation exchange capacity is determined at pH 7. The major grassland soils of the world have mollic epipedons (see Figure 17.1).

FIGURE 17.1   Soil with a mollic epipedon.

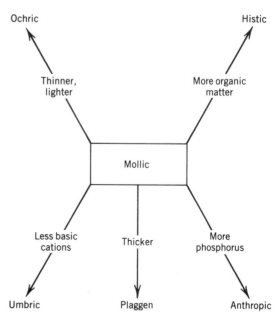

FIGURE 17.2   Relationships between the mollic epipedon and the umbric, ochric, histic, anthropic, and plaggen epipedons.

Umbric and Ochric Horizons

Other diagnostic epipedons can be related to the mollic horizon, as shown in Figure 17.2. The *umbric* epipedon is similar to the mollic in overall appearance of thickness and color; however, it is more leached and has a saturation with basic cations calcium, magnesium, potassium, and sodium that is less than 50 percent when the cation exchange capacity is determined at pH 7. Umbric horizons form in some very humid forests, as in the Coastal Range of the northwestern United States. In most forests, the epipedon is thinner than is that of the umbric, is lower in organic matter, and is lighter in color. This epipedon is called *ochric*.

Histic Horizon

Where soil development occurs under conditions of extreme wetness, as in swamps or lakes, the epipedon is organic in nature and is, typically, a *histic* epipedon. Histic epipedons are O horizons. A histic epipedon must be water saturated for at least 30 consecutive days during the year, unless the soil has been drained. The degree of decomposition of the organic matter in histic horizons is indicated with the following symbols: i for fibric, e for hemic, and a for sapric. Fibric (i) is the least decomposed and contains a large amount of recognizable plant fibers; sapric (a) is the most decomposed; and hemic (e) is of intermediate decomposition. The relationship of the histic, horizon to the mollic horizon is shown in Figure 17.2.

Melanic Horizon

Melanic epipedons are thick, black colored, and contain high concentrations of organic matter. The moist Munsell value and chroma are 2 or less, and the organic carbon content is 6 percent or more but less than 25 percent. The organic matter is thought to result from the supply of large amount of root residues from a graminaceous vegetation in contrast to organic matter that results from a forest vegetation.

Melanic horizons also have *andic soil properties*. Andic soil properties result mainly from the weathering of volcanic materials, which contain a significant amount of volcanic glass (noncrystalline ash, pumice, lava, etc). The dominant clay minerals include allophane and some oxides of iron and aluminum. Recent work with electron microscopy, and other techniques, has shown that these minerals are characterized by an orderly arrangement of atoms. But the orderly arrangement is short range and does not in a regular manner form a 3-dimensional pattern like that found in crystalline minerals such as vermiculite or kaolinite. The particles are very small with an enormous surface area, which results in high anion exchange capacity and a large capacity to complex organic matter. These minerals are considered to be amorphous or short-range minerals.

Melanic horizons are black and have high organic matter content similar to some histic horizons. Even though the organic matter content is high, the organic matter in melanic horizons is mostly associated or complexed with minerals, which results in properties dominated by the mineral fraction rather than the organic fraction. The horizon is also characterized by low bulk density and high anion exchange capacity.

Anthropic and Plaggen Horizons

The *anthropic* and *plaggen* horizons are formed by human activity. The anthropic epipedon resembles the mollic horizon in color, structure, and organic matter content; however, the phosphorus content is 250 or more ppm of P_2O_5 equivalent soluble in 1 percent citric acid solution. The anthropic horizon occurs where human activity resulted in the disposal of bones and other refuse near places of residence, or in agriculture at sites of long-continued use of soil for irrigated cropping. Measurement of the phosphorus en-

richment of soil is a standard technique used by anthropologists to determine where ancient campsites, villages, and roads were located. Many anthropic horizons have been found in Europe, and some have been found at old presettlement campsites in the United States. Some anthropic horizons in the Tombigee River valley in northern Mississippi are more than a meter thick, contain 0.6 to 3 percent organic carbon, and 1 to 5 percent charcoal fragments. In the deeper parts of the anthropic horizons the carbon had been dated as being 7,000 years old. The citrate soluble P_2O_5 content of the anthropic horizons exceeded 250 ppm compared with 80 and 20 ppm in the surface and subsoils, respectively, of adjacent soils.

The plaggen epipedon is found in Europe where long-continued manuring has produced a surface layer 50 centimeters or more thick. During the Middle Ages, farmers in northwestern Europe collected sods from forests and heath lands where soils were very sandy. These clumps of sod contained sand and were used for bedding in barns. The manure from the barn was applied to the land, which resulted in the slow accumulation of a thick, sandy epipedon enriched with organic matter. In some places, a 1-meter thick plaggen epipedon was created in 1,000 years. Plaggen horizons typically contain artifacts, such as bits of brick or pottery, and today these soils are valued for agricultural use.

The derivation of the the epipedon names and their major features are given in Table 17.1.

DIAGNOSTIC SUBSURFACE HORIZONS

Generally, diagnostic subsurface horizons are B horizons. They may form immediately below a layer of leaf litter, as in the case of the E horizon. More than 15 subsurface diagnostic horizons have been recognized.

Cambic Horizon

The *cambic* horizon can form quickly, relatively speaking, because changes in the color and struc-

TABLE 17.1   Derivations and Major Features of Diagnostic Surface Horizons

Horizon	Derivation	Major Features
Mollic	L. *mollis,* soft	Thick, dark-colored, high basic cation saturation ($> 50\%$), and strong structure so that the soil is not massive or hard when dry.
Umbric	L. *umbra,* shade	Same as mollic but low basic cation saturation $< 50\%$ and may be hard or massive when dry.
Ochric	Gr. *ochros,* pale	Thin, light-colored, and low in organic matter.
Histic	Gr. *histos,* tissue	Very high organic matter content and saturated with water at some time during the year unless artificially drained.
Melanic	Gr. *melas-anos,* black	Thick, black-colored with andic soil properties.
Anthropic	Gr. *anthropos,* man	Molliclike horizon that has a very high phosphate content, 250 ppm or more P_2O_5, resulting from longtime cultivation and fertilization.
Plaggen	Ger. *plaggen,* sod	Very thick, produced by long-continued manuring.

ture, and some leaching will convert the subsoil parent material into a cambic horizon. Cambic horizons are not illuvial horizons and, generally, they are not extremely weathered. A cambic horizon is a Bw horizon. A parent material with a texture of fine sand (or even finer) is required for the development of cambic horizons.

Agrillic and Natric Horizons

The gradual illuvial accumulation of clay in a cambic horizon converts a cambic horizon (Bw horizon) into a Bt horizon. Typically, clay skins occur on the ped surfaces of Bt horizons. When the ratio of clay in the Bt horizon, compared with that of the overlying eluvial horizon, is 1.2 or more, the Bt horizon becomes an *argillic* horizon, in which the overlying eluvial horizon contains 15 to 40 percent clay. If the eluvial horizon has less than 15 percent clay, there must be at least 3 percent more clay in the illuvial horizon to qualify as an argillic horizon. If the eluvial horizon has

more than 40 percent clay, the illuvial horizon must have at least 8 percent more clay to qualify as an argillic horizon. Argillic horizons usually require several thousands of years to form, as shown in Figure 2.7.

A *natric* horizon is a special kind of argillic horizon. These horizons have typically been affected by soluble salts. In addition to the properties of the argillic horizon, natric horizons commonly have a columnar structure and a sodium adsorption ratio (SAR) of 13 or more (or 15 percent or more exchangeable sodium saturation). The natric horizon is a Btn horizon; the n indicates the accumulation of exchangeable sodium. An example of a natric horizon is shown in Figure 17.3.

Kandic Horizon

In intensively weathered soils, the genesis of subsoil layers that have more clay than the overlying A horizon is obscure because of the very long time

FIGURE 17.3 Soil with a natric horizon. The upper part of the natric horizon is at 20 centimeters and is characterized by whitish-topped columnar peds.

lation of sesquioxides, humus, or humus plus sesquioxides, respectively. Evolution of the spodic horizon is summarized in Table 16.1.

Albic Horizon

The loss by eluviation of sesquioxides and clay, during the formation of spodic and argillic horizons, tends to leave behind a light-colored overlying eluvial horizon called the *albic* horizon. Albic is derived from the word *white*. The horizon is eluviated and is labeled an E horizon. The color of the albic horizon is due to the color of the primary sand and silt particles rather than to the particle coatings. Albic horizons are commonly underlain by spodic, argillic, or kandic horizons.

Oxic Horizon

The *oxic* horizon (Bo horizon) is a subsurface horizon at least 30 centimeters thick that is in an advanced stage of weathering. It is not dependent on a difference in the clay content of subsoil versus the topsoil horizons. Oxic is derived from the word *oxide*. Oxic horizons consist of a mixture of iron and/or aluminum oxides with variable amounts of kaolinite and highly insoluble accessory minerals such as quartz sand. There are few if any weatherable minerals to weather and supply plant nutrients, and the *clay* fraction has a small cation exchange capacity—16 cmol/kg (meq/100 g) or less as determined at pH 7 or 12 or less cmol/kg at the soil's natural pH. Soils with oxic horizons have essentially reached the end point of weathering.

of soil formation. Subsoils with much higher clay content than overlying horizons may show no evidence of clay skins. Therefore, many tropical soils have a subsoil horizon similar to the argillic horizon, except that the clay minerals have very low cation exchange capacity, which is indicative of *low activity clay*. The clay minerals are predominately kaolinite and oxidic clays. The cation exchange capacity of the *clay*, determined at pH 7, is 16 cmol/kg (16 meq/100 g) or less or the ECEC is 12 cmol/kg or less. These clay-rich horizons are called *kandic*.

Spodic Horizon

In sand-textured parent material there is little clay for illuviation and rapid water permeability. Under conditions of high precipitation, iron and aluminum oxides (sesquioxides) are released in weathering. These oxides complex or chelate with organic matter and are transported to the subsoil, resulting in the formation of an illuvial horizon. An illuvial subsoil horizon that is greatly enriched with these amorphous compounds is a *spodic* horizon. Spodic horizons can be Bs, Bh, or Bhs horizons, depending on the degree of accumu-

Calcic, Gypsic, and Salic Horizons

The *calcic* horizon is a horizon of calcium carbonate or calcium and magnesium carbonate accumulation. Calcic horizons develop in soils in which there is limited leaching and the carbonates are translocated downward. However, the carbonates are deposited within the soil profile because there is insufficient water to leach the carbonates to the water table. The symbol k indicates an accumulation of carbonates, as in Bwk or Ck. Gypsic horizons have an accumulation of secondary sulfates, and that is indicated with the symbol y. Cemented calcic and gypsic horizons are petrocalcic and petrogypsic horizons, respectively. Salic horizons contain a secondary enrichment of salts more soluble than gypsic and are indicated with the symbol z.

Subordinate Distinctions Within Horizons

Other symbols and meanings used for subordinate distinctions within soil horizons or layers include the following:

b   Buried genetic horizon
c   Concretions or hard nonconcretionary
       nodules
f   Permafrost
g   Strong gleying (indicating iron reduction)
m   Cementation or induration
q   Accumulation of silica
r   Weathered or soft rock
v   Plinthite (iron-rich reddish material that
       dries irreversibly)
x   Fragipan character (high density and
       brittleness)

SOIL MOISTURE REGIMES

The *soil moisture regime* is an expression of the changes in soil water over time. It refers to the presence or absence of either groundwater or of water held at a potential of less or more than -1,500 kPa (-15 bars) by periods of the year. Three soil water conditions are recognized: soil that is water saturated (wet soil), soil between saturation and the permanent wilting point (moist soil), and soil drier than the wilting point (dry soil).

The water content of the upper soil horizon changes frequently throughout the year because of repeated rainfall and subsequent drying. As a result, the soil moisture regime is based on a *moisture control section* that is represented by the difference between the depth of soil at the permanent wilting point that is wetted by 2.5 centimeters of water as compared with the depth wetted by 7.5 centimeters of water. The moisture control section is generally within the 10-centimeters and 90-centimeters depths, being thicker in sandy soils and thinner in clay soils. The amount of water in the moisture control section provides an indication of the sufficiency or deficiency of water for plant growth and the water status of the soil for other uses.

Aquic Moisture Regime

Soils with *aquic* moisture regime are wet. At sometime during the year, the lower soil horizons are saturated. In many instances, the entire soil is saturated part of the year. Water saturation occurs because of the location of the soil in a landscape position that is naturally poorly drained or where an impermeable layer causes saturation. Typically, the top of the saturated soil fluctuates during the year, being deeper during the driest part of the year. Artificial drainage is needed to grow plants that require an aerated root zone. Soils with an aquic moisture regime are unsuitable for building sites unless these soils are dewatered with a drainage system.

Subsoil colors in mineral soils with aquic regime are frequently gray, which indicates reducing conditions and low oxygen supply. Alternating reducing and oxidizing conditions are associated with a fluctuating water table that produces mottled colors. The high water content tends to be associated with a high content of organic matter and, when properly drained, many aquic soils become very productive agricultural soils.

Udic and Perudic Moisture Regimes

Udic means humid. Soils with a udic moisture regime are not dry in any part of the moisture control section as long as 90 *cumulative* days in most years. Udic moisture regimes are common in soils of humid climates that have well distributed precipitation. The amount of summer rainfall, plus stored water, is approximately equal to or exceeds the potential evapotranspiration (PE). In other words, if droughts occur, they are short and infrequent. Forests and tall grass prairies are typical native vegetation.

In most years there is some surplus water that percolates downward and produces leaching. Nutrient ions in the water are leached, removing a significant amount of plant nutrients from the soil. Leaching losses of calcium and magnesium may be greater than the amount required to produce an average crop. Udic soils are common from the East Coast of the United States to the western boundaries of Minnesota, Iowa, Missouri, Arkansas, and Louisiana. Old soils with udic moisture regime tend to be naturally acid and infertile as a result of leaching. An example of soil water changes over time in a soil with a udic soil moisture regime is shown in Figure 17.4 and the udic moisture regime diagram for Memphis, Tennessee is given in Figure 5.15.

If the regime is characterized by very wet conditions, the regime is *perudic*. In the perudic regime, precipitation exceeds evapotranspiration every month in most years, water moves through the soil every month unless the soil is frozen, and there are only occasional periods when some stored soil water is used.

Ustic Moisture Regime

The concept of the *ustic* moisture regime relates to soils with limited available water for plants, but water is present at a time when conditions are suitable for plant growth. Soils with ustic moisture regime are dry in some part of the moisture control section for as long as 90 *cumulative* days in most years. There is moisture available, stored soil water plus precipitation, for significant plant growth at some time during the year and a deficit part of the year, as shown in Figure 17.4. Most soils on the central and northern Great Plains

have an ustic moisture regime. Native vegetation was mainly mid, short, and bunch grasses. Surplus water is rare, leaching is limited, and soils tend to be fertile.

In areas where soils have an ustic moisture regime, there is enough rainfall in winter, early spring, and summer to produce a significant amount of soil water recharge or storage. This water, plus the precipitation during the spring and early summer, is sufficient for the production of a wheat crop that ripens before the soil becomes dry in mid- and late summer. Corn does not ripen until fall and generally cannot be produced unless irrigated. Droughts are common. Sorghum is a popular crop because it can interrupt growth when water is lacking and grow again if more rainfall occurs. Cattle grazing is an important land use.

Soils in monsoon climates have ustic moisture regimes because there is enough water during the rainy season for significant plant growth; however, there is a long dry season with a deficit.

Aridic Moisture Regime

The driest soils have an *aridic* moisture regime. The subsoil, or moisture control section, is dry (permanent wilting point or drier) in all parts for more than half the growing season, and is not moist (wetter than the permanent wilting point) in some parts for as long as 90 *consecutive* days during the growing season in most years. Most soils with an aridic moisture regime are found in arid or desert regions with widely spaced shrubs and cacti as the native vegetation. A crop cannot be matured without irrigation. Some climatic data and the soilwater balance typical of an aridic moisture regime are given in Figure 17.4. The precipitation is lower than the potential evapotranspiration (PE) during most months of the year. There is very little soil water storage or recharge, which means there is little water retained for plant growth. The result is a frequent lack of water for plants and a large water deficit. Many of the native plants have an unusual capacity to endure very

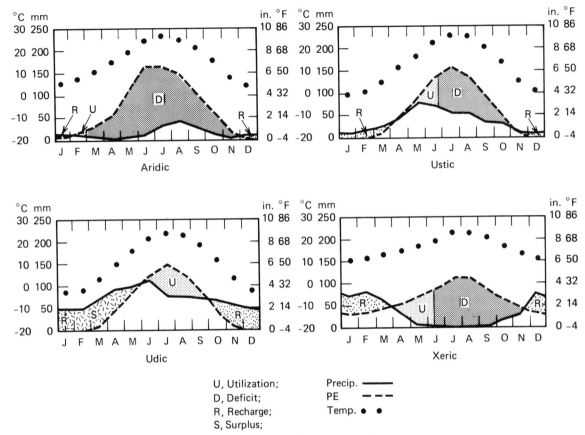

U, Utilization;
D, Deficit;
R, Recharge;
S, Surplus;

Precip. ——————
PE   – – – –
Temp. ● ●

FIGURE 17.4 Climatic data and water balance representing aridic, udic, ustic, and xeric soil moisture regimes. (From *Soil Taxonomy*, 1975.)

low water potentials in the soil and within their tissues. Grazing is the dominant land use except where irrigation permits crop production.

Xeric Moisture Regime

Soils in areas where winters are cool and moist and summers are hot and dry have a *xeric* soil moisture regime. This kind of climate is known as a Mediterranean climate. The precipitation occurs in the cool months when the evapotranspiration is low and surplus water may readily accumulate. This frequently results in considerable soil wet-ness during the winter. Leaching and weathering may occur during the winter even though summers are long, very dry, and hot. With limited water storage, and little if any summer rainfall, a large water deficit occurs in summer. Xeric soils are typical of the valleys of California where a large variety of crops, including vines, fruits, nuts, vegetables, and seeds are produced with irrigation. Xeric soils of the Palouse in Washington and Oregon are used for winter wheat. Here, there is a good match between the time when the wheat needs water and when soil water is available.

TABLE 17.2 Definitions and Features of Soil Temperature Regimes

Temperature Regime	Mean Annual Temperature in Root Zone, 5 to 100 centimeters		Characteristics and Some Locations
	C	F	
Pergelic	<0	<32	Permafrost and ice wedges common. Tundra of northern Alaska (and Canada) and high elevations of middle and northern Rocky Mountains.
Cryic and frigid[a]	0–8	32–47	Cool to cold soils of northern Great Plains of United States (and southern Canada) where spring wheat is the dominant crop. Forested regions of New England
Mesic	8–15	47–59	Midwestern and Great Plains regions where corn and winter wheat are common crops.
Thermic	15–22	59–72	Coastal plain of southeastern United States where temperatures are warm enough for cotton and Central valley of California.
Hyperthermic	Over 22	Over 72	Citrus areas of Florida peninsula, Rio Grande Valley of Texas, southern California, and low elevations in Puerto Rico and Hawaii. Tropical climates and crops.

[a] Frigid soils have warmer summers than cryic soils; both have same MAST. Frigid soils have more than 5° C temperature difference between mean winter and mean summer temperatures at a depth of 50 centimeters (or lithic or paralithic contact, if shallower).

SOIL TEMPERATURE REGIMES

Soil temperature regimes are defined according to the mean annual soil temperature (MAST) in the root zone (arbitrarily set at 5 to 100 cm). Regime definitions and some characteristics are given in Table 17.2. Both the soil moisture regimes and the temperature regimes are used to classify soils at various categories in Soil Taxonomy.

CATEGORIES OF SOIL TAXONOMY

The categories of Soil Taxonomy are order, suborder, great group, subgroup, family, and series. The order is the highest category and the series is the lowest category.

Soil Order

There are 11 orders, each ending in -sol; L. *solum*, meaning soil. The orders, together with their deri-

TABLE 17.3 Formative Syllables, Derivations, and Meanings of Soil Orders

Order	Formative Syllable	Derivation	Meaning
1. Alfisols	alf	Coined syllable	Pedalfer (Al-Fe)
2. Andisols	and	Ando	Volcanic ash soil
3. Aridisols	id	L. *aridus,* dry	Arid soil
4. Entisols	ent	Coined syllable	Recent soil
5. Histosols	ist	Gr. *histos,* tissue	Organic soil
6. Inceptisols	ept	L. *inceptum,* beginning	Inception, or young soil
7. Mollisols	oll	L. *mollis,* soft	Soft soil
8. Oxisols	ox	F. *oxide,* oxide	Oxide soil
9. Spodosols	od	Gr. *spodos,* wood ash	Ashy soil
10. Ultisols	ult	L. *ultimus,* last	Ultimate (of leaching)
11. Vertisols	ert	L. *verto,* turn	Inverted soil

vation and meaning are given in Table 17.3. For example, Histosols are organic soils composed of plant tissue, and Entisols are recent soils (see Table 17.3). Soils are keyed or placed into orders according to the following, very simplified, key:

1.	Soils that have organic materials that extend from the surface downward a significant distance:	Histosols

Other soils that

2.	Have andic soil properties:	Andisols
3.	Have a spodic horizon:	Spodosols
4.	Have an oxic horizon:	Oxisols
5.	Are clayey and develop deep cracks:	Vertisols
6.	Have aridic soil moisture regime:	Aridisols
7.	Have kandic or argillic horizon and <35% saturation of CEC with calcium, magnesium, potassium, and sodium:	Ultisols
8.	Have a mollic epipedon:	Mollisols
9.	Have argillic or kandic horizon and 35 percent or more saturation of CEC with calcium, magnesium, potassium, and sodium:	Alfisols
10.	Have a cambic (Bw) horizon:	Inceptisols
11.	Do not fit into one of the preceding orders:	Entisols

The Entisol order includes all soils that do not fit into one of the other 10 orders and is composed of soils with great heterogeneity. Extensive treatment of the soil orders can be found in the next chapter.

TABLE 17.4 Suborder Names, Formative Elements, Meaning, and Connotation

Suborder	Formative Element	Meaning and Connotation
		Alfisols
Aqualfs	aqu	Water, aquic moisture regime
Boralfs	bor	Northern, cool
Udalfs	ud	Humid, udic moisture regime
Ustalfs	ust	Hot summer, ustic moisture regime
Xeralfs	xer	Dry season, xeric moisture regime
		Andisols
Aquands	aqu	Water, aquic moisture regime
Cryands	cry	Cryic, cold
Torrands	torr	Hot and dry, aridic moisture regime
Udands	ud	Humid, udic moisture regime
Ustands	ust	Hot summer, ustic moisture regime
Vitrands	vitr	Vitreous, presence of glass
Xerands	xer	Dry season, xeric moisture regime
		Aridisols
Argids	arg	Clay, argillic horizon present
Orthids	orth	True, no argillic horizon, common ones
		Entisols
Aquents	aqu	Water, aquic moisture regime
Arents	ar	To plow, mixed horizons by deep plowing
Fluvents	fluv	River, floodplains parent material
Orthents	orth	True, common ones
Psamments	psamm	Sand, sand textured
		Histosols
Fibrists	fibr	Fiber, slightly decomposed
Folists	fol	Foliage, mass of leaves
Hemists	hem	Half, intermediate stage of decomposition
Saparists	sapr	Rotten, most decomposed stage
		Inceptisols
Aquepts	aqua	Water, aquic moisture regime
Ochrepts	ochr	Pale, presence of ochric epipedon
Plaggepts	plagg	Plaggen, presence of plaggen epipedon
Tropepts	trop	Tropical, continually warm
Umbrepts	umbr	Shade, presence of umbric epipedon

(*Continued*)

(Table 17.4, continued)

Suborder	Formative Element	Meaning and Connotation
		Mollisols
Albolls	alb	White, presence of albic horizon
Aquolls	aqu	Water, aquic moisture regime
Borolls	bor	Northern, cool
Rendolls	rend	Rendzina, high carbonate content
Udolls	ud	Humid, udic moisture regime
Ustolls	ust	Hot summer, ustic moisture regime
Xerolls	xer	Dry season, xeric moisture regime
		Oxisols
Aquox	aqu	Water, aquic moisture regime
Perox	per	Perudic moisture regime
Torrox	torr	Hot and dry, aridic moisture regime
Udox	ud	Humid, udic moisture regime
Ustox	ust	Hot summer, ustic moisture regime
		Spodosols
Aquods	aqu	Water, aquic moisture regime
Ferrods	ferr	Iron, presence of iron
Humods	hum	Humus, presence of organic matter
Orthods	orth	True, common ones
		Ultisols
Aquults	aqu	Water, aquic moisture regime
Humults	hum	Humus, presence of organic matter
Udults	ud	Humid, udic moisture regime
Ustults	ust	Hot summer, ustic moisture regime
Xerults	xer	Dry season, xeric moisture regime
		Vertisols
Torrerts	torr	Hot and dry, aridic moisture regime
Uderts	ud	Humid, udic moisture regime
Usterts	ust	Hot summer, ustic moisture regime
Xererts	xer	Dry season, xeric moisture regime

Suborder and Great Group

Orders are divided into suborders. The suborder name has two syllables. The first syllable connotes something of the diagnostic properties of the soil, and the second syllable is the formative element or syllable for the order. Suborder names and their formative elements, meaning, and connotation are given in Table 17.4.

For the suborder Aqualfs, aqu is derived from L. *aqua*, meaning water, and alf from Alfisol. Thus, Aqualfs are Alfisols that are wet; they have an aquic soil moisture regime. Boralfs are Alfisols of the cool regions. Udalfs have an udic moisture regime, Ustalfs have an ustic moisture regime, and Xeralfs have a xeric moisture regime. Note that many of the formative elements of the suborder names connote soil moisture or temperature regimes. The suborder is the most common category used in the discussion of soil geography and in the legends of the soil maps in the next chapter.

The legends for the soil maps also use some great group names. In the legend of Figure 18.1 (soil map of the United States), A1a refers to areas where soils are predominatly Aqualfs with significant amounts of other soils in decreasing order of importance. Udalfs are the second most extensive soils, followed by *Haplaquepts*, which is a great group name. Hapl means *minimal horizon* and refers to Aquepts that are weakly or minimally developed. Areas of A2a are dominated by Boralfs, with Udipsamments second in importance. Udipsamments are Psamments (sand-textured Entisols) with a udic moisture regime. Thus, some of the formative elements given in Table 17.4 for suborders are also used to form great group names. Prefixes and their connotations for great group names are given in Appendix 3.

Subgroup, Family, and Series

The taxonomic categories that follow great group are subgroup, family, and series. The subgroup denotes whether or not the soil is typical (typic) for the great group. Soils that are not typical of the great group are indicated as being intergrades. The family category provides groupings of soils with ranges in texture, mineralogy, temperature, and thickness. The primary use of the series name is to relate soils delineated on soil maps to the intrepretations for land use.

Series names are taken from the name of a town or landscape feature near the place where the soil series was first recognized. Many of the names are familar, such as Walla Walla in the state of Washington, Molokai in Hawaii, and Santa Fe in New Mexico. The Antigo soil is the *state* soil in Wisconsin. More than 16, 000 series have been recognized in the United States.

AN EXAMPLE OF CLASSIFICATION: The ABAC SOILS

Soils of the Abac series are Typic Ustorthents, meaning the soils are Entisols (ent), that occur on recent slopes, subject to erosion (orth), have an ustic moisture regime (ust), and are typical for the Ustorthent great group. The family is loamy, mixed, calcareous, frigid, and shallow. This means, for example, that the texture is finer than loamy sand (loamy), no one mineral dominates the mineralogy and the soil is calcareous (mixed, calcareous), the mean annual soil temperature is between 0 and 8° C (frigid), and there is rock within 50 centimeter of the soil surface (shallow). The Abac soil is a loamy, mixed, calcareous, frigid, shallow typic Ustorthent.

From this information it can be concluded that the soil is not likely to be cultivated and used for cropping. Such soils can produce a moderate amount of forage for grazing in the spring and early summer. The soils occur on steep slopes where it is shallow to rock in Montana. Such soils can be desirable locations for vacations and recreational activities, as shown in Figure 17.5.

FIGURE 17.5 Landscape in the mountains where many of the soils are—Ustorthents. The soils provide some pasture for grazing animals; however, the ideal summer climate makes such locations desirable for camping, horseback riding, and hiking.

THE PEDON

Soil bodies are large, and hence there is a need for a smaller soil unit that can be the object of scientific study. The *pedon* is this unit. A soil pedon is the smallest volume that can be called a soil, and it is roughly polygonal in shape. The lower boundary is the somewhat vague boundary between soil and nonsoil at the approximate depth of root penetration. Lateral dimensions are large enough to represent any horizon. The area of a pedon ranges from 1 to 10 square meters. The pedon is to the soil body what an oak tree is to an oak forest. A soil body is composed of many pedons, which is why a soil body is called a polypedon (see Figure 17.6).

FIGURE 17.6 Illustration of the relationship between the pedon and polypedon.

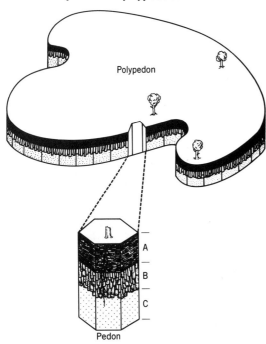

SUMMARY

Diagnostic surface and subsurface horizons, together with other diagnostic features, are used to classify soils in Soil Taxonomy.

Soil moisture regimes express the changes in soil water with time. Aquic soil moisture regimes are characterized by water-saturation. Aridic, ustic, and udic are three regimes with an increasing amount of water available for plant growth during the summer. Soils with perudic regime are continuously wet. Xeric soil moisture regime is characterized by cool, rainy winters and hot, dry summers.

Soil temperature regimes reflect the mean average soil temperature (MAST) in the root zone.

There are more than 16,000 soils (series) in the United States and they are classified into orders, suborders, great groups, subgroups, families, and series based on the presence of various diagnostic horizons and other criteria.

The pedon is the smallest volume that can be called a soil.

REFERENCES

Boul, S. W., F. D. Hole, and R. J. McCracken. 1980. *Soil Genesis and Classification*. 2nd ed. Iowa State University Press, Ames.

Fanning, D. S., and M. C. B. Fanning. 1989. *Soil Morphology, Genesis, and Classification*. Wiley, New York.

Foth, H. D. and J. W. Schafer. 1980. *Soil Geography and Land Use*. Wiley, New York.

Soil Survey Staff. 1975. *Soil Taxonomy*. USDA Agr. Handbook 436, Washington, D.C.

Soil Survey Staff. 1987. "Keys to Soil Taxonomy." *Soil Management Support Services Tech. Monograph* 6. USDA Washington, D.C.

Soil Survey Staff, 1989. *Amendments to Soil Taxonomy*. USDA, Washington, D.C.

Wilding, L. P., N. E. Smeck, and G. F. Hall. 1983. *Pedogenesis and Soil Taxonomy: I Concepts and Interactions*. Elsevier, New York

CHAPTER 18

SOIL GEOGRAPHY AND LAND USE

The extent and abundance of the orders for the United States and the world are shown in Table 18.1. Aridisols are the most abundant on a worldwide basis, occupying 19.2 percent of the land, whereas Mollisols are the most abundant soils in United States with 24.6 percent. Inceptisols and Alfisols are ranked 2 and 3, respectively, in both the United States and the world. Histosols and Oxisols are the least abundant soils in United States, and Histosols and Andisols are the least abundant soils worldwide.

The geographic distribution of soils in the United States is given in Figure 18.1 and for the world in Figure 18.2. The consideration of the soils is primarily at the suborder level.

ALFISOLS

The central concept of Alfisols is soils with an ochric epipedon and an argillic subsurface horizon, only moderately weathered and leached, and enough available water for the production of crops at least three months during the year. There is sufficient exchangeable calcium, magnesium, potassium, and sodium to saturate 35 percent or more of the CEC (at pH 8.3) at a depth of 1.25

meters below the upper boundary of the argillic horizon or 1.8 meters below the soil surface, whichever is greater.

In humid forested regions, many of the Alfisols have an albic (E) horizon between the ochric and argillic horizons. Alfisols that develop in regions or areas adjacent to grasslands lack E horizons and have properties very similar to Mollisols. The limited accumulation of organic matter and absence of a mollic horizon may result in a grassland soil being an Alfisol rather than a Mollisol. This is the situation in many areas in which the dominant soils have an ustic moisture regime. For these reasons, many Alfisols are quite similar to Mollisols in native fertility and ability to produce crops. Large areas of each of the suborders of Alfisols occur in United States.

Aqualfs

Aqualfs have an aquic soil moisture regime and are shown in Figure 18.1 as A1a. In the area west of Lake Erie in Ohio and Michigan, the Aqualfs have a naturally high water table. The Aqualfs formed mainly on glacial lake beds from fine-textured parent material. They occur on nearly 0 percent slopes and have high natural fertility.

TABLE 18.1 Extent and Rank of Orders in United States and the World

Order	Percent of Land		Ranking	
	United States	World	United States	World
Alfisols	13.4	14.7	3	3
Andisols	1.9	0.1	8	11
Aridisols	11.5	19.2	5	1
Entisols	7.9	12.5	6	4
Histosols	0.5	0.8	10	10
Inceptisols	16.3	15.8	2	2
Mollisols	24.6	9.0	1	6
Oxisols	<0.02	9.2	11	5
Spodosols	5.1	5.4	7	8
Ultisols	12.9	8.5	4	7
Vertisols	1.0	2.1	9	9

When artificially drained, the soils are very productive for corn, soybeans, wheat, and sugar beets. The soils are poorly suited to septic filter-field sewage disposal systems and other uses where natural soil wetness is undesirable.

Another area of Aqualfs (A1a in Figure 18.1) occurs in southern Illinois and west into Missouri. These soils are wet because of thick argillic horizons (claypans). The soils developed on nearly level upland areas on land surfaces that were formed by some of the earlier glaciations. Another area of claypan Aqualfs occurs in southwestern Louisiana, where wetland rice is a major crop.

Boralfs

Boralfs are the northern or cool Alfisols with cryic or frigid temperature regime. The mean annual soil temperature is warmer than 0° C (32° F) and colder than 8° C (47° F). Forestry rather than agriculture tends to be the major land use. The soils are too cold for winter wheat, corn, and soybeans.

Two kinds of Boralf areas are shown in Figure 18.1—A2a and A2s. The A2s areas occur on steep slopes in the Rocky Mountains. The soils of the Black Hills in western South Dakota are also mainly Boralfs. These areas are important for recreation and forestry uses, they provide runoff water for collection in reservoirs, and they are used for some grazing of cattle. The Boralfs, area A2a in Wisconsin and Minnesota, developed from recent glacial parent materials. Much of the land is found in forests. Areas in agriculture are devoted to dairying and specialty crops, such as potatoes. A very large area of Boralfs, similar to that in Minnesota and Wisconsin, occurs in the Soviet Union (area A1a of Figure 18.2). Moscow is located in this area; important crops are rye, oats, flax, sugar beets, and forage crops. Dairy cattle and poultry, together with pigs, are important livestock products. Boralfs represent some of the most northerly soils used for agriculture.

Udalfs

Udalfs are warmer than Boralfs and have a udic soil moisture regime. The major area of Udalfs in United States occurs east of the Prairie region in the north-central states. These areas (A3a in Figure 18.1) have trees for native vegetation because they are more humid than the major grasslands located to the west. The Udalfs have developed mainly from mid- to late-Pleistocene loess and till. In Indiana, Ohio, and southern Michigan, the soils are nearly as productive as are the grassland soils (Mollisols) in Illinois and Iowa. These soils are

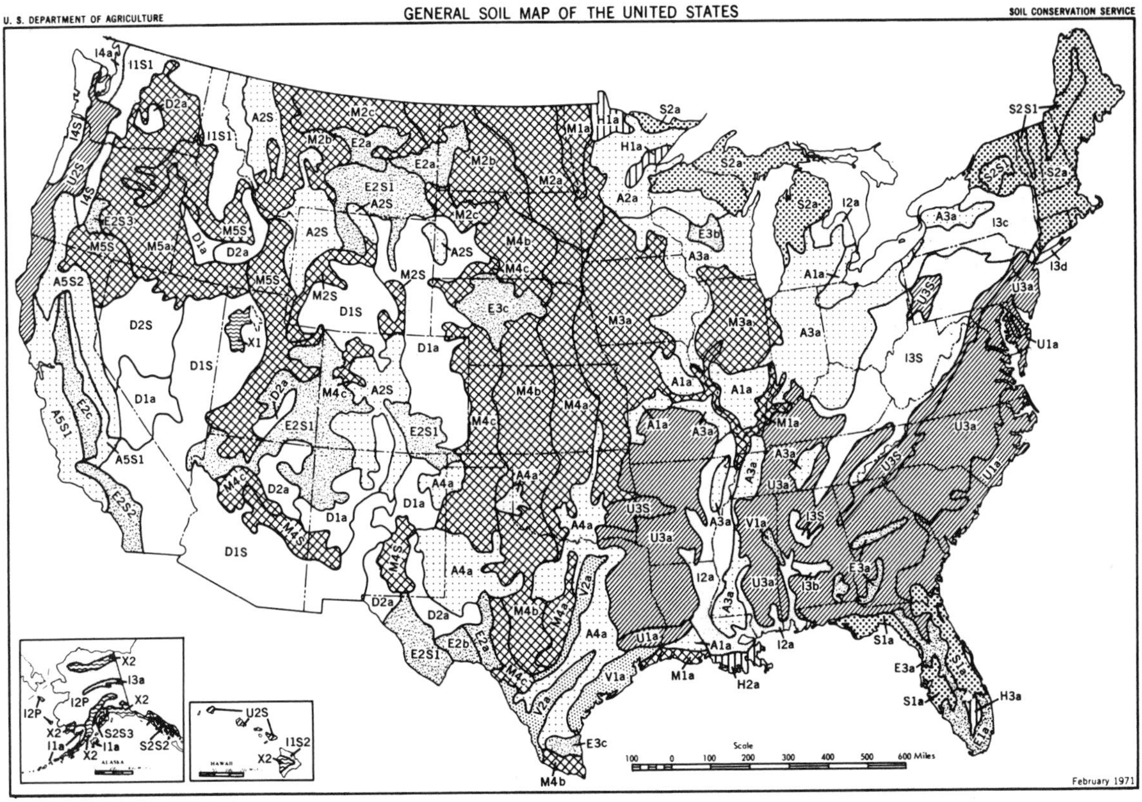

February 1971

FIGURE 18.1 General soil map of the United States.

Legend for general soil map of the United States.

ALFISOLS

Aqualfs
A1a—Aqualfs with Udalfs, Haplaquepts, Udolls; gently sloping.

Boralfs
A2a—Boralfs with Udipsamments and Histosols; gently and moderately sloping.
A2S—Cryoboralfs with Borolls, Cryochrepts, Cryorthods, and rock outcrops; steep.

Udalfs
A3a—Udalfs with Aqualfs, Aquolls, Rendolls, Udolls, and Udults; gently or moderately sloping.

Ustalfs
A4a—Ustalfs with Ustochrepts, Ustolls, Usterts, Ustipsamments, and Ustorthents; gently or moderately sloping.

Xeralfs
A5S1—Xeralfs with Xerolls, Xerothents, and Xererts; moderately sloping to steep.
A5S2—Ultic and lithic subgroups of Haploxeralfs with Andisols, Xerults, Xerolls, and Xerochrepts; steep.

ANDISOLS

Cryands
I1a-Cryands with Cryaquepts, Histosols, and rock land; gently and moderately sloping.

I1S1—Cryands with Cryochrepts, Cryumbrepts, and Cryorthods; steep.

Udands
I1S2—Udands and Ustands, Ustolls, and Tropofolists; moderately sloping to steep.

ARIDISOLS

Argids
D1a—Argids with Orthids, Orthents, Psamments, and Ustolls; gently and moderately sloping
D1S—Argids with Orthids, gently sloping; and Torriorthents, gently sloping to steep.

Orthids
D2a—Orthids with Argids, Orthents, and Xerolls; gently or moderately sloping.
D2S—Orthids, gently sloping to steep, with Argids, gently sloping; lithic subgroups of Torriorthents and Xerorthents, both steep.

ENTISOLS

Aquents
E1a—Aquents with Quartzipsamments, Aquepts, Aquolls, and Aquods; gently sloping.

Orthents
E2a—Torriorthents, steep, with borollic subgroups of Aridisols; Usterts and aridic and vertic subgroups of Borolls; gently or moderately sloping.

E2b—Torriorthents with Torrerts; gently or moderately sloping.

E2c—Xerothents with Xeralfs, Orthids, and Argids; gently sloping.

E2S1—Torriorthents; steep, and Argids, Torrifluvents, Ustolls, and Borolls; gently sloping.

E2S2—Xerorthents with Xeralfs and Xerolls; sleep.

E2S3—Cryorthents with Cryopsamments and Cryands; gently sloping to steep.

Pasmments

E3a—Quartzipsamments with Aquults and Udults; gently or moderately sloping.

E3b—Udipsamments with Aquolls and Udalfs; gently or moderately sloping.

E3c—Ustipsamments with Ustalfs and Aquolls; gently or moderately sloping.

HISTOSOLS

Histosols

H1a—Hemists with Psammaquents and Udipsamments; gently sloping.

H2a—Hemists and Saprists with Fluvaquents and Haplaquepts; gently sloping.

H3a—Fibrists, Hemists, and Saprists with Psammaquents; gently sloping.

INCEPTISOLS

Aquepts

I2a—Haplaquepts with Aqualfs, Aquolls, Udalfs, and Fluvaquents; gently sloping.

I2P—Cryaquepts with cryic great groups of Orthents, Histosols, and Ochrepts; gently sloping to steep.

Ochrepts

I3a—Cryochrepts with cryic great groups of Aquepts, Histosols, and Orthods; gently or moderately sloping.

I3b—Eutrochrepts with Uderts; gently sloping.

I3c—Fragiochrepts with Fragiaquepts, gently or moderately sloping; and Dystrochrepts, steep.

I3d—Dystrochrepts with Udipsamments and Haplorthods; gently sloping.

I3S—Dystrochrepts, steep, with Udalfs and Udults; gently or moderately sloping.

Umbrepts

14a—Haplumbrepts with Aquepts and Orthods; gently or moderately sloping.

I4S—Haplumbrepts and Orthods; steep, with Xerolls and Andisols; gently sloping.

MOLLISOLS

Aquolls

M1a—Aquolls with Udalfs, Fluvents, Udipsamments, Ustipsamments, Aquepts, Eutrochrepts, and Borolls; gently sloping.

Borolls

M2a—Udic subgroups of Borolls with Aquolls and Ustorthents; gently sloping.

M2b—Typic subgroups of Borolls with Ustipsamments, Ustorthents, and Boralfs; gently sloping.

M2c—Aridic subgroups of Borolls with Borollic subgroups of Argids and Orthids, and Torriorthents; gently sloping.

M2S—Borolls with Boralfs, Argids, Torriorthents, and Ustolls; moderately sloping or steep.

Udolls

M3a—Udolls, with Aquolls, Udalfs, Aqualfs, Fluvents,

Psamments, Ustorthents, Aquepts, and Albolls; gently or moderately sloping.

Ustolls

M4a—Udic subgroups of Ustolls with Orthents, Ustochrepts, Usterts, Aquents, Fluvents, and Udolls; gently or moderately sloping.

M4b—Typic subgroups of Ustolls with Ustalfs, Ustipsamments, Ustorthents, Ustochrepts, Aquolls, and Usterts; gently or moderately sloping.

M4c—Aridic subgroups of Ustolls with Ustalfs, Orthids, Ustipsamments, Ustorthents, Ustochrepts, Torriorthents, Borolls, Ustolls, and Usterts, gently or moderately sloping.

M4S—Ustolls with Argids and Torriorthents; moderately sloping or steep.

Xerolls

M5a—Xerolls with Argids, Orthids, Fluvents, Cryoboralfs, Cryoborolls, and Xerorthents; gently or moderately sloping.

M5S—Xerolls with Cryoboralfs, Xeralfs, Xerothents, and Xererts; moderately sloping or steep.

SPODOSOLS

Aquods

S1a—Aquods with Psammaquents, Aquolls, Humods, and Aquults; gently sloping

Orthods

S2a—Orthods with Boralfs, Aquents, Orthents, Psamments, Histosols, Aquepts, Fragiochrepts, and Dystrochrepts; gently or moderately sloping.

S2S1—Orthods with Histosols, Aquents, and Aquepts; moderately sloping or steep.

S2S2—Cryorthods with Histosols; moderately sloping or steep.

S2S3—Cryorthods with Histosols, Udands and Aquepts; gently sloping to steep.

ULTISOLS

Aquults

U1a—Aquults with Aquents, Histosols, Quartzipsamments, and Udults; gently sloping.

Humults

U2S—Humults with Andisols, Tropepts, Xerolls, Ustolls, Udox, Torrox, and rock land; gently sloping to steep.

Udults

U3a—Udults with Udalfs, Fluvents, Aquents, Quartzipsamments, Aquepts, Dystrochrepts, and Aquults; gently or moderately sloping.

U3S—Udults with Dystrochrepts; moderately sloping or steep.

VERTISOLS

Uderts

V1a—Uderts with Aqualfs, Eutrochrepts, Aquolls, and Ustolls; gently sloping.

Usterts

V2a—Usterts with Aqualfs, Orthids, Udifluvents, Aquolls, Ustolls, and Torrerts; gently sloping.

Areas With Little Soil

X1—Salt flats.

X2—Rock land (plus permanent snow fields and glaciers).

Slope Classes

Gently sloping—Slopes mainly less than 10 percent, including nearly level.

Moderately sloping—Slopes mainly between 10 and 25 percent.
Steep—Slopes mainly steeper than 25 percent.

SOILS OF THE WORLD

Distribution of Orders and Principal Suborders and Great Groups

A ALFISOLS—Soils with subsurface horizons of clay accumulation and medium to high base supply; either usually moist or moist for 90

consecutive days during a period when temperature is suitable for plant growth.

A1 Boralfs—cold.

    A1a with Histosols, cryic temperature regimes common

    A1b with Spodosols, cryic temperature regimes

A2 Udalfs—temperate to hot, usually moist.

    A2a with Aqualfs

    A2b with Aquolls

    A2c with Hapludults

FIGURE 18.2 Broad schematic map of the soil orders and suborders of the world.

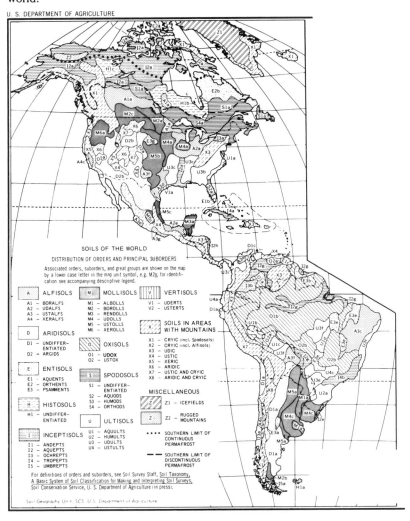

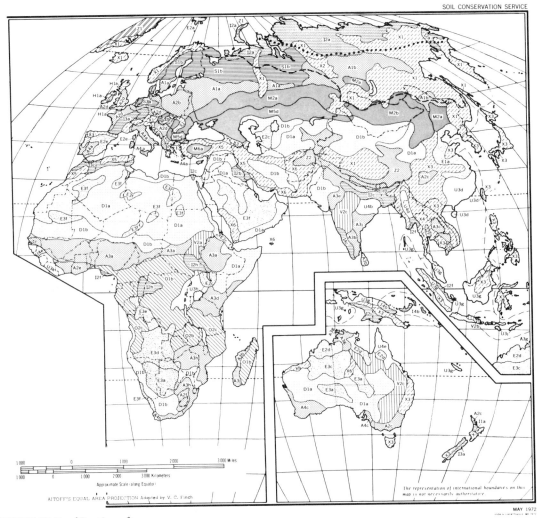

MAY 1972

FIGURE 18.2   (*Continued*)

A2d   with Ochrepts
A2e   with Troporthents
A2f   with Udorthents

A3   Ustalfs—temperate to hot, dry more than 90
cumulative days during periods when
temperature is suitable for plant growth.

    A3a   with Tropepts
    A3b   with Troporthents
    A3c   with Tropustults
    A3d   with Usterts
    A3e   with Ustochrepts

A3f   with Ustolls
A3g   with Ustorthents
A3h   with Ustox
A3j   Plinthustalfs with Ustorthents

A4   Xeralfs—temperate or warm, moist in
winter and dry more than 45 consecutive
days in summer.

    A4a   with Xerochrepts
    A4b   with Xerorthents
    A4c   with Xerults

D   ARIDISOLS—Soils with pedogenic horizons, usually dry in all horizons and never moist as long as 90 consecutive days during a period when temperature is suitable for plant growth.

   D1   Aridisols—undifferentiated.
      D1a   with Orthents
      D1b   with Psamments
      D1c   with Ustalfs

   D2   Argids—with horizons of clay accumulation.
      D2a   with Fluvents
      D2b   with Torriorthents

E   ENTISOLS—Soils without pedogenic horizons; either usually wet, usually moist, or usually dry.

   E1   Aquents—seasonally or perennially wet.
      E1a   Haplaquents with Udifluvents
      E1b   Psammaquents with Haplaquents
      E1c   Tropaquents with Hydraquents

   E2   Orthents—loamy or clayey textures, many shallow to rock.
      E2a   Cryothents
      E2b   Cryorthents with Orthods
      E2c   Torriorthents with Aridisols
      E2d   Torriorthents with Ustalfs
      E2e   Xerorthents with Xeralfs

   E3   Psamments—sand or loamy sand textures.
      E3a   with Aridisols
      E3b   with Udox
      E3c   with Torriorthents
      E3d   with Ustalfs
      E3e   with Ustox
      E3f   shifting sands
      E3g   Ustipsamments with Ustolls

H   HISTOSOLS—Organic soils.

   H1   Histosols—undifferentiated.
      H1a   with Aquods
      H1b   with Boralfs
      H1c   with Cryaquepts

I   INCEPTISOLS—Soils with pedogenic horizons of alteration or concentration but without accumulations of translocated materials other than carbonates or silica; usually moist or moist for 90 consecutive days during a period when temperature is suitable for plant growth.

   I1   Andepts—amorphous clay or vitric volcanic ash or pumice. The suborder Andepts was converted to the Andisol order in 1989.
      I1a   Dystrandepts with Ochrepts

   I2   Aquepts—seasonally wet.
      I2a   Cryaquepts with Orthents
      I2b   Haplaquepts with Salorthids
      I2c   Haplaquepts with Humaquepts
      I2d   Haplaquepts with Ochraqualfs
      I2e   Humaquepts with Psamments
      I2f   Tropaquepts with Hydraquents
      I2g   Tropaquepts with Plinthaquults
      I2h   Tropaquepts with Tropaquents
      I2j   Tropaquepts with Tropudults

   I3   Ochrepts—thin, light-colored surface horizons and little organic matter.
      I3a   Dystrochrepts with Fragiochrepts
      I3b   Dystrochrepts with Udox
      I3c   Xerochrepts with Xerolls

   I4   Tropepts—continuously warm or hot.
      I4a   with Ustalfs
      I4b   with Tropudults
      I4c   with Ustox

   I5   Umbrepts—dark-colored surface horizons with medium to low base supply.
      I5a   with Aqualfs

M   MOLLISOLS—Soils with nearly black, organic-rich surface horizons and high base supply; either usually moist or usually dry.

   M1   Albolls—light gray subsurface horizon over slowly permeable horizon; seasonally wet.
      M1a   with Aquepts

   M2   Borolls—cold.
      M2a   with Aquolls
      M2b   with Orthids
      M2c   with Torriorthents

   M3   Rendolls—subsurface horizons have much calcium carbonate but no accumulation of clay.
      M3a   with Usterts

   M4   Udolls—temperate or warm, usually moist.
      M4a   with Aquolls
      M4b   with Eutrochrepts
      M4c   with Humaquepts

   M5   Ustolls—temperate to hot, dry more than 90 cumulative days in year.
      M5a   with Argialbolls
      M5b   with Ustalfs
      M5c   with Usterts
      M5d   with Ustochrepts

   M6   Xerolls—cool to warm, moist in winter and dry more than 45 consecutive days in summer.
      M6a   with Xerorthents

O   OXISOLS—Soils with pedogenic horizons that are mixtures principally of kaolin, hydrated oxides, and quartz, and are low in weatherable minerals.
  O1   Udox—hot, nearly always moist.
    O1a   with Plinthaquults
    O1b   with Tropudults
  O2   Ustox—warm or hot, dry for long periods but moist more than 90 consecutive days in the year.
    O2a   with Plinthaquults
    O2b   with Tropustults
    O2c   with Ustalfs

S   SPODOSOLS—Soils with accumulation of amorphous materials in subsurface horizons; usually moist or wet.
  S1   Spodosols—undifferentiated.
    S1a   cryic temperature regimes; with Boralfs
    S1b   cryic temperature regimes; with Histosols
  S2   Aquods—seasonally wet.
    S2a   Haplaquods with Quartzipsamments
  S3   Humods—with accumulations of organic matter in subsurface horizons.
    S3a   with Hapludalfs
  S4   Orthods—with accumulations of organic matter, iron, and aluminum in subsurface horizons.
    S4a   Haplorthods with Boralfs

U   ULTISOLS—Soils with subsurface horizons of clay accumulation and low base supply; usually moist or moist for 90 consecutive days during a period when temperature is suitable for plant growth.
  U1   Aquults—seasonally wet.
    U1a   Ochraquults with Udults
    U1b   Plinthaquults with Udox
    U1c   Plinthaquults with Plinthaquox
    U1d   Plinthaquults with Tropaquepts
  U2   Humults—temperate or warm and moist all of year; high content of organic matter.
    U2a   with Umbrepts
  U3   Udults—temperate to hot; never dry more than 90 cumulative days in the year.
    U3a   with Andepts (Udands)
    U3b   with Dystrochrepts
    U3c   with Udalfs
    U3d   Hapludults with Dystrochrepts
    U3e   Rhodudults with Udalfs
    U3f   Tropudults with Aquults
    U3g   Tropudults with Hydraquents
    U3h   Tropudults with Udox
    U3j   Tropudults with Tropepts
    U3k   Tropudults with Tropudalfs
  U4   Ustults—warm or hot; dry more than 90 cumulative days in the year.
    U4a   with Ustochrepts
    U4b   Plinthustults with Ustorthents
    U4c   Rhodustults with Ustalfs
    U4d   Tropustults with Tropaquepts
    U4e   Tropustults with Ustalfs

V   VERTISOLS—Soils with high content of swelling clays; deep, wide cracks develop during dry periods.
  V1   Uderts—usually moist in some part in most years; cracks open less than 90 cumulative days in the year.
    V1a   with Usterts
  V2   Usterts—cracks open more than 90 cumulative days in the year.
    V2a   with Tropaquepts
    V2b   with Tropofluvents
    V2c   with Ustalfs

X   Soils in areas with mountains—Soils with various moisture and temperature regimes; many steep slopes; relief and total elevation vary greatly from place to place. Soils vary greatly within short distances and with changes in altitude; vertical zonation common.
  X1   Cryic great groups of Entisols, Inceptisols, and Spodosols.
  X2   Boralfs and cryic great groups of Entisols and Inceptisols.
  X3   Udic great groups of Alfisols, Entisols, and Ultisols; Inceptisols.
  X4   Ustic great groups of Alfisols, Inceptisols, Mollisols, and Ultisols.
  X5   Xeric great groups of Alfisols, Entisols, Inceptisols, Mollisols, and Ultisols.
  X6   Torric great groups of Entisols; Aridisols.
  X7   Ustic and cryic great groups of Alfisols, Entisols, Inceptisols, and Mollisols; ustic great groups of Ultisols; cryic great groups of Spodosols.
  X8   Aridisols, torric and cryic great groups of Entisols, and cryic great groups of Spodosols and Inceptisols.

Z   MISCELLANEOUS
  Z1   Icefields.
  Z2   Rugged mountains—mostly devoid of soil (includes glaciers, permanent snow fields, and, in some places, areas of soil).
. . . Southern limit of continuous permafrost.
- -Southern limit of discontinuous permafrost.

FIGURE 18.3 Many Udalfs occur on gentle slopes of glaciated areas in the north-central part of the United States and New York, where a wide variety of crops is grown and dairying is a major industry.

well suited to the production of a wide variety of crops, especially corn, soybeans, wheat, and forages (see Figure 18.3). Udalfs have developed along the lower Mississippi River in loess and are similar to Udalfs that developed from loess in the Midwest. A Udalf is shown in Color Plate 5.

Many of the best agricultural soils in northwestern Europe are Udalfs comparable to those of the north-central United States in that they developed in glacial parent materials of similar age. In both areas, dairying is a major industry.

Soils of the Blue Grass region in central Kentucky are predominantly Udalfs that developed from a high phosphorus limestone capped by a thin layer of loess. Early settlers prized the soils for growing food and, today, they are used for raising racehorses in addition to being used for general agriculture.

Ustalfs

Ustalfs have an ustic moisture regime and tend to occur where temperatures are warm or hot and the vegetation is grassland or savanna. The large areas dominated by Ustalfs in Texas and New Mexico (A4a in Figure 18.1) occur south of areas where the soils are mainly Mollisols. In these Ustalf areas, higher temperature and sandy parent materials have contributed to the development of soils with ochric epipedons under grass or sa-

vanna vegetation. Much of the land is devoted to cattle grazing, and warm temperatures allow cotton to be grown.

The largest area in the world in which Alfisols are the dominant soils occurs south of the Sahara Desert in Africa. The soils are Ustalfs (see area A3a in Figure 18.2). The soils have undergone a long period of weathering; many of them have kandic horizons, and would normally be Ultisols. However, calcareous dust from the Sahara Desert is carried south by seasonal winds and is deposited on the soils. This has kept the base saturation at 35 percent or more. Considerable cattle raising, subsistence farming, and shifting cultivation occur in this region. The large size of raindrops of the tropical thunderstorms contributes to high rates of water runoff and soil erosion on exposed land.

Xeralfs

Xeralfs have a zeric moisture regime and are the dominant soils in the coastal ranges and inland mountains in California and southern Oregon (see areas A5s1 and A5s2 in Figure 18.1). Natural

FIGURE 18.4 Grazing of cattle is important on Xeralfs on the lower slopes of the Sierra Nevada Mountains in California. On the cooler and more humid upper slopes are Xerults, Ultisols with a xeric moisture regime.

vegetation is a mixture of annual grasses, forbs, and shrubs. Many of these sloping lands are used for animal grazing, and some areas are irrigated for the production of horticultural crops. In the mountains at highest elevations with greater precipitation, Ultisols occur as shown in Figure 18.4.

ANDISOLS

Andisols are weakly developed soils produced by the weathering of volcanic ejecta. Although Andisols occupy less than 1 percent of the world's land surface, they are important soils of the Pacific Ring of Fire—that ring of active volcanoes along the western coast of the American continents, across the Aleutian Islands, down Japan, the Philippine Islands, other Pacific Islands, and down to New Zealand. The most important areas of Andisols in the United States occur in the Pacific northwest and Hawaiian Islands.

Genesis and Properties

An important feature of Andisols is the presence of volcanic glass. The dominant process in most Andisols is one of weathering and mineral transformation. Volcanic glass weathers rapidly producing allophane and oxides of aluminum and iron. Translocation and accumulation within the soil are minimal. Repeated ash falls maintain the weakly developed nature of the soils and buried horizons are common.

Andisols have andic soil properties. The weathered minerals, including allophane and oxides of aluminum and iron, are amorphous or short-range minerals with enormous surface area and great capacity to complex with organic matter. Andisols are characterized by high organic matter content, ion exchange capacity, and low bulk density. The capacity to hold available water for plant growth is high. Soils are generally considered fertile, although high fixation and low availability of phosphorus are sometimes a problem. Many Andisols have melanic epipedons. Large areas of intensively cropped black soils,

Andisols with melanic epipedons, are visible from the air in the Andes of South America.

Suborders

Most Andisols suborders are based on moisture and temperature regimes. Aquands have an aquic moisture regime, Cryands are cold (cryic temperature regime or colder), Torrands have an aridic moisture regime, and Xerands have a xeric moisture regime. Vitrands have the greatest content of glassy or vitreous materials. Sufficient weathering of Vitrands and disappearance of glass results in Udands with udic moisture regime or Ustands with ustic moisture regime.

The wide range in temperature and moisture regimes is associated with great variation in native vegetation from bamboo, grass, sedges, and forest. Xerands are the dominant Andisols of the Pacific northwest in the United States. Udands and Ustands are the dominant Andisols of the Island of Hawaii where they are important soils for sugarcane production and cattle grazing.

Locations where soils are likely to be Andisols are shown on the world soil map as mountainous areas, X areas. One area of volcanic ash soils is located in northern New Zealand. The soils are Andepts in the legend of Figure 18.2. In 1989 the Andept suborder was converted to the Andisol order.

ARIDISOLS

Aridisols have an aridic moisture regime and are the dominant soils of deserts. Large areas of desert occur in northern Africa (Sahara Desert), central Eurasia, central and western Australia, in southern South America east of the Andes Mountains, in northern Mexico, and in the southwestern United States. Small deserts exist on many islands in the rain shadow of mountains, as in the Hawaiian and Caribbean islands. Aridisols are the most abundant soils worldwide and the fifth most abundant soils in the United States (see Table 18.1).

FIGURE 18.5 Vegetation on the Sonoran Desert in Arizona. Note gravel pavement in the foreground because of removal of fine particles by the wind.

Shrubs, the dominant vegetation, give way to bunch grasses with increasing precipitation. Plants are widely spaced and use the limited water supply quite effectively. Many desert plants grow and function during the wetter seasons and go dormant during the driest seasons. Suprisingly, there can be great plant diversity and a considerable amount of vegetation, as seen in Figure 18.5. On the more humid or eastern edge of the arid region in southwestern United States, the desert bunch grasses give way to taller and more vigorous grasses; Aridisols merge with Alfisols and Mollisols.

Genesis and Properties
In arid regions, there is a slow rate of plant growth, resulting in low organic matter accumu-lation and limited weathering, leaching, and eluviation. A striking feature of most Aridisols is a zone below the surface in which calcium carbonate has accumulated to form a calcic (k) horizone as a result of limited leaching by water. Winds play a role by removing the finer particles from exposed areas and, where gravel occurs in the soil, a surface layer of gravel (desert pavement) is formed (see Figure 18.5).

Some Aridisols, however, have a well-developed argillic horizon (Bt), which indicates considerable weathering and clay movement. The occurrence of argillic horizons suggests that a more humid climate existed many years ago. The Mohave is a common Aridisol in the Sonoran Desert of Arizona; and some data are given in Table 18.2 to illustrate properties commmonly found in Aridisols. Note the Bt horizon (argillic) at the 10- to 69- centimeter depth. This is underlain by Bk and Ck horizons that contain 10 and 22 percent calcium carbonate, respectively. The horizons have low organic matter contents and low carbon-nitrogen ratios. The argillic horizon has high CEC because of clay accumulation. Significant exchangeable sodium and high pH values reflect limited leaching, at least, in recent years. A photograph of an Aridisol with A, Bt, and Ck horizons is shown in Color Plate 6.

Aridisol Suborders
Aridisols are placed in suborders on the basis of the presence or absence of an argillic horizon. Orthids do not have argillic horizons, whereas Argids do. As shown in Table 18.2, the Mohave soil has an argillic horizon and is an Argid. The general distribution of Orthids and Argids in the United States is shown in Figure 18.1 and for the world in Figure 18.2.

It is believed that Orthids are young Aridisols and that Argids, by comparison, are much older Aridisols. There is evidence that the Orthids in the United States developed largely within the past 25,000 years in an arid climate. Orthids are usually located where recent alluvium has been deposited. Argids are common on old land surfaces

TABLE 18.2 Some Properties of Mohave Sandy Clay Loam–an Aridisol

Horizon	Depth, cm	Clay %	Organic Matter, %	C/N	CEC cmol (+)/kg	Exch. Na, %	pH	$CaCO_3$, %
A	0–10	11	0.25	6	8	1.2	7.8	—
Bt1	10–25	14	0.19	6	15	2.0	7.4	—
Bt2	25–69	25	0.24	7	22	2.5	8.5	—
Bk	69–94	21	0.25	8	17	4.1	8.9	10
Ck	94–137	17	0.08	—	6	12.7	9.2	22

Adapted from profile 62 of *Soil Classification, a Comprehensive System*, USDA, 1960.

in any landscape where there has been enough time for genesis of an argillic horizon during a humid period that existed more than 25,000 years ago. About 75 percent of the Aridisols in United States are Argids. Entisols occur on the most recent land surfaces. The desert landscape shown in Figure 18.6 has land surfaces with greatly different ages.

Land Use on Aridisols

Aridisols of the western United States occur almost entirely within a region called the "western range and irrigated region." As the name implies, grazing of cattle and sheep and production of crops by irrigation are the two major agricultural land uses. The use of land for grazing is closely related to precipitation, which largely determines the amount of forage produced. Some areas are too dry for grazing, whereas other areas receive more favorable precipitation and thus may be used for summer grazing on mountain meadows. As many as 35 hectares (75 acres) or more in the drier areas are required per head of cattle, thus making large farms or ranches a necessity. Most ranchers supplement range forage by producing some crops on a small irrigated acreage as shown in the left foreground of Figure 18.6. Overgrazing,

FIGURE 18.6 Grazing lands dominated by Aridisols. Soils in the area are closely related to age of land surface with Aridisols on the older surfaces and Entisols on the youngest surfaces. Note the small irrigation reservoir and irrigated cropland near ranch headquarters. (USDA photograph by B. W. Muir.)

which results in the invasion of less desirable plant species and increased soil erosion, is the major hazard in the use of these grazing lands.

In Arizona, for example, 1 or 2 percent of the land is irrigated. Most of the irrigated land is located on soils developed from alluvial parent materials along streams and rivers. Here, the land is nearly level, irrigation water is obtained from the river, and water is distributed over the field by gravity. The alkaline nature of Aridisols may cause deficiencies of various micronutrients on certain crops. Major crops include alfalfa, cotton, citrus fruits, vegetables, and grain crops. In Arizona, crop production on only 1 or 2 percent of the land that is irrigated accounts for 60 percent of the total farm income. Grazing, by contrast, uses 80 percent of the land and accounts for only 40 percent of the total farm income.

ENTISOLS

Entisols are soils of recent origin. The central concept is soils developed in unconsolidated earth, or parent material, with no genetic horizons except an A horizon. Some Entisols have rock close to the surface and have A and R horizons as shown in Figure 18.7. All soils that do not fit into one of the other 10 orders are Entisols. The result is that the Entisol order is characterized by great diversity.

Aquents

Small areas of Aquents are widely distributed. In the United States, however, the only major area shown in Figure 18.1 is in southern Florida (area E1a). Parent materials are mainly marine sands and marl. The soils are wet as a result of a naturally high water table. Although they are drained for cropping, water for irrigation is pumped from a shallow water table when needed. The frost-free weather in southern Florida makes these soils valuable for production of winter vegetables and fruits.

FIGURE 18.7 An Entisol developing in parent material formed by the weathering of granite in the U.S. Rocky Mountains.

The largest area in which soils are mainly Aquents worldwide occurs on the North China Plain near the mouths of Yellow, Yangtze, and several other rivers (area E1a in Figure 18.2). It is a vast floodplain where inhabitants must cope with occasional catastrophic floods. This large floodplain is one of the most densely populated and intensively cropped areas of the world.

Fluvents

Fluvents develop in alluvium of recent origin. They are, in large part, soils having properties that have been inherited from the parent material. For a soil to be a Fluvent it must have, with increasing depth, an irregular distribution of organic matter that reflects the organic matter content of different layers of recently deposited alluvium. The organic matter in soils formed from alluvium decomposes over time. Therefore, in the absence of continued flooding and deposition of alluvium, a normal gradual decrease in organic matter content with increasing soil depth is produced with time. In this way, a Fluvent may develop in the absence of continued flooding and deposition of alluvium into an Inceptisol, an Aquept. For this reason many soils on large river valley floodplains, such as the Mississippi River floodplains, are not Fluvents.

Alluvial sediments were deposited in the Imperial Valley in southeastern California by several years of uncontrolled flooding of the Colorado River when the valley underwent widespread development of irrigation. Some Fluvents have developed in these recent sediments, however, most of the soils are Orthents. The uncontrolled flooding created the Salton Sea in a depression below sea level north of the valley. Now, the Salton Sea serves as a sink for the salt in the drainage water from the irrigated fields.

FIGURE 18.8   Cow-calf production is the major land use on the Psamments of the Nebraska Sandhills. Maintenance of a grass cover is essential to prevent serious wind erosion.

maintain a vegetative cover to control wind erosion.

Psamments occupy most of the central ridge of the Florida peninsula. The hyperthermic temperature regime provides frost-free winters and makes the soils prized for citrus production. Outside the United States many large areas in north and south Africa, Australia, and Saudi Arabia are dominated by Psamments.

Psamments

Psamments develop in sand parent materials and do not have an aquic moisture regime. Psamments develop when unstablized sand dunes become stablized by vegetation. The texture must be loamy fine-sand or coarser to a depth of 1 meter or more, or to a hard rock layer. The Sandhills of north-central Nebraska (see area E3c of Figure 18.1) are dominated by Psamments with an ustic moisture regime. These soils are Ustipsamments. This area consists of sand dunes stabilized by mid- and tall grasses and used for cattle grazing. The major product is feeder stock that is sold for fattening in feedlots. A typical Sandhills landscape is shown in Figure 18.8. Numerous small lakes in the area provide watering places. The major soil management challenge is to control grazing so that productive forage species

Orthents

The suborder Orthents includes all Entisols that do not fit into one of the other suborders. As a result, Orthents are very diverse and form in parent materials finer in texture than sands and without an aquic moisture regime. Delayed genesis is caused by active erosion on steep slopes, short period of time, and/or hardness of rock close to the soil surface. Where the parent materials are derived from unconsolidated rock, A–C soils occur. A–R soils occur where rock is near the soil surface (see Figure 18.7). Active erosion on steep slopes accounts for many Orthents in the Badlands of the western United States, and active erosion, together with rock close to the soil surface, account for many of the Orthents in the Rocky Mountains. Orthents are common in the valleys of California, including the large Central

Valley (area E2c of Figure 18.1). The soils have a xeric moisture regime, are mainly Xerorthents, and were formed in sediments carried to the valley floor as a result of erosion of the surrounding hills and mountains. The soils are intensively irrigated, and many high value crops are produced. As a result, the value of agricultural crops produced in California is the largest of any state (see Figure 18.9).

HISTOSOLS

Histosols are organic soils, and they have been called bogs, peats, mucks, and moors. Most Histosols are water saturated for most of the year unless they have been drained. Histosols contain organic soil materials to a significant depth. Organic soil material contains a variable amount of organic carbon, depending on the clay content. Soil material containing 60 or more percent clay has an organic carbon content of 18 or more percent. Material containing no clay has an organic carbon content of 12 or more percent. Interme-

FIGURE 18.9 Orthents are the most important agricultural soils in California in large, broad, and nearly level valleys. A wide variety of crops is grown, including fruits, nuts, vegetables, and farm crops. Cotton is shown in the foreground and large stacks of baled alfalfa in the rear.

diate clay contents have intermediate amounts of organic carbon (see Figure 9.5).

Many Histosols have developed in organic materials of great depth (many meters deep). Others have developed in parent materials containing organic layers of varying thickness. Where the organic material overlies rock at a shallow depth, Histosols can be thin.

Histosol Suborders

During water saturation, organic matter resists decomposition but decomposes when the soil drains and desaturates. The degree of decomposition of the organic fibers or plant materials increases with increasing time of drainage or desaturation. Three degrees of decomposition are recognized and reflected in the suborders *Fibrists*, *Hemists*, and *Saprists*. Fibrists consist largely of plant remains so little decomposed that their botanical origin can be determined readily. Hemists are decomposed sufficiently so that the botanical origin of as much as two-thirds of the materials cannot be determined. Saprists consist of almost completely decomposed plant remains and they usually have a black color.

Histosol horizons are O horizons: Oi, Oe, and Oa refer to O horizons with fibric, hemic, and sapric organic materials, respectively. Folists are the other suborder and consist of organic soil materials resulting from the deposition of leaf or foliar material. Folists may be thin over hard rock, and it is likely that they may never be water saturated.

One eighth of the soils in Michigan are Histosols, which occur in all the counties, with only a few areas being quite extensive. In Figure 18.1, two large areas of Hemists (H1a) are shown in northern Minnesota, and one area is shown in the lower Mississippi River delta in Louisiana. The Everglades in Florida contain Fibrists, Hemists, and Saprists (see area H3a in Figure 18.1). An Everglades landscape is shown in Figure 18.10. Large acreages of Histosols occur in northern Canada (see H1b and H1c in Figure 18.2).

FIGURE 18.10 Histosol landscape of the Florida Everglades.

FIGURE 18.11 In 1924 the bottom of this 9-foot concrete post was placed on the underlying limestone at Belle Glade, Florida, and the top of the post was at ground level. This photograph, which was taken in 1972, shows that subsidence of the organic soil (Histosol) has been about 1 inch per year.

Land Use on Histosols

A major limitation of northern Histosols for agricultural use is frost hazard. The Histosols tend to occupy the lowest position in the landscape and are the first areas to freeze. Use for crop production requires water drainage, which enhances decomposition of the organic matter. The subsidence and loss of soil by oxidation and disappearance of the soil is a major concern, as shown in Figure 18.11. A large acreage in the Florida Everglades is used primarily for the production of sugarcane. During 50 years of agriculture the soils subsided an amount equal to an amount of soil that required 1,200 years to form. Since the Histosols of the Everglades overlie limestone, many areas now cultivated will be too thin for continued cultivation past the year 2000. The likely use for such soils will then be pasture, which requires no tillage.

Other problems in the use of Histosols are fire hazards and wind erosion during dry seasons. The soils contain few minerals for supplying many of the plant nutrients; however, the organic matter contributes to high nitrogen-supplying power. Despite the problems in their use, many Histosols are intensively managed for the production of a wide range of field and vegetable crops, including sod.

INCEPTISOLS

Inceptisols typically have an ochric or umbric epipedon that overlies a cambic (Bw) horizon. They typically show little evidence of eluviation, illuviation, or extreme weathering. They are too old to be Entisols and lack sufficient diagnostic features to be placed in one of the other orders. Inceptisols occur in all climatic zones where there is some leaching in most years, and are the second most abundant soils both in the United States and worldwide. In the United States about 70 percent of the Inceptisols are Aquepts, 26 percent are Ochrepts, and 4 percent are Umbrepts.

Aquepts

Aquepts are wet Inceptisols that have an aquic moisture regime, unless they are artificially drained. Mottled rusty and gray colors are common at depths of 50 centimeters or less. Aquepts occur in many depressions in geologically recent landscapes where soils are water saturated at some time during the year, and are surrounded by better drained soils that are Alfisols or Mollisols. Water saturation inhibits eluviation and illuviation. With better drainage, Aquepts may develop into soils that are older or more mature. Aquepts can have any temperature regime and almost any vegetation.

The largest areas of Aquepts occur in the Tundra regions of the Northern Hemisphere in Alaska, Canada, Europe, and Asia (areas I2a in Figure 18.2). Most of these cold Aquepts (Cryaquepts) have permafrost, which inhibits the downward movement of water and contributes to soil wetness when frozen soil melts during the summer. Trees grow poorly on soils with permafrost, so forestry is not important. Locally, above timberline in many mountains, there are Cryaquepts with permafrost (see Figure 18.12).

Few people inhabit tundra regions; some of those who do work in the mining and oil industries, and a few people in northern Europe and Asia raise reindeer. The reindeer browse on the tundra vegetation and may be readily contaminated with radioactive elements that fall out of the atmosphere. Reindeer farmers in Finland found that their animals were unfit for human consumption because of the radioactive contamination following the Chernobyl (USSR) reactor explosion.

Other large and important areas of Aquepts occur in the world's major river valleys. Note the large I2a area in Figure 18.1 that includes the lower Mississippi River valley. The soils have formed in alluvial materials, but most of the materials are not recent and the soils are not Entisols, because the river has wandered back and forth over a wide floodplain over a long period of time. These Aquepts tend to be fine textured and very fertile. Good drainage and management have made the Aquepts of the Mississippi River valley among the most productive soils in the United States. Sugarcane is an important crop in southern Louisiana where temperatures are warm. A landscape in which most of the soils are Aquepts in the lower Mississippi River valley is shown in Figure 18.13.

The Ganges-Brahmaputra River valley in northern India and Bangladesh is the largest area in the world in which Aquepts are the dominant soils (see Area I2c of Figure 18.2). The presence of Aquepts in the river valleys of Asia, and the large rice consumption of the inhabitants makes Aquepts the world's most important soil for rice production. Aquepts are also the dominant soils in the Amazon River valley, (see Figure 18.2).

FIGURE 18.12 Dominant soils on the tundra above timberline in the mountains are Inceptisols, Aquepts. Soils have permafrost, vegetation is sparse, and the environment is fragile.

Ochrepts

Ochrepts are the more or less light-colored and freely drained Inceptisols of the mid- to high latitudes. They tend to occur on geologically young land surfaces and have ochric and cambic horizons. Most of them had, or now have, a forest vegetation. Ochrepts are the dominant soils of the Appalachian Mountains (area I3s of Figure 18.1). Erosion on steep slopes has contributed to the formation and maintenance of these Inceptisols.

Umbrepts

Umbrepts are Inceptisols with umbric epipedons. Most of them occur in hilly or mountainous regions of high precipitation and under coniferous forest. Umbrepts in the United States are the most extensive near the Pacific Ocean in the states of

FIGURE 18.13 Intensive agriculture on Aquepts in the lower Mississippi River valley.

Oregon and Washington (see areas I4S and I4a of Figure 18.1). These Umbrepts have a mesic temperature regime and udic or xeric moisture regimes. Many of the soils are on steep slopes that are densely covered with coniferous forest and lumbering is the principal industry.

MOLLISOLS

Mollisols are the dominant soils of the temperate grasslands or steppes. These regions are frequently bordered on the drier side by deserts with Aridisols and on the wetter side by forests in which the dominant soils are Alfisols. There is sufficient precipitation to support perennial grasses that contribute a large amount of organic matter to the soil annually by deeply pentrating roots. Precipitation, however, is somewhat limited so the soils are only minimally or moderately weathered and leached. Mollisols typically have a mollic epipedon. Subsurface horizons are typically cambic or argillic, and a few have natric horizons. Mollisols are characterized by a moderate or high organic matter content and a high content of weatherable minerals. Many of the soils are calcareous at the surface and the acid Mollisols tend to be only mildly acid and without significant exchangeable aluminum. Two Mollisols, both with k (calcic) horizons, are shown in Color Plate 6.

Mollisols are considered to be some of the most naturally fertile soils for agriculture. Where temperature and water are favorable, high grain yields are easily obtained. The most abundant soils in the United States are Mollisols, being 25 percent of the total (see Table 18.1). Few Mollisols occur in the tropics because of the generally long period of weathering and leaching.

The world's major grasslands lacked trees for lumber, readily available water supplies, and natural sites for protection against invaders. As a result, these areas dominated with Mollisols were inhabited mainly by nomadic people until about 150 years ago. The farmers who settled in the central part of United States tended to avoid the Mollisols because they were used to forest soils, and they found the sod of the Udolls difficult to plow and convert into cropland. The Virgin Lands of the Soviet Union were opened to settled agriculture in 1954. Today, these areas are characterized by their excellent agricultural soil and a low population density. Land use is related mainly to water supply and temperature.

Aquolls

Aquolls are the wet Mollisols that have an aquic moisture regime or are artificially drained. They occur as inclusions in many grassland landscapes owing to water saturation of the soil in local depressions. Drummer silty clay loam is an

Aquoll and is the most abundant soil in Illinois. Much of the Corn Belt's reputation for corn production is due to large acreages of Aquolls that developed on nearly level and depressional areas under poor drainage.

Aquolls are the dominant soils of the Red River Valley bordering Minnesota and North Dakota (area M1a in Figure 18.1). These Aquolls have formed from lacustrine sediments deposited in glacial Lake Agassiz about 9,000 to 12,000 years ago. The soils tend to be fine-textured, to have a relatively high organic matter content, and to be well suited for grain crops. The Red River Valley is an area of large-scale cash crop farming. Spring wheat, barley, sugar beets, corn, and sunflowers are the major crops.

Borolls

Borolls are Mollisols that do not have an aquic moisture regime but are northerly or cool with cryic temperature regime (have mean annual soil temperature less than 8° C or 47° F). Borolls are dominant on the northern Great Plains in the United States and Canada and in the Soviet Union and northern China (areas M2a, b, and c in Figure 18.2). Spring wheat is the major crop and much of the land is used for cattle grazing. Low temperature is frequently coupled with low precipitation so that much of the wheat is produced in a wheat-fallow system.

Ustolls and Udolls

Ustolls and Udolls have warmer temperatures than do Borolls, and have ustic and udic moisture regimes, respectively. Ustolls typically have k horizons and a neutral or alkaline reaction. Udolls are without k layers. Udolls are typically acid, and lime is frequently applied. Cambic and argillic horizons are common in both.

In both the Great Plains and the north-central states, Ustolls are dominant west of the approximate boundary between Iowa and Nebraska and Udolls are dominant east of the boundary. (Contrast areas M4a, b, and c with the M3a areas in

Figure 18.1). Winter wheat, produced in a wheat-fallow system, is the major crop grown on Ustolls. Kansas ranks first in wheat production in the United States. The winter wheat is planted in the fall. The milder winters on Ustolls (as compared with those on Borolls) allow the wheat to become established in the fall and to begin growing in early spring. Consequently, winter wheat yields are greater than those for spring wheat. Spring wheat is grown primarily on Borolls of the northern Great Plains in the United States and Canada, and in the Soviet Union. Cattle grazing is also a major enterprise on Ustolls, especially on land not suitable for wheat (see Figure 18.14). Irrigation water pumped from the Ogallala formation is used for intensive cropping on the High Plains of Texas. Common crops include sorghum, winter wheat, and peanuts.

Corn and soybeans are the major crops grown on Udolls, as shown in Figure 18.15. The two most famous areas of Udolls are the Corn Belt in the United States and the humid part of the Pampa in Argentina, which are areas of intensive corn production. There is limited production of corn in the Soviet Union, which is a reflection of the cooler and drier climate and the dominance of Ustolls and Borolls rather than Udolls on the grasslands. Note in Figure 18.2 that there is a west to east distribution of Aridisols, Ustolls, and Udolls in

FIGURE 18.14 Beef cattle grazing on Ustolls where the slopes are unfavorable for wheat production. (Courtesy USDA.)

FIGURE 18.15 Typical Aquoll-Udoll landscape in central Illinois with corn and soybeans as the major crops.

Argentina, similar to that which occurs in the United States. Land uses in the two areas are also similar, with wheat and cattle grazing on the Ustolls and corn production concentrated on the Udolls. Most of the Mollisols of Argentina formed in loess that was contaminated with volcanic ash, and therefore the soils have a greater inherent fertility than do comparable Mollisols in the United States and the Soviet Union.

Xerolls

Xerolls are Mollisols with warmer temperatures than are associated with Borolls and a xeric moisture regime. Most of the Xerolls in the United States occur in Washington, Oregon, and Idaho, where summers are dry and winters are cool and rainy. Water availability is suited for winter wheat production, and Whitford County in the state of Washington produces more wheat than any other county in the United States. Whitford County is located in the Palouse of eastern Washington and western Idaho. The Xerolls developed in loess

that overlies basalt in a rolling landscape (see Figure 18.16). The soils are very similar to Mollisols developed from loess in the central United States. The xeric moisture regime precludes corn production because the summers are very dry. Much of the land with Xeroll soils is used for cattle grazing.

OXISOLS

Oxisols are the most intensively weathered soils, consisting largely of mixtures of quartz, kaolinite, oxides of iron and aluminum, and some organic matter. Many Oxisols contain little silt because the silt sized mineral particles have essentially weathered leaving a clay fraction composed mainly of oxidic clays plus kaolinte and varying amounts of sand because of the great weathering resistance of large quartz grains. Most Oxisols have oxic horizons. Some have *plinthite*, which forms a continuous layer within 30 centimeters of the soil surface and is water saturated at this

COLOR PLATE 5

Vertisol (Ustert)

Spodosol (Orthod)

Alfisol (Udalf)

Ultisol (Udult)

Developed from clay parent material where climate has distinct wet and dry seasons. Scale in centimeters.

Developed from sand parent material under forest in humid climate. Scale in feet.

Developed from loamy parent material under forest in humid climate. Scale in feet.

Developed from loamy parent material under forest in a warm and humid climate for very long time. Scale in centimeters.

COLOR PLATE 6

Oxisol (Torrox)

Mollisol (Boroll)

Mollisol (Boroll)

Aridisol (Argid)

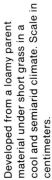

Developed over a long period of time from parent material weathered from basalt in Hawaii. Scale in feet.

Developed from fine-loamy parent material under short grass in a cool and subhumid climate. Scale in centimeters.

Developed from a loamy parent material under short grass in a cool and semiarid climate. Scale in centimeters.

Developed from loamy parent material under short grass and shrubs in arid climate. Scale in feet.

FIGURE 18.16 Landscape of Xerolls in the Palouse in Washington composed mainly of loessial dunes. Winter wheat is the major crop. (Photograph courtesy Dr. H. W. Smith.)

depth sometime during the year. Plinthite is an iron-rich material that forms some distance below the soil surface. When plinthite is exposed to drying, it dries irreversibly, forming a rocklike material.

Differences in properties with depth in Oxisols are so gradual that horizon boundaries are difficult to determine. A photograph of an Oxisol is shown in Color Plate 6.

Oxisols are characterized by a very low amount of exchangeable nutrient elements, very low content of weatherable minerals, high aluminum saturation of the cation exchange capacity, and moderate acidity. Although aluminum saturation may be high, the crops do not generally develop aluminum toxicity because of the low CEC. The mineralogy contributes to very stable peds or structure that results in qualities associated with sandy, quartzitic soils in the temperate regions. Generally, Oxisols represent the most naturally infertile soils for agriculture. Some are, however, the world's most productive soils because of inten-

sive management, as in the case of pineapple production in Hawaii, shown in Figure 18.17.

There are no Oxisols within the continental United States; however, they are found in Hawaii and Puerto Rico. As a result, Oxisols are the least

FIGURE 18.17 Pineapple production on Oxisols in Hawaii.

abundant soils in United States. Worldwide, Oxisols are the sixth most abundant soils; however, Oxisols are the most abundant soils of the tropics, where they occur on about 22 percent of the land.

Oxisol Suborders

Oxisols tend to occur along or parallel to the equator. Near the equator, where there is significant rainfall each month, many Udox and Perox soils have developed. Udox have udic moisture regime and Perox have perudic moisture regime. North and south of the equator are regions in which there is abundant rain in summer but none in winter. In these areas, the Oxisols that develop have an ustic moisture regime and are Ustox. About two thirds of the Oxisols are Udox (and Perox) and about one third are Ustox. Oxisols are dominant soils in two large areas. One is in the Amazon basin in South America and the other is centered on the Congo River or Zaire basin in central Africa. Note the distribution of Udox and the Ustox soils by comparing the location of the Udox areas, O1b, with the Ustox areas, O2b, in Figure 18.2. Oxisols of the Aquox and Torrox suborders are of minor extent.

Land Use on Udox Soils

The Oxisols in the year-round rainy climates, Udox, are mainly in the tropical rain forests where a small human population is supported by shifting cultivation or slash-and-burn agriculture. Trees are killed by girdling and/or burning, and the ashes and decomposing plant remains provide nutrients for crops for a year or two. Then, soil fertility exhaustion and weed invasion result in abandonment of the land, which is gradually reforested by invasion of trees and shrubs. After a period of about 20 years, sufficient nutrients will have accumulated within living and dead plant materials and in soil organic matter to allow for another short period of cropping. Land prepared for growing crops is shown in Figure 18.18. Note the large amount of organic residues on the soil

FIGURE 18.18 Preparation of land in the foreground by slash and burn. Note stumps left in field and the considerable amount of organic matter on the soil surface that will mineralize and provide plant nutrients.

surface and the dense rain forest growth in the background.

Frequently, a village or tribal chief allocates land to the families. Vegetables grown in small gardens near houses, and fish and game animals supplement the diet. There are no farm animals to provide manure for the maintenance of soil fertility, and only a sparse population can be supported. Where population growth occurs, shorter fallow periods (periods when the forest is regrowing and accumulating nutrients within plant and soil organic matter) are used and it may become impossible to regenerate soil fertility. Then, the system becomes ineffective. The accumulation of nutrients in the vegetation during forest fallow in Zaire (Congo) is shown in Table 18.3. The 18- to

TABLE 18.3 Nutrient Accumulation in Forest Fallow in the Zaire Basin

Age of Forest Fallow	Nutrients in Vegetation, kg/ha.				
	N	P	S	K	Ca + Mg
2 years	188	22	37	185	160
5 years	566	33	103	455	420
8 years	578	35	101	839	667
18–19 years	701	108	196	600	820

Adapted from C.E., Kellogg, "Shifting Cultivation," *Soil Sci.,* 95:221–230, 1963. Used by permission of Williams and Watkins Co.

19-year-old forest vegetation contained five times more nutrients than were contained in the 2-year-old forest fallow. Forests grow vigorously on Oxisols in the humid tropics because water is generally available and (1) plant nutrients are efficiently cycled and (2) deeply penetrating roots absorb nutrients from soil layers that are much less weathered and more fertile than the A and B horizons.

Land Use on Ustox Soils

The native vegetation of the savanna areas, where Ustox occur, is not suited for the type of slash-and-burn agriculture that occurs in the rain forests on Udox soils (see Figure 18.19). There are various kinds of fallow systems used. However, the opportunity to accumulate nutrients in plant material is much less in the savanna than in the rain forest. Farmers in these areas tend to be small subsistence growers who cultivate a small acreage permanently and give greater attention to the use of manures, plant residues, and fertilizers to maintain soil fertility. Periods of 6 months with little if any rainfall are common, and each family tends to farm a small acreage of wetland, (if it is available), where vegetables are harvested during the dry season. Many fruit trees are scattered

FIGURE 18.20 Land preparation of Ustox soils on the African savanna. Organic refuse, including any animal manure, is placed in the furrows between the ridges. Then, each ridge is split and the soil is pulled into the furrow to form a new ridge where the furrow previously existed. Ridged land increases water infiltration and reduces erosion from torrential summer storms.

throughout the savanna and provide some food with minimal care. Corn, beans, cowpeas, and cassava are important crops. Some farmers have large herds of cattle. A farmer is shown preparing land for cropping on Ustox in Africa in Figure 18.20.

Extremely Weathered Oxisols

The maximum effective cation exchange capacity (ECEC) of the clay fraction of oxic horizons is 12 or less cmol/kg. By contrast, the ECEC of the most weathered horizons in Oxisols is 1.5 or less cmol/kg. On this basis, soils with a horizon having an ECEC of 12.5 to 1.5 cmol/kg are considered to be intensively weathered and soils with a horizon having an ECEC of 1.5 or less cmol/kg are considered to be extremely weathered. The prefix *acr* is used to indicate this low ECEC as in the case of Acrudox, extremely weathered Udox. The horizon with acricity must be 18 or more centimeters thick.

FIGURE 18.19 Natural vegetation on savannas in Brazil, consisting of small trees, shrubs, and grass in an area dominated by Ustox soils.

The pH of soils is closely related to the kinds of exchangeable cations. Extremely weathered soils with very low ECEC have a small capacity to develop low pH values even when nearly 100 percent aluminum saturated and tend to have a pH that is greater than that for intensively weathered soils. Even though extremely weathered soils may be essentially 100 percent aluminum saturated, there may be insufficient Al^{3+} in solution to produce aluminum toxicity. Such soils have very small amounts of exchangeable calcium and magnesium. Lime is used in very small quantities to supply calcium and magnesium with minimal effect on soil pH.

Horizons in extremely weathered soils that contain very little organic matter sometimes have a net charge of zero or a net positive charge. Soils with a horizon 18 or more centimeters thick that has a net charge of zero or a net positive charge are considered *anionic*. Thus, an Acrudox with an anionic horizon is an Anionic Acrudox.

Plinthite or Laterite

A soil horizon associated with the tropics and many Oxisols, and also with Ultisols and Alfisols, is *plinthite*. It is an iron-rich mixture containing quartz and other dilutents that remains soft as long as it is not dried. On exposure at the soil surface, due to erosion, the material dries and hardens irreversibly with repeated wetting and drying. The hard, rocklike material is hardened plinthite and is commonly called ironstone or laterite.

Plinthite can, perhaps, form from the concentration of iron in *situ*, but it more likely forms via the influx of iron in moving groundwater. Plinthite formation is characteristically associated with a fluctuating water table in areas having a short dry season. Reduced iron (ferrous), produced in water-saturated soil, is much more mobile than oxidized iron (ferric) and moves in water and precipitates on contact with a good supply of oxygen. Plinthite normally forms as a subsurface layer that is saturated at some season, but it can

also form at the base of slopes where water seeps out at the soil surface. Plinthite is soft to the extent that it can be cut with a spade and readily mined. On exposure to the sun, the material hardens to a bricklike substance, which is used as a building material (see Figure 18.21).

There is great diversity in plinthite, ranging from continuous thick layers to thin, discontinuous nodular forms. Small amounts of nodular plinthite or deeply buried, continuous layers below the root zone have little influence on plant growth. Where continuous plinthite is exposed by erosion, drying and hardening can seriously limit plant growth. As long as the layer remains moist and soft, and is below the root zone, soils can be effectively used for a wide variety of crops. In some places holes are dug in hardened plinthite and then filled with soil to grow high-value crops. It is estimated that hardened plinthite, ironstone or laterite, occurs on 2 percent of the land in tropical America, 5 percent in Brazil, 7 percent in the tropical part of the Indian subcontinent, and 15 percent for Sub-Saharan West Africa.

SPODOSOLS

Spodosols contain spodic horizons in which amorphous illuvial materials, including organic matter and oxides of aluminum and iron, have accumulated. Quartzitic sand-textured parent material and a humid climate with intense leaching promote development of spodic horizons. As a result, Spodosols are widely distributed from the tropics to the tundra. Trees are the common vegetation.

Spodosol Suborders

In the United States, about 86 percent of the Spodosols are Orthods and the remainder are Aquods. Aquods have aquic moisture regime and are the dominant soils in the state of Florida, which has abundant sand-textured parent material, abundant rainfall, and shallow water tables

FIGURE 18.21 Mining plinthite in Orissa state, India. On drying, the bricks harden and are used for construction.

that are close to the soil surface at least part of the year (see the S1a areas for Florida in Figure 18.1). The two major kinds of soils in Florida are both very sandy and naturally infertile—Aquods and Psamments. The warm climate, together with intensive soil management, makes these soils productive for winter vegetables, citrus crops, and year-round pastures.

Orthods are the most common Spodosols; an Orthod is shown in Color Plate 5. They are abundant in the northeastern United States where they have developed in late Pleistocene sandy parent materials (see areas S2a in Figure 18.1). They occur in association with Alfisols in the Great Lake states where the Spodosols have developed on sandy parent materials and the Alfisols have developed on the loamy parent materials. The soil boundary line across lower Michigan and northern Wisconsin is, in essence, a parent material boundary. The boundary separates an area with loamy Alfisols to the south from an area to the north dominated by sandy Spodosols. This pattern is repeated across northern Europe and Asia, as shown in Figure 18.2.

Suborders of minor extent include Humods and Ferrods with a relatively high organic matter and iron content in the spodic horizons, respectively.

Spodosol Properties and Land Use

Spodosols typically have a very low clay content in all horizons; less than 5 percent for the soil horizons cited in Table 18.4. Sola are quite acid, with a pH less than 5 being very common. There is a small cation exchange capacity, ranging from 1.3 cmol/kg (meq/100 g) for the C horizon to 5.2 cmol/kg (meq/100 g) for the Bhs horizon. The cation exchange capacity of each horizon is closely related to the amount of amorphous material—greater in the spodic horizon (Bhs) than in the E or lower B and C horizons. There is a modest aluminum saturation of the cation exchange capacity which, coupled with a low cation exchange capacity, results in a very low quantity of exchangeable calcium, magnesium, and potassium (see Table 18.4). These properties contribute to soils that are too infertile for general agriculture and have a low water-holding capacity.

TABLE 18.4   Selected Properties of a Spodosol (Orthod)

Depth cm	Horizon	Clay %	pH	Ca	Mg	K	Na	Al	Total
								Exchangeable cations, cmol (+)/kg	
0–25	Ap	3.2	4.8	3.2	0.2	tr[a]	tr	1.1	4.5
25–36	E	3.3	4.1	1.9	0.2	tr	tr	1.0	3.1
36–53	Bhs	2.5	4.2	2.1	0.2	tr	tr	2.9	5.2
53–84	BC	1.7	4.6	0.5	0.1	tr	tr	1.7	2.3
84–105	C	1.3	4.7	0.5	0.1	tr	tr	0.7	1.3

[a] Trace amount. Selected data for pedon 25 in *Soil Taxonomy*, 1975.

The natural infertility and droughtiness of Spodosols has been cited as an example of a case where virgin soils are not necessarily fertile and suitable for agriculture. The following quote is from an article that appeared in *Time* magazine in 1954:

The virgin soil under a long-established forest is not always good. When the settlers cleared New England forests 300 years ago, the topsoil they found was only 2 to 3 inches thick. Below this was sterile subsoil and when the plow mixed the two together, the blend was low in nearly everything a good soil should have. It was not the lavish virgin soil of popular fancy. Such a soil could not sup- *port extractive agriculture which takes nutrients out of the soil and does not replace them. Many New England lands that were treated in this way soon went back to forest.*

In 1870, the widespread conversion of forest land in New Hampshire peaked to about 75 percent improved farmland and then rapidly decreased to only about 10 percent improved farmland by 1930, as shown in Figure 18.22. As settlement moved westward, the plow followed the cutting of the forest and a similar land-use pattern also occurred on the Spodosols in the Great Lake states. The first bulletin published by the Michigan Agricultural Experiment Station in 1885 was devoted to solving problems on the *sand plains* of north-central Michigan. The abandonment of land for agricultural use where soils are Spodosols is shown in Figure 18.23.

The cool summers of the northern Spodosol region attract many tourists. Many cities in this area owe their existence to mining and lumbering. Although agriculture is of minor importance, there are localized areas of intense fruit, vegetable, and dairy farming. This farming occurs on two kinds of soils. In one kind the agriculture occurs on inclusions of Alfisols within the Spodosol region. These are situations where the parent materials are loam-textured and the Alfisols are Boralfs. In the other kind, the soils have a sandy upper Spodosol solum that overlies a horizon of increased clay content—a Bt (argillic) horizon. These are bisequum soils that contain the E

FIGURE 18.22   Percent of improved farmland in New Hampshire from 1790 to 1970. (Courtesy Steven Hamburg, Yale University.)

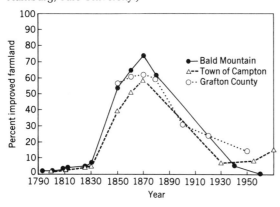

FIGURE 18.23 These disintegrating buildings tell the story of forest removal, farming, and then abandonment of land dominated by Spodosols in the Great Lake States.

and Bhs horizons of a Spodosol that typically overlie some E and the argillic horizon of an Alfisol. These are the dominant soils in the well-known potato-growing area in Aroostook County, Maine and in adjoining New Brunswick, Canada.

ULTISOLS

The central concept of Ultisols is *ultimately* leached soils of the mid- to low latitudes that have a kandic or an argillic horizon but few weatherable minerals. In parent materials with almost no weatherable minerals, a Ultisol may develop in a few thousand years. Perhaps, 100,000 years or more are required if the Ultisol develops from the weathering and leaching of a former Alfisol. Ultisols are the dominant soils of the Piedmont and coastal plain of the southeastern United States beyond the area of glaciation and young parent materials. The large Ultisol region extends from Washington, D.C., south and westward into eastern Texas, as shown by the U3a areas in Figure 18.1. Ultisols are the fourth most abundant soils in the United States (see Table 18.1) and the fourth most abundant soils in the tropics, following Oxisols, Aridisols, and Alfisols.

Ultisol Suborders

Aquults have an aquic moisture regime and occur extensively along the lower coastal plain of the southeastern United States. The water tables are naturally close to the soil surface at some time during the year. Further inland, at higher elevations on the middle and upper coastal plains, the Ultisols are mainly Udults. Contrast areas U1a (Aquults) with U3a (Udults) in Figure 18.1. Humults are found in the coastal ranges of northern California, Oregon, and Washington (area U2s in Figure 18.1). Humults have more than the usual amount of organic matter and develop under conditions of high precipitation and cool temperatures, associated with mountain slopes.

A large region of mostly Udult soils is found in southeastern China at a latitude and continental position similar to that of the southeastern United States (see U3d in Figure 18.2). Many smaller areas of Ultisols occur in Africa and South America, usually in association with Oxisols.

Properties of Ultisols

A common horizon sequence for many Ultisols is A, E, Bt and C. Ultisols that develop in iron-rich parent materials may have intense red colors and lack an obvious E horizon. A photograph of a Udult is shown in Color Plate 5. The symbol, v, as in Btv, indicates plinthite, which in this instance is not continuous and has mixed red and yellow color (see the Ultisol in Color Plate 5).

In the United States, 76 percent of the Ultisols are Udults. The horizons of Udults are quite acid to a great depth and they contain a very small amount of exchangeable calcium, magnesium, potassium, and sodium. The very low amounts of these exchangeable cations reflect the low content of weatherable minerals—less than 10 percent in the coarse silt and fine sand fractions. Leaching removes cations from the soil more rapidly than they are released by weathering. Ultisols are required to have less than 35 percent of the cation exchange capacity, which is determined at pH 8.2, saturated with calcium, magnesium, potassium, and sodium in the lower part of the root zone (at a depth of about 125 centimeters below the upper boundary of the argillic horizon or 180 centimeters below the soil surface). By contrast, the exchangeable aluminum is high and roots may grow poorly in subsoils because of an aluminum toxicity. Many subsoils contain too little calcium for good root growth. Most of the plant nutrient cations are bound within the plants and/or soil organic matter. Ultisols are naturally too infertile for the production of many agricultural crops because of their acidity, low content of exchangeable basic cations, and the potential for aluminum toxicity.

Land Use on Ultisols

During the early settlement of United States, agriculture on Ultisols in Virginia and the surrounding states consisted mainly of tobacco and cotton production for the European market. After clearing the forest and cropping for only a few years, the naturally infertile soils became exhausted. Abandonment of land and a shifting type of cultivation became widespread as the frontier moved westward. This was the plight of early Virginians. One farmer in particular, Edmund Ruffin, described the situation as follows:

Virginians by the thousands, seeing only a dismal future at home, emigrated to the newer states of the South and Middle West. The population growth dropped from 38 percent in 1820 to less than 14 percent in 1830, and then to a mere two percent in 1840. All wished to sell, none to buy.

So poor and exhausted were Ruffin's lands and those of his neighbors that they averaged only 10 bushels of corn per acre and even the better lands a mere 6 bushels of wheat. Ruffin happened to read *Elements of Agricultural Chemistry* by Sir Humphrey Davy, from which he learned that acidity was making the soils sterile, and liming was needed to neutralize soil acidity. The application of marl to the land was very successful. Now, it is known that the marl (or liming) added calcium and magnesium and reduced aluminum toxicity. Liming and the increase in soil pH made the application of manure and use of legumes in rotations more successful. For this discovery, Ruffin was credited with bringing about an agricultural revival in the South. With the long, year-round growing season and abundant rainfall, this region that once proved so troublesome is now one of the most productive agricultural and forestry regions in the United States (see Figure 18.24).

Land use on Ultisols in the tropics is quite similar to land use on nearby Oxisols. Slash-and-burn and subsistence agriculture are common. Many plantations or estates produce tea, rubber, bananas, tobacco, and other valuable crops.

VERTISOLS

Vertisols are mineral soils that are more than 50 centimeters thick, contain 30 percent or more clay in all horizons, and have cracks at least 1 centimeter wide to a depth of 50 centimeters, unless irrigated at some time, in most years. The typical vegetation in natural areas is grass or herbaceous annuals, although some Vertisols support drought-tolerant woody plants.

Worldwide, and in United States, Vertisols are the ninth most abundant soils. The three largest areas are in Australia (70 million acres), India (60 million acres), and the Sudan (40 million acres). These soils are mainly Usterts, and are shown as areas of V2a and V2c in Figure 18.2.

Vertisols in the United States have formed mainly in marine clay sediments on the coastal plains of Texas, Alabama, and Mississippi. Very

FIGURE 18.24 Udults on the coastal plain in the southeastern United States have nearly level slopes and sandy surface soils that are easily tilled.

sharp boundaries between soils of different orders may occur in the area because Vertisols developed on clay sediments, whereas Ultisols developed on adjacent sandy sediments.

Vertisol Suborders

Approximately 40 percent of the Vertisols in United States are Uderts and 60 percent are Usterts. Usterts are sufficiently dry so that cracks remain open more than 90 cumulative days in a year. There are two areas in Texas where Usterts are the dominant soils (areas V2a in Figure 18.1). Uderts have a wetter climate than do Usterts and have cracks that are open 90 or fewer cumulative days in a year. A large area of Uderts occurs along the Gulf Coast of Texas (V1a in Figure 18.1). A smaller area occurs along the border of Alabama and Mississippi. Torrerts and Xererts occur to a lesser extent.

Vertisol Genesis

Conditions that give rise to the formation of Vertisols are: (1) parent materials high in, or that weather to form, a large amount of expanding clay, usually smectites; and (2) a climate with alternating wet and dry seasons. After the cracks develop in the dry season, surface soil material and material along the crack faces, fall to the bottom of the cracks. The soil rewets at the beginning of the rainy season from water that quickly runs into and to the bottom of the cracks, which causes the soil to wet from the bottom up during the early part of the wet season. After the cracks at the surface close, the soil wets normally from the surface downward. This results in a *dry* soil layer sandwiched between two layers that are being wetted and undergoing expansion. Thus, there is considerable moving and sliding of adjacent soil masses and the formation of shiny ped surfaces—*slickensides*. A large slickenside is shown near the AC label of the Vertisol in Color Plate 5.

Expansion or swelling of soil in the vicinity of cracks results in great lateral and upward pressures, which cause a slow gradual movement of soil upward between areas that crack. This soil movement produces a microrelief called *gilgai* (see Figure 18.25). The major features of Vertisols are illustrated in Figure 18.26.

FIGURE 18.25   Gilgai microrelief on Vertisols in Texas. (Photograph Soil Conservation Service, USDA.)

Vertisol Properties

The high content of swelling clay and movement of soil by expansion and contraction retard the development of B horizons. Typically, the soil remains quite undeveloped and often has an A-C profile. Some properties of the Houston Black clay from the Blackland Prairie region of Texas are presented in Table 18.5. There is a high and consistent clay content in each horizon. Organic matter content is typical for soils with high clay content, but the organic matter decreases gradually with increasing soil depth. This pattern of organic matter content is similar to that which occurs in many soils; in Vertisols, the change in organic matter content may be associated with no change in color (see Vertisol in Color Plate 5). This Vertisol contains some $CaCO_3$ and has a pH of about 8.3 in all horizons. Some Vertisols, Uderts, are sufficiently leached to be acid in some horizons.

The soil has high fertility for many agricultural

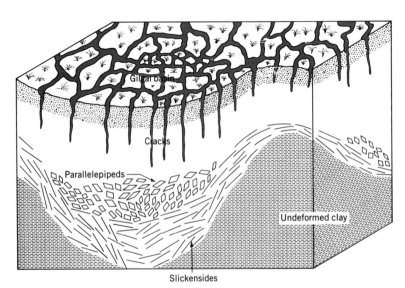

FIGURE 18.26   Schematic drawing showing a soil profile of Vertisols. Major features include deep cracks, parallelepiped structure (rhombohedral aggregates), slickensides, and gilgai microrelief basin. (Adapted from Dudal and Eswaran, 1988.)

TABLE 18.5 Some Properties of Houston Black Clay—a Vertisol

Horizon	Depth, cm	Clay %	Organic Matter, %	CaCO₃ %	Cation Exchange Capacity cmol (+)/kg
A1	0–46	58	4.1	17	64
A2	46–100	58	2.1	20	58
AC	100–152	58	1.0	26	53
C	152–198	59	0.4	32	47

Table is based on data of Kunze and Templin, 1956. Average of 5 profiles.

crops and the soils can retain a large amount of water. Conversely, when the soils are wet the infiltration of water is essentially zero, which results in excessive water runoff and soil erosion on sloping cultivated land.

Land Use on Vertisols

It is difficult for rainwater to move through Vertisols and contribute to a groundwater aquifer. Those who settled on Vertisols had difficulty drilling wells and obtaining water. For this reason, and because many Vertisols occur in regions of limited rainfall where moisture regimes are ustic, many Vertisols have been minimally or only moderately leached. Thus, Vertisols are considered fertile for many agricultural crops (see Figure 18.27).

Most soil management problems are the results of the high content of expanding clay and accompanying physical properties. In Texas, the Vertisols were referred to as "dinner-pail-land," being too wet and sticky to till before the noontime meal (dinner) and too dry and hard to till after the noontime meal.

Farmers in India found the Usterts too hard to till with their small plows and oxen at the end of a long dry season. They were unable to plant crops before the monsoon rains started, which meant that crops were not produced during the season when water is most available. Then, crops were planted after the moonsoon rains stopped, and

the crops depended primarily on stored soil water. Some creative research done at the International Crops Research Institute for the Semiarid Tropics (ICRISAT) in Hyderabad, India, resulted in recommendations about some different tillage methods. Tillage was to be done immediately after harvest of the winter crops before the soils became too dry to till. Some occasional tillage during the dry season was also recommended. Then, farmers could establish the crop in loose, dry soil before the monsoon rains began. The

FIGURE 18.27 Post-monsoon crop of cotton on Usterts in India. Cotton is a major crop, which is why the Usterts are called Black-Cotton soils. Note the many slickensides surfaces and black color to a depth of more than a meter.

FIGURE 18.28 Structural failure caused by landslide on Vertisols produced by wet soil sliding downhill. Note that pipes are on top of the ground to prevent breaking caused by expansion and contraction of soil.

result was high-yielding crops during the monsoon season (in addition to the post-monsoon crops), and more than 100 percent more total production per year as compared with the traditional system.

Open cracks are a hazard for grazing animals during dry seasons. The expansion and contraction of soil cause a misalignment of fence and telephone posts, breaks pipelines, and destroys road and building foundations as shown in Figure 18.28.

SUMMARY

Histosols are soils formed from organic soil materials. Typically, they are water saturated for part of the year. Drainage is required for cropping, but subsequently, decomposition of the organic matter results in a gradual disappearence of the soil and subsidence of the soil surface.

Andisols, Spodosols and Vertisols owe their existence, largely to the parent material factor of soil formation. Andisols develop in volcanic ejecta. Spodosols develop from sand parent material in humid regions and are droughty, acid, and infertile. Vertisols develop in clay materials that expand greatly with wetting and shrink with drying. They are frequently minimally weathered and leached, and are fertile soils for cropping. Vertisols have unique use problems associated with their clayey nature.

Aridisols, Entisols, and Inceptisols typically are minimally weathered soils that tend to be rich in primary minerals and to be fertile for cropping. Some Inceptisols in tropical regions, however, have developed in intensively weathered parent material and are acid and infertile.

Moderately weathered and generally fertile soils include Alfisols and Mollisols. These soils are common in the temperate regions and are generally good for agricultural use.

Ultisols and Oxisols are intensively weathered soils. They are generally acid and infertile, and aluminum is toxic for many plants. Many are used for slash-and-burn agriculture and subsistence farming. Intensive management can result in productive soils with high crop yields.

REFERENCES

Buol, S. W. 1966. "Soils of Arizona." *Ariz. Agr. Exp. Sta. Tech. Bull*. 117, Tuscon.

Dudal, R. and H. Eswaran. 1988. "Distribution, Properties, and Classification of Vertisols," in *Vertisols: Their Distribution, Properties, Classification, and Management*. L. P. Wilding and R. Puentes (eds.) Texas A & M University Printing Center, College Station.

Foth, H. D. and J. W. Schafer. 1980. *Soil Geography and Land Use*. Wiley, New York.

Kanwar, J. S., J. Kampen, and S. M. Virmani. 1982. "Management of Vertisols for Maximizing Crop Production-ICRISAT Experience." *Vertisols and Rice Soils of The Tropics, Symposia Papers II*. pp. 94–118. 12th Int. Congress Soil Sci., New Delhi.

Kedzie, R. C. 1885. "Early Amber Cane as a Forage Crop." *Agr. College of Michigan Bull*. 1. East Lansing.

Kellogg. C. E. 1963. "Shifting Cultivation." *Soil Sci*. 95:221–230.

Kunze, G. W. and E. H. Templin. 1956. "Houston Black Clay, the Type Grumusol: II Mineralogical and Chemical Characterization." *Soil Sci. Soc. Am. Proc*. 20: 91–96.

Ruffin, Edmund. 1961. *An Essay on Calcareous Manures*. Cambridge, Mass. (Reprint of original book that was published in 1832).

Sanchez, P. A. 1976. *Properties and Management of Soils in the Tropics*. Wiley, New York.

Soil Survey Staff. 1975. "Soil Taxonomy." *USDA Agr. Handbook* 436. Washington, D.C.

Soil Survey Staff. 1987. "Keys to Soil Taxonomy." *Soil Management Support Services Tech. Monograph* 6. Cornell University, Ithaca, New York.

Swanson, C. L. W. 1954. "The Road to Fertility." *Time*, January 18.

Tuan, Yi-Fu. 1969. *China*. Aldine, Chicago.

Young, A. 1976. *Tropical Soils and Soil Survey*. Cambridge, London.

CHAPTER 19

SOIL SURVEYS AND LAND USE INTERPRETATIONS

A *soil survey* is the systematic examination, description, classification, and mapping of soils in an area. The kinds of soil in the survey area are identified and their extent is shown on a map. The soil map, together with an accompanying report that describes, defines, classifies, and makes interpretations for various uses of the soils, is published as a Soil Survey Report.

MAKING A SOIL SURVEY

A soil survey requires making soil maps of the area and writing an informational and interpretative report. In many states, a soil survey is a cooperative activity between the U. S. Department of Agriculture and state Agricultural Experiment Stations. The size of the area surveyed and the intensity of mapping depend on the amount of detail required to make the necessary land-use interpretations.

Making a Soil Map

Soil maps are made by scientists who are familiar with the local soil-forming factors, an understanding of soil genesis, and the ability to identify the soils. An aerial photograph is used as the map base. The soil surveyor walks over the land and simultaneously studies the aerial photograph and the landscape. This includes observations of ground cover or vegetation, slope, and landforms. In a glacial landscape, outwash plains will have a nearly level surface and typically contain considerable sand and gravel. Lake beds are nearly level with fine-textured soils. Ground moraines have an irregular surface with much variation in slope and texture. These three different landforms will likely have different plant species in the native vegetation or, if in a cultivated field, have growth differences of crop plants. These three different situations (outwash plain, lacustrine plain, ground moraine)in a field will result in at least three different kinds of soil. At selected locations, the soil surveyor uses a soil auger to bore a hole and observe the sequential horizons of the soil profile as the bore hole samples of horizons are extracted (see Figure 19.1).

Observations of each horizon include: thickness, color, texture, structure, pH, presence or absence of carbonates, and so on. In addition, the slope and degree of erosion are noted. This information is then recorded directly on the aerial photograph, and lines are drawn to separate different

FIGURE 19.1 A soil surveyor indentifies the soil by inspecting the soil horizons. Lines are then drawn around areas of similar soil on an aerial photograph to produce a soil map. (Photograph courtesy USDA.)

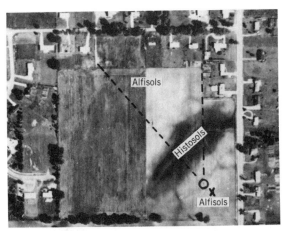

FIGURE 19.2 Aerial photograph of a field in Ingham County, Michigan, where soils are Alfisols and Histosols. Standing at point X and looking north a person would observe a black area of Histosols sandwiched between areas of light-colored soils that are Alfisols.

soil areas (mapping units) to create the soil map. Other information is also recorded, such as location of gravel pits, farmsteads, cemeteries, dams, and lakes.

Areas of different kinds of soil typically appear on the aerial photograph as areas with different intensity of black and white, as shown in Figure 19.2. A large black area of Histosols is sandwiched between two light-colored areas of Alfisols in a recently cultivated field. A soil surveyor standing at point X in Figure 19.2 would see the landscape in late summer after grain harvest, as shown in Figure 19.3. It is obvious in Figure 19.3 that at point X (Figures 19.2 and 19.3) the soil is on a slope. The soils near X are well-drained mineral soils that have been subjected to erosion. These soils have a grayish-brown surface color and appear as a light-colored area on the aerial photograph. These soils are Alfisols. The slope ends in a depression in which the soils are very poorly drained. These black-colored soils are or-

ganic soils (Histosols) and appear as a dark area on the aerial photograph. Beyond the depression, the soils are again well-drained Alfisols on a slope similar to that at point X. From observations of the aerial photograph and landscape, the soil surveyor constructs a mental image of what kind of soil is located in each part of the landscape. Then, at selected locations, soil borings are made to verify and identify the soil. The result is that the soil surveyor does not randomly bore holes and make observations, but instead uses the auger to confirm the hypotheses. This results in the boring of holes at strategic locations and the construction of the soil map with minimum physical input.

The soil map made for the recently cultivated field (without a vegetative cover in Figure 19.2), is shown in Figure 19.4. The symbols on the map refer to the mapping units. Typically, the mapping unit contains three symbols such as 3B1—3 for soil, B for slope, and 1 for erosion. In this case it means: 3 for Marlette sandy loam, B for a 2- to 6-percent slope, and 1 for slightly eroded. Some soils are always located on level or nearly level areas that are not eroded, and the slope and ero-

FIGURE 19.3 On-the-ground view of the landscape from X of Figure 19.2 and looking north-northwest. The soil differences shown on the aerial photograph are also apparent after wheat harvest. The dark-colored area of Histosols in the depression area can be seen to be sandwiched between the Alfisol areas that are sloping.

FIGURE 19.4 Soil map of cultivated filed shown in Figure 19.2.

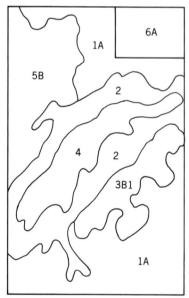

1A Capac loam, 0–3 percent slopes
2   Colwood-Brookston loams
3B1 Marlette sandy loam, 2–6 percent slopes
4   Palms muck
5B Spinks lomy sand, 0–6 percent slopes
6A Urban land, Capac-Colwood complex, 0–4 percent slopes

sion notations are omitted. The mapping symbols for the soil map in Figure 19.4 are as follows:

1A Capac loam, 0 to 3 percent slopes
2   Colwood-Brookston loam
3B1 Marlette sandy loam, 2- to 6-percent slope, slight erosion
4   Palms muck
5B Spinks loamy sand, 0- to 6-percent slope
6A Urban land, Capac-Colwood complex, 0- to 4-percent slope

The mapping symbol at point X in Figures 19.2 and 19.3—3B1—is for the Marlette soil, which is a Udalf. In the center of the depressional area, the soil is Palms muck (4)—a Saprist. These two soils have very different properties, including slope, erosion, and drainage characteristics and would demand very different management regardless of whether they are used for agriculture, forestry, or for a home-building site.

Writing the Soil Survey Report

Typically, soil survey reports are for a county, parish, or another governmental unit. The soil maps usually form the latter part of the report. Anyone desiring information on a specific parcel of land can refer to the appropriate soil map sheet and determine the kinds of soils. The first part of the report is written and consists of information about the genesis and classification of the soils, descriptions and properties of each soil, and interpretative tables for various soil uses.

Using the Soil Survey Report

Imagine that you are driving down the road along the eastern side of the cultivated field shown in Figure 14.2, because you are looking for a site on which to build a house in the country. Would you be able to find a suitable parcel of land in that field to use as a home site? How could the soil survey report help you in making your decision?

To find the answers to these questions you would need the Soil Survey Report for Ingham County, Michigan. First, locate in the soil survey report the appropriate soil map in the back section and identify the soils within the parcel of land that interests you. Then, refer to the descriptions

of the soils for the mapping units in the front portion of the report to learn about the general features of these soils. This includes the kinds of horizons and their properties and the conditions under which the soils developed. For example, the 4 mapping unit includes Palms muck, which consists of very poorly drained organic soil about 36 inches thick that overlies moderately coarse to fine-textured deposits.

Factors that are important for building sites include natural drainage, depth to water table, septic-field suitability (where municipal sewage disposal is not available), supporting ability, shrink-swell potential, slope, and lawn and landscaping suitability. Some hazards and suitabilities of the various mapping units, relative to constructing a building with a basement, are given in Table 19.1. The Palms soils (4) have severe problems related to flooding and wetness. The water table is at or within a foot of the soil surface from November to May. These conditions make the use of a septic filter field for sewage effluent disposal impossible because of the high water table. A dry

TABLE 19.1 Land-Use Interpretations

| Mapping Unit | Suitability and/or Hazard | | | Water Table | |
	Dwelling with Basement	Lawn and Landscaping	Septic Field	Depth, Feet	Months
1A	Severe: wetness	Moderate: wetness	Severe: wetness, percs slowly	1–2	November to May
2	Severe: floods, wetness	Severe: wetness, floods	Severe: wetness, floods	0–1	October
3B1	Moderate: low strength, wetness	Slight:	Severe: wetness, percs slowly	2.5–6	December to April
4	Severe: wetness, low strength, floods	Severe: wetness, floods, excess humus	Severe: wetness, floods, subsides	0–1	November to May
5B	Slight	Moderate: too sandy	Slight	>6	—

Data from Soil Survey Report of Ingham County, Michigan.

basement would be a virtual impossibility. Driveways and sidewalks, without footings or support on the mineral soil under the muck, would subside as the muck dries out and decomposes. Palms soil is obviously poorly suited for a building site.

From Table 19.1, the soil with the best overall suitability is 5B (Spinks loamy sand). It is a well-drained and moderately to rapidly permeable soil. The water table is below 6 feet throughout the year, and the soil has sufficient water permeability for the septic-filter field and only moderate limitations for lawn and landscaping. Spinks soils would also have a low shrink-swell potential.

A precaution is in order. Soils do not occur as uniform bodies. Mapping units typically contain 15 percent or more inclusions. That is, an area labeled 5B on a soil map could contain a significant amount of similar but different soils. Thus, to establish whether or not a septic-sewage disposal system will operate successfully in a particular soil mapping area, a percolation test is needed for that small area within the mapping unit where the filter field will be located.

SOIL SURVEY INTERPRETATIONS AND LAND USE PLANNING

When a soil survey is made, there are specific needs to be fullfilled by the survey. The writers of the soil survey report include many tables of soil properties and characteristics, together with tables that contain predictions of soil behavior for many uses. We have used this type of soil survey information in regard to selection of a building site for a house in the country. Estimates of crop yields and rates of forest growth are useful for prospective land buyers and land tax assessors. Highway planners need information about flooding and wetness conditions and the ability of soils to support roads. Landfills for waste disposal require underlying layers that are impermeable to water (see Figure 19.5). Thus, land-use interpretation maps are needed to satisfy the requirements of various land users.

FIGURE 19.5 Sites selected for solid waste disposal (landfills) need water-impermeable underlying strata to minimize groundwater pollution.

Examples of Interpretative Land-Use Maps

All soils that are predicted to perform similarly for a particular use can be grouped together, and an interpretative land-use map can be constructed.

FIGURE 19.6 Soil suitablility map for buildings with basements of the recently cultivated field shown in Figure 19.2 (also in Figure 19.4).

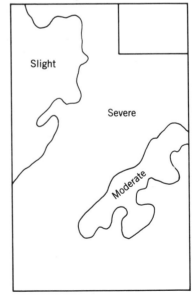

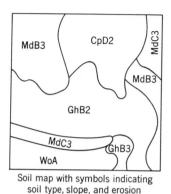

Soil map with symbols indicating soil type, slope, and erosion

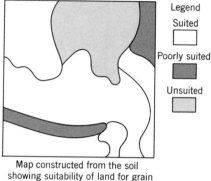

Map constructed from the soil showing suitability of land for grain and seed crops for wildlife management

Legend

Suited

Poorly suited

Unsuited

FIGURE 19.7 Soil survey map on the left was used to construct a wildlife suitability map on the right for a 40-acre tract of land. (Based on information from Allan, et al., 1963.)

The suitability map for construction of buildings with basements in Figure 19.6 is based on the soil map of Figure 19.4 and the interpretations given in Table 19.1.

Soils directly affect the kind and amount of vegetation available to wildlife as food and cover, and the soils affect the construction of water impoundments. The kind and abundance of wildlife that populate an area depend largely on the amount and distribution of food, cover, and water. Figure 19.7 shows a soil map and an accompanying suitability map for the production of grain and seed crops for wildlife use. The requirements for grain and seed crops are much like those of most agricultural crops. Similar maps could be constructed for suitability of land for wetland habitat for wildfowl, open-land habitat for pheasant and bobwhite quail, or woodland habitat for ruffed grouse, squirrel, or deer.

Land Capability Class Maps

The Soil Conservation Service of the U.S. Department of Agriculture developed eight land-use capability classes based on limitations due to erosion, wetness, rooting zone, and climate. The limitations and restrictions for these land use capability classes are given in Table 19.2. Classes I through IV apply to land suited for agriculture and classes V through VIII to land generally not suited to cultivation. Class I land has few use restrictions

in terms of erosion, wetness, rooting depth, and climate. Class VIII land, by contrast, precludes normal agricultural production and restricts use to recreation, wildlife, water supply, or esthetic purposes. A land-use capability class map for a 200-acre dairy farm is Wisconsin in shown in Figure 19.8.

Computers and Soil Survey Interpretations

One of the major problems with the use of Soil Survey Reports is the large amount of available information and the difficulty of locating the specific material needed by a particular user. In Minnesota the most frequent nonfarm users wanted soil survey information to determine land values for purchase, sale, rent, or as collateral for loans by financial institutions. Other nonfarm uses were made by local government officials, zoning administrators, sanitarians, and tax assessors. Many persons make frequent use of the Soil Survey Report. To increase the usefulness and the speed with which information can be extracted from Soil Survey Reports, computers are being used to develop digitized records (files) of soil maps and soil data bases. Computers help to solve the problem of locating information by locating only the particular part of the record that the user needs. Another big advantage of computerized soil survey informtion is that the soils' data bases can be

TABLE 19.2 Land Capabilty Classes

Land Suited to Cultivation and Other Uses

Class I.   Soils have few limitations that restrict their uses.

Class II.   Soils have some limitations that reduce the choice of plants or require moderate conservation practices.

Class III. Soils have severe limitations that reduce the choice of plants, require special conservation practices, or both.

Class IV.   Soils have very severe limitations that restrict the choice of plants, require very careful management, or both.

Land Limited in Use—Generally Not Suited for Cultivation

Class V.   Soils have little or no erosion hazard, but have other limitations impractical to remove that limit their use largely to pasture, range, woodland, or wildlife food and cover.

Class VI.   Soils have severe limitations that make them generally unsuited to cultivation and limit their use largely to pasture or range, woodland, or wildlife food and cover.

Class VII. Soils have very severe limitations that make them unsuited to cultivation and that restrict their use largely to grazing, woodland, or wildlife.

Class VIII. Soils and landforms have limitations that preclude their use for commercial plant production and restrict their use to recreation, wildlife, or water supply, or to esthetic purposes.

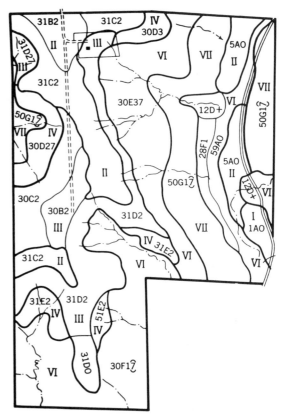

FIGURE 19.8   Land capability map for a 200-acre dairy farm in Wisconsin. Roman numerals designate land capability classes; other symbols indicate land characteristics. For example, 30E37, 30 designates type of soil, E indicates slope, and 37 means that more than 75 percent of the topsoil has been lost and that there are occasional gullies. (Courtesy USDA, Soil conservation Service.)

combined with systems to produce a Geographic Information System (GIS). Within these computer files information is stored, such as land ownership, elevation-topographic information, vegetation, and land use. Systems are being developed cooperatively with university and government personnel so that natural resources (soil, water, wetland, etc.) can be better protected through improved land-use planning.

SOIL SURVEYS AND AGROTECHNOLGY TRANSFER

Field experiments are conducted to test hypotheses for increasing the amount and efficiency of agricultural production. Suppose an experiment indicated that a 3-ton application of lime was the most profitable rate of application for production of alfalfa in a particular crop rotation system. The question that arises is: Where can this information be used or applied? If the experiment was conducted in the state of Arkansas, where in Arkansas are the results valid? Can the experimental results be used outside of Arkansas, and if so, where?

The transfer of this kind of information is valid only for locations with very similar soils, climate, and soil management practices. Thus, it is very important to (1) establish field experiments on the kinds of soils or mapping units for which interpretations or recommendations are needed, and (2) restrict the recommendations or predictions to those soils or very similar soils and experimental conditions. Because experiments cannot be conducted for every mapping unit or soil, it becomes necessary to extrapolate and interpret the information for situations other than those represented by the experiments.

SUMMARY

A soil survey is a systematic examination and mapping of the soils of an area. The soil map, together with information about the soils and land-use interpretations, is published as a Soil Survey Report.

Soil Survey Reports are used by a wide variety of persons, including farmers, governmental officals, land appraisers and tax assessors, land-use and zoning personnel, and ordinary citizens.

Soil survey information and interpretations are being computerized to facilitate access by users and to allow for better planning by various governmental agencies.

Because research information can only be obtained on a limited number of soils, soils that are expected to perform similarly must be grouped together for making interpretations.

REFERENCES

Allan, P. F., L. E. Garland, and R. F. Dugan. 1963. "Rating Northeastern Soils for Their Suitability for Wildlife Habitat." *Trans. 28th Nor. Am. Wildlife and Nat. Res. Con.*

Anderson, J. 1981. "A News Letter for Minnesota Cooperative Soil Survey." *Soil Survey Scene* **3,** (4):1–3. University of Minnesota, St. Paul.

Bartelli, L. J., A. A. Klingbiel, J. V. Baird, and M.K.R. Heddleson. 1966. *Soil Surveys and Land Use Planning.* Soil Sci. Soc. Am., Madison, Wis.

Soil Conservation Service, 1961. Land-Capability Classification. USDA, Washington, D.C.

Soil Conservation Service. 1981. *Soil Survey Manual.* 430 V-SSM, USDA, Washington, D.C.

CHAPTER 20

LAND AND THE WORLD FOOD SUPPLY

Since the beginning of time, increased food production was mainly the result of using more land. Now, new cropland supplies are diminishing as the human population continues to increase. This chapter considers the adequacy of the world's land, or soil, resources for the future.

POPULATION AND FOOD TRENDS

Human population growth and food supply are ancient problems. For several million years, humans migrated onto new lands and slowly improved their hunting and gathering techniques. The population increased slowly; about 10,000 years ago there were an estimated 5 million people in the world. At this time, people had just begun to colonize the last remaining continents, the Americas, by crossing over the land bridge between Asia and Alaska during the last ice age.

Development of Agriculture
Settled agriculture was just getting started in the Fertile Crescent of the Middle East when the migration to the Americas began. Agriculture may have developed at the same time in other isolated places in the world. It is uncertain why early people gave up the more leisurely hunting and gathering activities and took up a more rigorous way of living to produce crops and animals. Development of agriculture resulted in both an increased food supply and a population explosion. After reaching a population of 5 million over several million years, the human population increased 16 times to 86 million within the next 4,000 years (10,000 to 6,000 years before present). Population growth tended to level off, and the world population increased only six to seven times to 545 million over the next 6,000 years. This was followed by the Industrial Revolution that began about 1650.

The Industrial Revolution
The Industrial Revolution brought energy and machines into the food production process, resulting in an increased food supply. However, a new dimension was added to population growth—a reduced death rate because of the development of medicine and public sanitation. Now, the world population increases annually at the rate of 1.7 percent and doubles every 40 years. The most rapid rate of increase is in Africa, where the an-

nual increase is 2.9 percent, with a doubling every 24 years. Western Europe has a growth rate of 0.2 percent and a population doubling about every 400 years. The slowest population growth is in Europe, where some countries have negative population growth rates.

Estimates of world population exceeding 6 billion by the year 2000 will require about 40 percent more food just to maintain the status quo. Considering that more than 500 million people are currently suffering from malnutrition, and increased food consumption rates are occurring in the richest parts of the world, it appears that food needs, shortly beyond the year 2000, will be double today's needs. It is predicted that there will be a world population of 12 billion or more by the twenty-second century.

Recent Trends in Food Production

As recently as the period from 1950 to 1975, there was a 98 percent increase in world cereal production and a 22 percent increase in cropland. In recent decades, during a period of rapid population growth, world food production was able to keep pace with population growth. In fact, the rate of food production has exceeded the population growth rate. For the period from 1971 to 1980, both the developing and developed countries increased food production faster than population, as shown in Figure 20.1. For Africa, however, population increased faster than food production, which caused a 14 percent decrease in per capita food production during the 1970's.

There has been a declining per capita grain production trend for the past 20 years in Africa

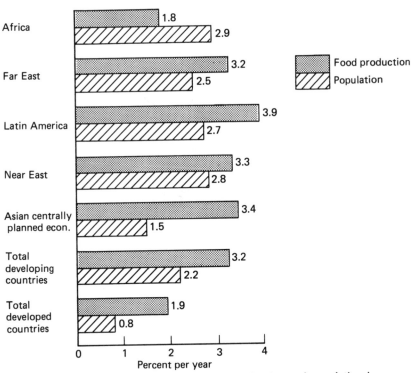

FIGURE 20.1 Annual rates of change of food production and population in developing and developed countries and regions, 1971–1980. (Data from the State of Food and Agriculture 1980, FAO, 1981.)

and in the Andean countries of Bolivia, Chile, Ecuador, and Peru. Although total world food production is increasing, the rate of increase is slowing. Since the late 1970s, there appears to be a declining trend in world per capita food production and, now, 50 to 60 nations have declining per capita food production, as shown in Figure 20.2.

Recent Trends in Per Capita Cropland

Expansion onto new lands has been the traditional means to increase the food supply. One reason for the present declining world per capita food production is the slower rate of bringing new land into production. The rate of cropland increase has declined from an annual increase of 1.0 percent in the 1950s to 0.3 percent in the 1970s, and is expected to be 0.15 percent in the 1990s. New settlement projects are underway in

the Amazon Basin and in the central uplands of Brazil and in the outer islands of Indonesia. Conversely, the area of cropland peaked in the 1950s and 1960s in many Western nations, with declines from a maximum in 1980 of 29.4 percent for Ireland, 21 percent for Sweden, and 19.6 percent for Japan.

About 2 billion people will be added to the world in another two or three decades, which means that a great deal of land will be needed for nonagricultural use. When new families are formed in many developing nations, houses are built on cropland with fields immediately adjacent to them. The use of local building materials (such as bamboo) make multistoried housing impractical. Much land will also be used for roads, strip mining, and transportation facilities, as shown in Figure 20.3.

At the 1978 International Soil Science Congress

FIGURE 20.2 The geographic distribution of declining annual rates of change in per capita food production, 1971–1980. (Data from *The State of Food and Agriculture,* 1980, FAO, 1981.)

FIGURE 20.3 Farmland in the United States is converted to urban use at the annual rate of about 400,000 hectares (1 million acres).

held in Edmonton, Canada, concern was expressed about Canada's ability to remain self-sufficient in food supply by the year 2000 and beyond. A population growth of 20 to 45 percent is expected, and cities continue to expand onto prime farmland. The amount of land for cereal grains per capita was predicted to decline from 0.184 hectares in 1978 to 0.128 hectares by 2000. Only three of Canada's Prairie provinces are surplus food producers. Only 10 percent of the food produced is now exported.

Land requires considerable inputs to remain productive, and it is subject to degradation by erosion, salinity, water-logging, and desertification. A recent United Nations report estimated that 10 percent of the world's irrigated land has had a significant decline in productivity due to water-logging, and another 10 percent due to an increase in salinity. In some irrigated regions, geologic groundwater will be exhausted. In other cases, the high cost of pumping water and competition with urban water users will result in conversion of irrigated land to range land. Desertification is a problem, especially in subSaharan Africa. As the need for land grows, farmers are forced onto lands with greater production hazards and limitations as shown in Figure 20.4.

Summary Statement

The past few decades have brought an end to the traditional expansion onto new lands as a major means to increase food production. For the foreseeable future, it appears that an expanding world population will have to produce its food on a declining amount of land per person.

POTENTIALLY AVAILABLE LAND AND SOIL RESOURCES

About 65 percent of the ice-free land has a climate suitable for some cropping. Only 10 or 11 percent of the world's land is cultivated, which suggests a considerable possibility for increasing the world's cropland.

World's Potential Arable Land

The distribution of the world's cropland is shown in Figure 20.5. The five countries with the most

FIGURE 20.4 Farmers are pushed higher and higher on slopes in the Himalayan foothills due to population growth. Many such lands are used only a few years before it is no longer profitable to crop the land due to severe soil erosion and landslides.

FIGURE 20.5 Distribution of the approximate cropland area of the world. (Data of USDA.)

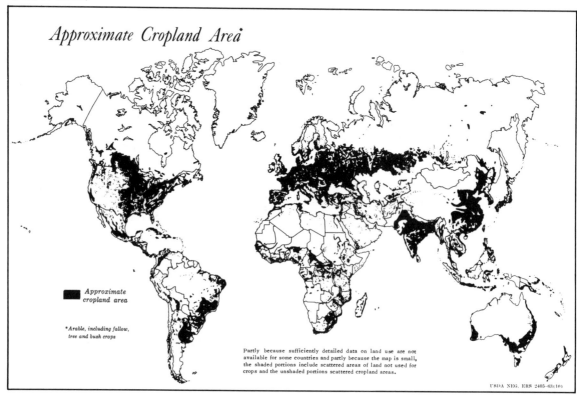

Approximate Cropland Area

■ *Approximate cropland area*

*Arable, including fallow, tree and bush crops

Partly because sufficiently detailed data on land use are not available for some countries and partly because the map is small, the shaded portions include scattered areas of land not used for crops and the unshaded portions scattered cropland areas.

USDA NEG. ERS 2405-63(10)

cropland, in decreasing order, are: USSR, United States, India, China, and Canada. The cropland tends to exist either north or south of a large central band where there is only a small percentage of cropland, in northern South America and Africa and through central Asia. Much of the cropland in the USSR, United States, and Canada is in areas where the dominant soils are Mollisols. Few new cropland areas are available in western Europe and India, as can be seen by examining Figure 20.5.

Many estimates of potential cropland, or arable land, have been made. Several well-documented studies concluded that the present cultivated land area could at least be doubled. The data from one of the studies, (see Table 20.1) show the potentially arable land and currently cultivated land by continents. The data support the following conclusions:

1. The world's arable land can be increased 100 percent or more.
2. Of the potentially arable land that can be developed, 66 percent is located in Africa and South America.

The 66 percent potentially arable land that can be developed is located mainly in tropical and subtropical areas of Africa and South America. China and India are the two largest and most populated countries in Asia. At present in Asia, there is only 0.7 acre of cultivated land per person, and little additional unused and potentially arable land exists. India is unique as a large country, since it has one of the the greatest agricultural potentials of any country, because of its soils and climate. This is reflected in the fact that about 50 percent of India's land is currently cultivated, as compared to only 10 percent for the world and 20 percent for the United States.

The developing countries, with 77 percent of the world's population, have 54 percent of the current cultivated land area and 78 percent of the undeveloped, potentially arable land. By contrast, the developed countries have 23 percent of the world's population, 46 percent of the cultivated land, and 28 percent of the undeveloped and potentially arable land. Currently, the developed countries have about twice as much cultivated land on a per capita basis than do the developing nations. Note that Europe has about as much cultivated land per capita as does Asia, which has essentially no unused, potentially arable land. In fact, the actual cropland area in many European countries is declining.

TABLE 20.1 Present Cultivated Land on Each Continent Compared with Potentially Arable Land

Continent	Area, Billions of Acres[a]				Acres of Cultivated Land per Person
	Total	Potentially Arable	Cultivated	Potentially Arable minus Cultivated	
Africa	7.46	1.81	0.39	1.42	1.3
Asia	6.67	1.55	1.28	0.27	0.7
Australia and New Zealand	2.03	0.38	0.08	0.30	2.9
Europe	1.18	0.43	0.38	0.05	0.9
North America	5.21	1.15	0.59	0.56	2.3
South America	4.33	1.68	0.19	1.49	1.0
USSR	5.52	0.88	0.56	0.32	2.4
Total	32.49	7.88	3.47	4.41	1.0

From *The World Food Problem, Vol. 2,* The White House, May 1967.
[a] To convert to hectares multiply by 0.405.

Limitations of World Soil Resources

On a global basis, the main limitations for using the world's soil resources for agricultural production are drought (28%), mineral stress or infertility (23%), shallow depth (22%), excess water (10%), and permafrost (6%). Only 11 percent of the world's soils are without serious limitations, as shown in Table 20.2. Because the currently used agricultural land represents the world's best land, development of the unused, potentially arable land would be expected to have more serious limitations than those shown in Table 20.2.

Future improvements in the world food situation, in terms of new land for cultivation, will require developing land in Africa and South America. Almost 50 percent of the South American continent is centered on the Amazon Basin and central uplands of Brazil, where soils are mainly Oxisols and Ultisols. Not surprisingly, 47 percent of soil-use limitations are related to mineral stress and only 17 percent are due to drought. In Africa, 44 percent of the area has an arid or semiarid climate. Many Oxisols and Ultisols occupy the Zaire (Congo) basin, but an enormous area of Ustalfs occurs along the subSaharan region. This reflects the 44 percent drought and only 18 percent mineral stress limitations in Africa.

Summary Statement

The world's arable (or cultivated) land can be doubled with reasonable inputs. Increasing food production by bringing more land into cultivation in South America is limited mainly by soil infertility, but also by available water or drought. In Africa, the major limitations to bringing new land into cultivation are first, lack of water or drought and, second, soil infertility.

FUTURE OUTLOOK

In New York state, wheat yields increased only 3.1 bushels per acre (209 kg per hectare) over the 70-year period from 1865 to 1935. This also typifies the nature of crop yield increases in Europe during the same period. Beginning in 1935, improved varieties and better management practices increased wheat yields 20.1 bushels per acre during the next 40 years, as shown in Figure 20.6. Fifty-one percent of the increase was due to new technology and 49 percent of the increase was due to improved varieties of wheat. Low energy costs coupled with the adoption of improved varieties and other technology have resulted in a continuing overproduction of food in the United States.

TABLE 20.2   World Soil Resources and their Major Limitations for Agricultural Use

	Drought	Mineral Stress[a]	Shallow Depth	Water Excess	Permafrost	No Serious Limitations
	% of total land area					
North America	20	22	10	10	16	22
Central America	32	16	17	10	—	25
South America	17	47	11	10	—	15
Europe	8	33	12	8	3	36
Africa	44	18	13	9	—	16
South Asia	43	5	23	11	—	18
North and Central Asia	17	9	38	13	13	10
Southeast Asia	2	59	6	19	—	14
Australia	55	6	8	16	—	15
WORLD	28	23	22	10	6	11

From *The State of Food and Agriculture 1977*, FAO, 1978.
[a] Nutritional deficiencies or toxicities related to chemical composition or mode of origin.

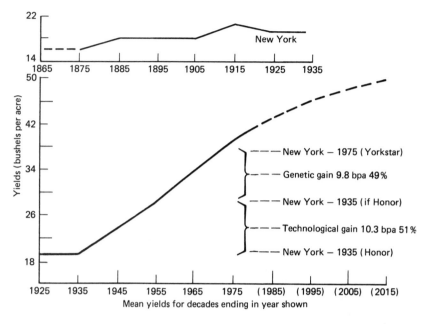

FIGURE 20.6 Wheat yields in New York State from 1866–1975. Between 1935 and 1975, yield increase was the result of 51 percent to technology and 49 percent to genetic gain. (Adapted from Jensen, 1978, and used by permission of *Science.*)

Large yield increases are common when the base yield is very low, which accounts for much of the relatively large increase in food production in many developing nations in recent years (see Figure 20.1). In the future, yield increases will become more difficult as the yield base increases. Thus, with very little new land coming into use and the need to increase yields on current cropland, increases in food production are becoming more difficult.

Beyond Technology

From a technological point of view, a large potential for increasing food production could be realized by greater photosynthetic efficiency, greater use of biological nitrogen fixation, more efficient water and nutrient use by plants, and greater plant resistance to disease and environmental stress. In many developing countries, technology has bought about large and rapid increases in food production, only to be overwhelmed by population growth. Without a suitable social, economic, and political environment, the benefits of technology have, in some cases, been temporal.

The World Grain Trade

The extent of food production sufficiency in various parts of the world can be deduced in part from an observation of the world grain trade. Western Europe, after the Industrial Revolution, became dependent on other countries for food grains and remained a consistent importer for a very long period of time. We have noted that Europe has little more cropland per capita than does Asia, and Europe has little unused potentially arable land to bring into production. The large and consistent import of grain by Europe since the 1930s is shown in Table 20.3.

Many nations that are now importers of food grains were exporters during the 1930s (see Table 20.3). India was a wheat exporter until population growth overtook food production growth. Latin America and even Africa were food grain exporters. The USSR was an exporter before the 1917 Russian revolution. Its inability to provide adequate diets, and the resulting riots, caused the USSR to enter the world market and buy a large amount of grain in 1973. This produced the so-called great grain robbery as the grain was secretly purchased without allowing the increased

TABLE 20.3   World Grain Trade for Selected Years

Region	1934–38	1950	1960	1970	1980	1984
	(million metric tons)					
North America	+5[a]	+23	+39	+56	+131	+126
Latin America	+9	+1	0	+4	−10	−4
Western Europe	−24	−22	−25	−30	−16	+13
Eastern Europe and Soviet Union	+5	0	0	0	−46	−51
Africa	+1	0	−2	−5	−15	−24
Asia	+2	−6	−17	−37	−63	−80
Australia and New Zealand	+3	+3	+6	+12	+19	+20

United Nations Food and Agriculture Organization, *Production Yearbook* (Rome: various years); U.S. Department of Agriculture, *Foreign Agriculture Circulars,* August 1983 and November 1984; adjustments by Worldwatch Institute.
[a] Plus sign indicates net exports; minus sign, net imports.

demand to affect prices. Note from Table 20.3 that the USSR is a major grain importer in spite of a near stagnant rate of population growth. The USSR has more cultivated land per person than does North America (see Table 20.1). Production problems stem from their unique economic-political situation and less than ideal climatic conditions. The large imports of grain into the USSR from the United States, which began in the early 1970s, are continuing. Much of the imported grain is fed to farm animals (see feed grains in Figure 20.7).

FIGURE 20.7   Grain-exports to the Soviet Union from the United States. The increase in exports has continued thorough the 1980s. (Data USDA.)

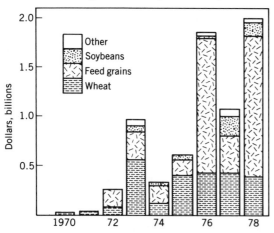

North America—the United States and Canada—together with Argentina, Australia, and New Zealand, have become the world's exporters of food grains. The data in Table 20.3 show that despite great efforts to increase food production in the developing countries, the food balance situation has deteriorated, a trend that is expected to continue.

Population Control and Politics

A population growth rate of zero is a mathematical certainty. Even a low growth rate, over a long enough period of time, results in more people than there is standing room available on earth. At the extreme of estimates for the earth's carrying capacity, deWit (1967) calculated that if all of the land surface produced calories equivalent to the maximum photosynthetic potential, there would be enough calories for 1,000 billion people. Such a world would have no place for people to live and work and no room for animals.

Population control as government policy is difficult because of long-time family traditions, religious beliefs, and a diminished sense of importance as the country's population declines. A one-child family policy was adopted by China to curb population growth. Today, the slowest rates of population growth are in the areas with the greatest amount of resources to invest in food production. Population control programs warrant more support.

In Africa, south of the Sahara, some countries are contributing to a food reserve while other countries are food deficient. Drought has been used to explain the food shortage in some of these African countries. It is becoming more and more apparent, however, that a lack of appropriate infrastructures is a problem. In Zambia, there are few people within a large area. In spite of low population, there has been considerable erosion and degradation of the most fertile soils. Overgrazing also contributes to soil erosion and land degradation. Yet, in Zambia there is no soil conservation service and little effort has been made at the national level to control land use.

In Nepal, population growth has forced farmers to plant crops on higher and steeper land in the hills. The result is extensive soil erosion and eventual abandonment of the land as a result of landslides. Erosion is further promoted by the removal of the forest cover for fuel wood, which has resulted in severe flooding downstream. Depletion of the forest (fuel wood) and subsequent conversion to kerosene to cook food may absorb 15 to 20 percent of a poor family's income. Erosion is an ever present and pressing problem for small farmers and results in higher fuel costs for the urban poor. The political power, however, rests with those who are not directly affected. Some of these include wealthy people who own rental properties in the capital city, the large landowners with estates on the fertile strip of land along the Indian border (the Terai), and merchants who import goods for resale. These groups wield the most political power in Nepal, so the needs of the urban and rural poor are ignored.

SUMMARY

In 1948, the late Dr. Charles E. Kellogg pointed out that sufficient land and soil resources (and other physical resources) are available to feed the world, provided, that economic, social, and polit-

ical problems are solved. Today, the statement is still true. Looking at the food-population problem from a technological point of view, one can be optimistic. Looking at the problem in its totality, it is hard to be optimistic. The difficult task ahead is to provide the economic, social, and political environment in which hundreds of millions of small peasant farmers, and with simple tools, can produce enough food for their families and some surplus to sell to the urban sector. The problem is an ancient one, and the challenge for its solution continues.

REFERENCES

Bentley, C. F. 1978. "Canada's Agricultural Land Resources and the World Food Problem," in *Trans. 11th Int. Cong. Soil Sci.* **2:**1–26, Edmonton, 1978.

Blaikie, Piers. 1985. *The Political Economy of Soil Erosion in Developing Countries*. Longman, London.

Brown, H. 1954. *The Challenge of Man's Future*. Viking, New York.

Brown, L. R. 1978. "Worldwide Loss of Cropland." *Worldwatch Paper* 24, Worldwatch Institute, Washington, D.C.

Brown, L. R. 1985. "Reducing Hunger." *State of the World*, pp. 23–41, Norton, New York.

Brown, L. R. 1989. *State of the World*. Norton, New York.

deWit, C. T. 1967. "Photosynthesis: Its Relationship to Overpopulation." *Harvesting the Sun*, A. J. Pietro, F. A Greer, and T. J. Army, Eds, pp. 315–332, Academic, New York.

FAO. 1978. *The State of Food and Agriculture 1977*. Rome.

FAO. 1981. *The State of Food and Agriculture 1980*. Rome.

Foth, H. D. 1982. Soil Resources and Food: A Global View, in *Principles and Applications of Soil Geography*. E. M. Bridges and D. A. Davidson, eds., Longman, London.

Jensen, N. F. 1978. "Limits to Growth in World Food Production." *Science*. **210:**317–320.

Kellogg, C. E. 1948. "Modern Soil Science." *Am. Scientist*. **38:**517–536.

Kellogg, C. E. and A. C. Orvedal. 1969. "Potentially

Arable Soils of the World and Critical Measures for Their Use," in *Advances in Agronomy*. Vol. 21. Academic Press, New York.

Morgan, Dan. 1979. *Merchants of Grain*. Viking, New York.

The White House. 1967. *The World Food Problem*. U.S. Government Printing Office, Washington, D.C.

USDA. 1964. "A Graphic Summary of World Agriculture." *Misc. Pub. No.* 705. Washington, D.C.

SOIL TEXTURE BY THE FIELD METHOD

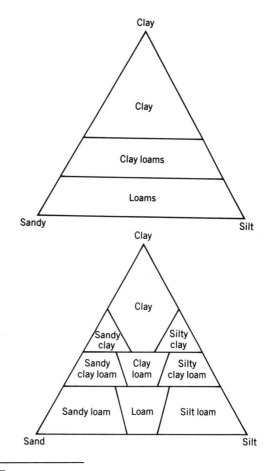

1. Moisten a sample of soil the size of a golf ball but don't get it very wet. Work it until it is uniformly moist; then squeeze it out between the thumb and forefinger to try to form a ribbon.

2. First decision: If the moist soil is:
   (a) *Extremely* sticky and stiff — One of the CLAYS
   (b) Sticky and stiff to squeeze — One of the CLAY LOAMS
   (c) Soft, easy to squeeze, only slightly sticky — One of the LOAMS

3. The second decision: Add an adjective to refine our description?
   (a) The soil feels very smooth — Use adjective SILT OR SILTY
   (b) The soils feels somewhat gritty — Use no adjective
   (c) The soil feels very, very gritty — Use adjective SANDY

4. The final refinement: The true texture triangle has two small additional changes.
    (a) The lines jog a little.
    (b) There are three additional (and less common) classes: sand, loamy sand, and silt.

5. Beware, the feel of a soil is modified by:
    (a) The amount of moisture present. Compare soils at like moisture contents.
    (b) The amount of organic matter. This especially affects clayey soils. Very high amounts of organic matter cause the soil to be "smooth," causing an overestimation of silt content.
    (c) The kind of clay. In tropic and subtropic regions different types of clay, particularly kaolinite, predominate and give a less sticky feel.

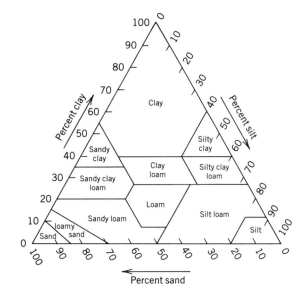

APPENDIX II

TYPES AND CLASSES OF SOIL STRUCTURE

Type (Shape and Arrangement of Peds)

Class	Platelike with one dimension (the vertical) limited and greatly less than the other two; arranged around a horizontal plane; faces mostly horizontal	Prismlike with two dimensions (the horizontal) limited and considerably less than the vertical; arranged around a vertical line; vertical faces well defined; vertices angular		Blocklike; polyhedronlike, or spheroidal, with three dimensions of the same order of magnitude, arranged around a point			Spheroids of polyhedrons having plane or curved surfaces which have slight or no accommodation to the faces of surrounding peds
		Without rounded caps	With rounded caps	Blocklike; blocks or polyhedrons having plane or curved surfaces that are casts of the molds formed by the faces of the surrounding peds			
				Faces flattened; most vertices sharply angular	Mixed rounded and flattened faces with many rounded vertices	
	Platy	Prismatic	Columnar	(Angular) Blocky [a]	Subangular Blocky [b]	Granular
Very fine or very thin	Very thin platy; <1 mm	Very fine prismatic; <10 mm	Very fine columnar; <10 mm	Very fine angular blocky; <5 mm	Very fine subangular blocky; <5 mm	Very fine granular; <1 mm
Fine or thin	Thin platy; 1–2 mm	Fine prismatic; 10–20 mm	Fine columnar; 10–20 mm	Fine angular blocky; 5–10 mm	Fine subangular blocky; 5–10 mm	Fine granular; 1–2 mm
Medium	Medium platy; 2–5 mm	Medium prismatic; 20–50 mm	Medium columnar; 20–50 mm	Medium angular blocky; 10–20 mm	Medium subangular blocky; 10–20 mm	Medium granular; 2–5 mm
Coarse or thick	Thick platy; 5–10 mm	Coarse prismatic; 50–100 mm	Coarse columnar; 50–100 mm	Coarse angular blocky; 20–50 mm	Coarse subangular blocky; 20–50 mm	Coarse granular, 5–10 mm
Very coarse or very thick	Very thick platy; >10 mm	Very coarse prismatic; >100 mm	Very coarse columnar; >100 mm	Very coarse angular blocky; >50 mm	Very coarse subangular blocky; >50 mm	Very coarse granular; >10 mm

From: Soil Survey Staff, SCS, USDA. 1951. *Soil Survey Manual.* Agric. Handbook 18, p. 228. U.S. Government Printing Office, Washington, D.C.

[a] Sometimes called nut. (b) The word *angular* in the name ordinarily can be omitted.
[b] Sometimes called nuciform, nut, or subangular nut. Since the size connotation of these terms is a source of great confusion to many, they are not recommended.

APPENDIX III

PREFIXES AND THEIR CONNOTATIONS FOR NAMES OF GREAT GROUPS IN THE U.S. SOIL CLASSIFICATION SYSTEM (SOIL TAXONOMY)

Prefix	Connotation of Prefix	Prefix	Connotation of Prefix
acr	Extreme weathering	luv	Illuvial
agr	An agric horizon	med	Of temperate climates
alb	An albic horizon	nadur	See the formative elements *natr* and *dur*
and	Ando-like	natr	Presence of natric horizon
anthr	An anthropic epipedon	ochr	Presence of ochric epipedon
arg	An argillic horizon	pale	Old development
bor	Cool	pell	Low chroma
calc	A calcic horizon	plac	Presence of a thin cemented layer
camb	A cambic horizon	plag	Presence of plaggen horizon
chrom	High chroma	plinth	Presence of plinthite
cry	Cold	psamm	Sand textures
dur	A duripan	quartz	High quartz content
dystr, dys	Low basic cation saturation	rhod	Dark red color
eutr, eu	High basic cation saturation	sal	Presence of salic horizon
ferr	Presence of iron	sider	Presence of free iron oxides
fluv	Floodplain	sombr	A dark horizon
frag	Presence of fragipan	sphagn	Presence of sphagnum moss
fragloss	See the formative elements *frag* and *gloss*	sulf	Presence of sulfides or their oxidation products
gibbs	Presence of gibbsite	torr	Aridic moisture regime
gloss	Tongued	trop	Continually warm and humid
gyps	Presence of gypsic horizon	ud	Udic moisture regime
hal	Salty	umbr	Presence of umbric epipedon
hapl	Minimum horizon	ust	Ustic moisture regime
hum	Presence of humus	verm	Wormy, or mixed by animals
hydr	Presence of water	vitr	Presence of glass
kand	Presence of low activity clay	xer	Xeric moisture regime

GLOSSARY*

A horizons.   Mineral horizons that formed at the surface, or below an O horizon, and are characterized by an accumulation of humified organic matter intimately mixed with the mineral fraction.

Acid soil.   Soil with a pH value < 7.0.

Acidic cations.   Hydrogen ions or cations that, on being added to water, undergo hydrolysis, resulting in an acidic solution. Examples in soils are H^+, Al^{3+}, and Fe^{3+}.

Acidity, residual.   Soil acidity that is neutralized by lime or other alkaline materials, but that cannot be replaced by an unbuffered salt solution.

Acidity, salt—replaceable.   The aluminum and hydrogen that can be replaced from an acid soil by an unbuffered salt solution such as KCl. Essentially, the sum of the exchangeable Al and H.

Acidity, total.   Sum of salt—replaceable and residual acidity.

Actinomycetes.   A nontaxonomic term applied to a group of gram-positive bacteria that have a superficial resemblance to fungi.

Activity.   Informally, may be taken as the effective concentration of a substance in a solution.

Aerate.   To allow or promote exchange of soil gases with atmospheric gases.

Aeration porosity.   The fraction of the bulk soil volume that is filled with air at any given time, or under a given condition, such as a specified soil-water matric potential.

Aerobic.   (1) Having molecular oxygen as a part of the environment. (2) Growing only in the presence of molecular oxygen, as aerobic organisms. (3) Occurring only in the presence of molecular oxygen, such as aerobic decomposition.

Aggregate.   A unit of soil structure, usually formed by natural processes, and generally below 10 millimeters in diameter.

* Adapted from *Glossary of Soil Science Terms*, published by the Soil Science Society of America, July 1987, and reprinted by permission of the Soil Science Society of America.

Air dry.   The state of dryness at equilibrium with the water content in the surrounding atmosphere.

Albic horizon.   A mineral soil horizon from which clay and free iron oxides have been removed or segregated. The color of the horizon is determined primarily by the color of primary sand and silt particles rather than coatings on these particles. An E horizon.

Alfisols.   Mineral soils that have umbric or ochric epipedons, agrillic or kandic horizons, and that have plant-available water during at least 90 days when the soil is warm enough for plants to grow. Alfisols have a mean annual soil temperature below 8° C or a base saturation in the lower part of the argillic horizon of 35 percent or more when measured at pH 8.2. A soil order.

Alkaline soil.   Any soil having a pH above 7.0.

Ammonification.   The biological process leading to the formation of ammoniacal nitrogen from nitrogen-containing organic compounds.

Ammonium fixation.   The process of converting exchangeable or soluble ammonium ions to those occupying interlayer positions similar to K^+ in micas.

Amorphous material.   Noncrystalline soil constitutents.

Anaerobic.   (1) The absence of molecular oxygen. (2) Growing in the absence of molecular oxygen, such as anaerobic bacteria. (3) Occurring in the absence of oxygen, such as a biochemical process.

Andisols.   Mineral soils developed in volcanic ejecta that have andic soil properties. A soil order.

Anion exchange capacity.   The sum total of exchangeable anions that a soil can adsorb.

Aquic.   A mostly reducing soil moisture regime nearly free of dissolved oxygen due to saturation by groundwater or its capillary fringe and occurring at periods when the soil temperature at 50 centimeters below the surface is above 5° C.

Argillic horizon.   A mineral soil horizon that is characterized by the illuvial accumulation of layer-silicate clays.

Aridic.   A soil moisture regime that has no plant-available water for more than half the cumulative time that the soil temperature at 50 centimeters below the surface is above 5° C and has no period as long as 90 consecutive days when there is water for plants, while the soil temperature at 50 centimeters is continuously above 8° C.

Aridisols. Mineral soils that have an aridic moisture regime but no oxic horizon. A soil order.

Autotroph. An organism capable of utilizing CO_2 or carbonates as a sole source of carbon and obtaining energy for carbon reduction and biosynthetic processes from radiant energy (photoautotroph) or oxidation of inorganic substances (chemoautotroph).

Available nutrients. Nutrient ions or compounds in forms that plants can absorb and utilize in growth.

Available water. The portion of water in a soil that can be absorbed by plant roots; the amount of water between in situ field capacity and the wilting point.

B horizons. Horizons that formed below an A, E, or O horizon that have properties different from the overlying and underlying horizons owing to the soil forming processes.

Bacteroid. An altered form of cells of certain bacteria. Refers particularly to the swollen, irregular vacuolated cells of *Rhizobium* in nodules of legumes.

Bar. A unit of pressure equal to 1 million dynes per square centimeter.

Basic cation saturation percentage. The extent to which the cation exchange capacity is saturated with alkali (sodium and potassium) and alkaline earth (calcium and magnesium) cations expressed as a percentage of the cation exchange capacity, as measured at a particular pH, such as 7 or 8.2.

Biomass. The total mass of living microorganisms in a given volume or mass of soil.

Biosequence. A sequence of related soils, differing one from the other, primarily because of differences in kinds and numbers of plants and soil organisms as a soil-forming factor.

Biuret. A toxic product formed at high temperature during the manufacturing of urea.

Buffer power. The ability of ions associated with the solid phase to buffer changes in ion concentration in the solution phase.

Bulk density, soil. The mass of dry soil per unit bulk volume, expressed as grams per cubic centimeter,

g/cm^3. The bulk volume is determined before drying to constant weight at 105° C.

C horizons. Horizons or layers, excluding hard rock, that are little affected by the soil-forming processes. C horizons typically underly A, B, E, or O horizons.

Calcareous soil. Soil containing sufficient free $CaCO_3$, and/or $MgCO_3$, to effervesce visibly when treated with cold 0.1 molar HCl. These soils usually contain from as little as 1 to 20 percent $CaCO_3$ equivalent.

Calcic horizon. A mineral soil horizon of secondary carbonate enrichment that is more than 15 centimeters thick, has a $CaCO_3$ equivalent above 150 g/kg, and has at least 50 g/kg more $CaCO_3$ equivalent than the underlying C horizon.

Caliche. A zone near the surface, more or less cemented by secondary carbonates of calcium or magnesium precipitated from the soil solution. Caliche may occur as a soft thin soil horizon, as a hard thick bed, or as a surface layer exposed by erosion.

Cambic horizon. A mineral soil horizon that has a texture of loamy very fine sand or finer, has soil structure rather than rock structure, contains some weatherable minerals, and is characterized by alteration or removal of mineral material, or the removal of carbonates.

Capillary fringe. A zone in the soil just above the water table that remains saturated or almost saturated with water.

Carbon—organic nitrogen ratio. The ratio of the mass of organic carbon to the mass of organic nitrogen in soil, organic material, plants, or the cells of microorganisms.

Cat clay. Wet clay soils containing ferrous sulfide, which become highly acidic when drained.

Catena. A sequence of soils of about the same age, derived from similar parent material, and occurring under similar climatic conditions, but having different characteristics because of variations in relief and in drainage.

Cation exchange. The interchange between a cation in solution and another cation on the surface of any negatively charged material such as clay colloid or organic colloid.

Cation exchange capacity (CEC). The sum of exchangeable cations that a soil, soil constitutent, or other material can adsorb at a specific pH; commonly expressed as milliequivalents per 100 grams or centimoles per kilogram.

Chemical potential. Informally, it is the capacity of a solution or other substance to do work by virtue of its chemical composition.

Chlorosis. Failure of plants to develop chlorophyll caused by a deficiency of an essential element. Chlorotic leaves range in color from light green through yellow to almost white.

Chroma. The relative purity, strength, or saturation of a color; directly related to the dominance of the determining wavelength of the light and inversely related to grayness; one of the three variables of color.

Chronosequence. A sequence of related soils that differ one from the other, in certain properties primarily as a result of time as a soil-forming factor.

Citrate soluble phosphorus. That part of the total phosphorus in fertilizer that is insoluble in water but soluble in neutral 0.33 M ammonium citrate and which, together with water-soluble phosphorus represents the readily available phosphorus content of the fertilizer.

Clay. (1) A soil separate consisting of particles <0.002 millimeters in equivalent diameter. (2) A textural class.

Clay films. Coatings of clay on the surfaces of soil peds, mineral grains, and in soil pores. (Also called clay skins, clay flows, or agrillans.)

Clay mineral. Any crystalline inorganic substance of clay size.

Claypan. A dense, compact layer in the subsoil having a much higher clay content than the overlying material, from which it is separated by a sharply defined boundary. Claypans usually impede the movement of water and air and the growth of plant roots.

Colluvium. A general term applied to deposits on a slope or at the foot of a slope or cliff that were moved there chiefly by gravity.

Concretion. A local concentration of a chemical compound, such as calcium carbonate or iron oxide, in the form of a grain or nodule of varying size, shape, hardness, and color.

Consistency. The manifestations of the forces of cohesion and adhesion acting within the soil at various water contents, as expressed by the relative ease with which a soil can be deformed or ruptured.

Consumptive water use. The water used by plants in transpiration and growth, plus water vapor loss from adjacent soil or snow, or from intercepted precipitation in any specified time. Usually expressed as equivalent depth of free water per unit of time.

Creep. Slow mass movement of soil and soil material down relatively steep slopes primarily under the influence of gravity.

Crust. A soil-surface layer, ranging in thickness from a few millimeters to a few tens of millimeters, which, when dry, is much more compact, hard, and brittle than the material immediately beneath it.

Cryic. A soil temperature regime that has mean annual soil temperatures of above 0° C but below 8° C at 50 centimeters. There is more than a 5° C difference between mean summer and mean winter soil temperatures.

Darcy's law. A law describing the rate of flow of water through porous media. Named for Henry Darcy of Paris, who formulated it in 1856 from extensive work on the flow of water through sand filter beds.

Denitrification. Reduction of nitrate or nitrite to molecular nitrogen or nitrogen oxides by microbial activity or by chemical reactions involving nitrite.

Desert pavement. The layer of gravel or stones remaining on the land surface in desert regions after the removal of the fine material by wind erosion.

Diffuse double layer. A heterogeneous system that consists of a solid surface layer having a net electrical charge, together with an ionic swarm under the influence of the solid and a solution phase that is in direct contact with the surface.

Duripan. A mineral soil horizon that is cemented by silica to the point that air-dry fragments will not slake in water or HCl.

E horizons. Mineral horizons in which the main feature is loss of silicate clay, iron, aluminum, or some combination of these, leaving a concentration of sand and silt particles of quartz or other resistant minerals.

EC$_e$. The electrolytic conductivity of an extract from saturated soil, normally expressed in units of decisiemens per meter at 25° C.

Ecology. The science that deals with the interrelationships between organisms and between organisms and their environment.

Ecosystem. A community of organisms and the environment in which they live.

Ectomycorrhiza. A mycorrhizal association in which the fungal mycelia extend inward, between root cortical cells, to form a network ("Hartig net") and outward into the surrounding soil.

Edaphology. The science that deals with the influence of soils on living things, particularly plants.

Eluviation. The removal of soil material in suspension (or in solution) from a layer or layers of soil.

Endomycorrhiza. A mycorrhizal association with intracellular penetration of the host root cortical cells by the fungus as well as outward extension into the surrounding soil.

Entisols. Mineral soils that have no distinct subsurface diagnostic horizons within 1 meter of the soil surface. A soil order.

Erosion. (1) The wearing away of the land surface by running water, wind, ice, and other geological agents, including such processes as gravitational creep. (2) Detachment and movement of soil or rock by water, wind, ice, or gravity.

Erosion potential. A numerical value expressing the inherent erodibility of a soil.

Eutrophic. Having concentrations of nutrients optimal, or nearly so, for plant or animal growth.

Evapotranspiration. The combined loss of water from a given area, and during a specified period of time, by evaporation from the soil surface and by transpiration from plants.

Exchangeable cation percentage. The extent to which the adsorption complex of a soil is occupied by a particular cation.

Exchangeable ion. A cation or anion held on or near the surface of a solid particle, which may be replaced by other ions of similar charge that are in solution.

Fallowing. The practice of leaving land uncropped for periods of time to accumulate and retain water and mineralized nutrient elements.

Family, soil. In soil classification one of the categories intermediate between the great soil group and the soil series. Families provide groupings of soils with ranges in texture, mineralogy, temperature, and thickness.

Fertigation. Application of plant nutrients in irrigation water.

Fertility, soil. The ability of a soil to supply elements essential for plant growth without a toxic concentration of any element.

Fertilizer. Any organic or inorganic material of natural or synthetic origin (other than liming materials) that is added to a soil to supply one or more elements essential to the growth of plants.

Fibric material. Mostly undecomposed plant remains that contain large amounts of well-preserved and recognizable fibers.

Field capacity, in situ (field water capacity). The content of water, on a mass or volume basis, remaining in a soil 2 or 3 days after having been wetted with water and after free drainage is negligible.

Film water. A thin layer of water, in close proximity to soil-particle surfaces, that varies in thickness from 1 or 2 to perhaps 100 or more molecular layers.

Fixation. The process by which available plant nutrients are rendered less available or unavailable in the soil.

Floodplain. The land bordering a stream, built up of sediments from overflow of the stream, and subject to inundation when the stream is at flood stage.

Flux. The time rate of transport of a quantity across a given area.

Free iron oxides. A general term for those iron oxides that can be reduced and dissolved by a dithionite treatment. Often includes goethite and hematite.

Friable. A consistency term pertaining to the ease of crumbling of soils.

Fluvic acid. The *colored* material that remains in solution after the removal of humic acid by acidification.

Gibbiste. A mineral with a platy habit that occurs in highly weathered soils. $Al(OH)_3$.

Gilagi. The microrelief of soils produced by expansion and contraction with changes in water content, a common feature of Vertisols.

Glacial drift. Rock debris that has been transported by glaciers and deposited, either directly from the ice or from the meltwater.

Goethite. A yellow-brown iron oxide mineral that is very common and is responsible for the brown color in many soils. $FeOOH$.

Great soil group. One of the categories in the system of soil classification that has been used in the United States for many years. Great groups place soils according to soil moisture and temperature, basic cation saturation status, and expression of soil horizons.

Green manure. Plant material incorporated into soil while green or at maturity; for soil improvement.

Groundwater. That portion of the water below the surface of the ground at a pressure equal to or greater than atmospheric.

Guano. The decomposed dried excrement of birds and bats, used for fertilizer purposes.

Gypsic horizon. A mineral soil horizon of secondary calcium sulfate enrichment that is more than 15 centimeters thick.

Gypsum. The common name for calcium sulfate (Ca-SO$_4$·2H$_2$O), used to supply calcium and sulfur to ameliorate sodic soils.

Hardpan. A hardened soil layer, in the lower A or in the B horizon, caused by cementation of soil particles with organic matter or with materials such as silica, sesquioxides, or calcium carbonate.

Heavy metals. Those metals that have high density; in agronomic usuage includes Cu, Fe, Mn, Mo, Co, Zn, Cd, Hg, Ni, and Pb.

Hematite. A red iron oxide mineral that contributes red color to many soils. Fe$_2$O$_3$.

Hemic material. An intermediate degree of decomposition, such as two-thirds of the organic material cannot be recognized.

Heterotroph. An organism capable of deriving carbon and energy for growth and cell synthesis by the utilization of organic compounds.

Histosols. Organic soils that have organic soil material in more than half of the upper 80 centimeters, or that are of any thickness overlying rock or fragmental materials, which have interstices filled with organic soil materials. A soil order.

Hue. One of the three variables of color.

Humic acid. The dark-colored organic material that can be extracted from soil by various agents and that is precipitated by acidification to pH 1 or 2.

Humification. The process whereby the carbon of organic residues is transformed and converted to humic substances through biochemical and/or chemical processes.

Humus. All of the organic compounds in soil exclusive of undecayed plant and animal tissues, their partial decomposition products, and the soil biomass. Resistant to further alteration.

Hydraulic conductivity. The proportionality factor in Darcy's Law as applied to the viscous flow of water in soil.

Hydroxy-aluminum. Aluminum hyroxide compounds of varying composition.

Hyperthermic. A soil temperature regime that has mean annual soil temperatures of 22° C or more and a higher than 5° C difference between mean summer and mean winter soil temperatures at 50 centimeters below the surface.

Illuvial horizon. A soil layer or horizon in which material carried from the overlying layer has been precipitated from solution or deposited from suspension. The layer of accumulation.

Illuviation. The process of deposition of soil material removed from one horizon to another in the soil, usually from an upper to a lower horizon in the soil profile.

Immobilization. The conversion of an element from the inorganic to the organic form in microbial or plant tissues.

Inceptisols. Mineral soils having one or more pedogenic horizons in which mineral materials, other than carbonates or amorphorus silica, have been altered or removed but not accumulated to a significant degree. Water is available to plants more than half of the year or more than 90 consecutive days during a warm season. A soil order.

Indicator plants. Plants characteristic of specific soil or site conditions.

Infiltration. The downward entry of water through the soil surface.

Ion activity. Informally, the effective concentration of an ion in solution.

Iron oxides. Group name for the oxides and hydroxides of iron. Includes the minerals goethite, hematite, lepidocrocite, ferrihydrite, maghemite, and magnetite. Sometimes referred to as sesquioxides or hydrous oxides.

Iron pan. An indurated soil horizon in which iron oxide is the principal cementing agent, as in plinthite or laterite.

Ironstone. Hardened plinthite materials often occurring as nodules and concretions.

Isomorphous substitution. The replacement of one atom by another of similar size in a crystal structure without disrupting or seriously changing the structure.

Jarosite. A yellow potassium iron sulfate mineral.

Kandic horizon. Subsoil diagnostic horizon having a clay increase relative to overlying horizons and having low-activity clays, below 16 meq/100 g or below 16 cmol(+)/kg of clay.

Kaolinite. A clay mineral of the kaolin subgroup. It has a 1 : 1 layer structure and is a nonexpanding clay mineral.

Labile. A substance that is readily transformed by microorganisms or is readily available to plants.

Landscape. All the natural features such as fields, hills, forests, water, and such, which distinguish one part of the earth's surface from another part. Usually that portion of the land that the eye can comprehend in a single view.

Lattice. A regular geometric arrangement of points in a plane or in space. Lattice is used to represent the distribution of repeating atoms or groups of atoms in a crystalline substance.

Leaching. The removal of materials in solution from the soil.

Lepidocrocite. An orange iron oxide mineral that is found in mottles and concretions of wet soils. FeOOH.

Lime, agricultural. A soil amendment containing calcium carbonate, magnesium carbonate and other materials; used to neutralize soil acidity and furnish calcium and magnesium for plant growth.

Lithosequence. A group of related soils that differ, one from the other, in certain properties primarily as a result of differences in the parent rock as a soil-forming factor.

Loam. A soil textural class.

Loess. Material transported and deposited by wind and consisting of predominantly silt-sized particles.

Luxury uptake. The absorption by plants of nutrients in excess of their need for growth.

Macronutrient. A plant nutrient usually attaining a concentration of more than 500 mg/kg in mature plants.

Maghemite. A dark, reddish-brown, magnetic iron oxide mineral chemically similar to hematite, but structurally similar to magnetite. Fe_2O_3. Often found in well-drained, highly weathered soils of the tropical regions.

Magnetite. A black, magnetic iron oxide mineral usually inherited from igneous rocks. Often found in soils as black magnetic sand grains.

Manganese oxides. A group term for oxides of manganese. They are typically black and frequently occur as nodules and coatings on ped faces, usually in association with iron oxides.

Manure. The excreta of animals, with or without an admixture of bedding or litter, fresh or at various stages of further decomposition or composting.

Mass flow (nutrient). The movement of solutes associated with the net movement of water.

Melanic horizon. A thick, dark colored surface horizon having andic soil properties.

Mesic. A soil temperature regime that has mean annual soil temperatures of 8° C or more but less than 15° C, and more than 5° C difference between mean summer and mean winter temperatures at 50 centimeters below the surface.

Mica. A layer-structured aluminosilicate mineral group of the 2 : 1 type that is characterized by high layer charge, which is usually satisfied by potassium.

Microclimate. The sequence of atmospheric changes within a very small region.

Microfauna. Protozoa, nematodes, and arthropods of microscopic size.

Microflora. Bacteria (including actinomycetes), fungi, algae, and viruses.

Micronutrient. A chemical element necessary for plant growth found in small amounts, usually less than 100 mg/kg in the plant. These elements consist of B, Cl, Cu, Fe, Mn, Mo, and Zn.

Mineralization. The conversion of an element from an organic form to an inorganic state as a result of microbial activity.

Mineral soil. A soil consisting predominantly of, and having properties determined predominantly by, mineral matter. Usually contains less than 200 g/kg of organic carbon.

Mollisols. Mineral soils that have a mollic epipedon overlying mineral material with a basic cation saturation of 50 percent or more when measured at pH 7.0. A soil order.

Montmorillonite. An aluminum silicate (smectite)

with a layer structure composed of two silica tetrahedral sheets and a shared aluminum and magnesium octahedral sheet.

Mor. A type of forest humus in which the Oa horizon is present and in which there is almost no mixing of surface organic matter with mineral soil.

Mottled zone. A layer that is marked with spots or blotches of different color or shades of color (mottles).

Muck soil. An organic soil in which the plant residues have been altered beyond recognition.

Mulch. Any material such as straw, sawdust, leaves, plastic film, and loose soil, that is spread upon the soil surface to protect soil and plant roots from the effects of raindrops, soil crusting, freezing, evaporation, and such.

Mulch farming. A system of tillage and planting operations resulting in minimum incorporation of plant residues or other mulch into the soil surface.

Mull. A type of forest humus in which the Oe horizon may or may not be present and in which there is no Oa horizon. The A horizon consists of an intimate mixture of organic matter and mineral soil with gradual transition between the A horizon and the horizon underneath.

Munsell color system. A color designation system that specifies the relative degrees of the three simple variables of color: hue, value, and chroma. For example: 10YR 6/4 is a color with a hue = 10YR (yellow - red), value = 6, and chroma = 4.

Mycorrhiza. Literally "fungus root." The association, usually symbiotic, of specific fungi with the roots of higher plants.

N value. The relationship between the percentage of water under field conditions and the percentages of inorganic clay and of humus.

Natric horizon. A mineral soil horizon that satisfies the requirements of an argillic horizon, but that also has prismatic, columnar, or blocky structure and a subhorizon having 15 percent or more saturation with exchangeable sodium.

Nitrification. Biological oxidation of ammonium to nitrite and nitrate, or a biologically induced increase in the oxidation state of nitrogen.

Nitrogenase. The specific enzyme required for biological dinitrogen fixation.

No-tillage system. A procedure whereby a crop is planted directly into the soil with no preparatory tillage since harvest of the previous crop; usually a special planter is necessary to prepare a narrow, shallow seedbed immediately surrounding the seed being planted.

Nutrient antagonism. The depressing effect caused by one or more plant nutrients on the uptake and availability of another.

Nutrient interaction. A statistical term used when two or more nutrients are applied together to denote a departure from additive responses occurring when they are applied separately.

O horizon. Layers dominated by organic material, except limnic layers that are organic.

Ochric epipedon. A thin, light colored surface horizon of mineral soil.

Organic soil. A soil that contains a high percentage of organic carbon (>200 g/kg or >120 - 180 g/kg if saturated with water) throughout the solum.

Outwash. Stratified glacial drift deposited by meltwater streams beyond active glacier ice.

Ovendry soil. Soil that has been dried at 105° C until it reaches constant mass.

Oxisols. Mineral soils that have an oxic horizon within 2 meters of the surface or plinthite as a continuous phase within 30 centimeters of the surface, and that do not have a spodic or argillic horizon above the oxic horizon. A soil order.

Paleosol, buried. A soil formed on a landscape during the geological past and subsequently buried by sedimentation.

Pans. Horizons or layers in soils that are strongly compacted, indurated, or having very high clay content.

Paralithic contact. Similar to lithic contact except that it is softer, and is difficult to dig with a spade.

Parent material. The unconsolidated and more or less chemically weathered mineral or organic matter from which the solum of soils is developed by pedogenic processes.

Particle density. The density of soil particles, the dry mass of the particles being divided by the solid volume of the particles.

Pascal. A unit of pressure equal to 1 Newton per square meter.

Peat. Unconsolidated soil material consisting largely of undecomposed, or only slightly decomposed, organic matter accumulated under conditions of excessive moisture.

Ped. A unit of soil structure such as an aggregate, crumb, prism, block, or granule, formed by natural processes.

Pedon. A three-dimensional body of soil with lateral dimensions large enough to permit the study of horizon shapes and relations. Its area ranges from 1 to 10 square meters.

Penetrability. The ease with which a probe can be pushed into the soil.

Percolation. The downward movement of water through the soil.

Pergelic. A soil temperature regime that has mean annual soil temperatures of less than 0° C. Permafrost is present.

Permafrost. A perennially frozen soil horizon.

Permafrost table. The upper boundary of the permafrost, coincident with the lower limit of seasonal thaw.

Permanent wilting point. The largest water content of a soil at which indicator plants, growing in that soil, wilt and fail to recover when placed in a humid chamber. Often estimated by the water content at -15 bars, -1,500 kilopascals, or -1.5 megapascals soil matric potential.

Petrocalcic horizon. A continuous, indurated calcic horizon that is cemented by calcium carbonate and, in some places, with magnesium carbonate.

Petrogypsic horizon. A continuous, strongly cemented, massive gypsic horizon that is cemented with calcium sulfate.

pH-dependent charge. The portion of the cation or anion exchange capacity which varies with pH.

pH, soil. The negative logarithm of the hydrogen ion activity of a soil.

Phase. A utilitarian grouping of soils defined by soil or environmental features that are not class differentia used in the U.S. system of soil taxonomy, for example, surface texture, surficial rock fragments, salinity, erosion, thickness, and such.

Phosphate. In fertilizer trade terminology, phosphate is used to express the sum of the water-soluble and citrate-soluble phosphoric acid (P_2O_5); also referred to as the available phosphoric acid (P_2O_5).

Phosphoric acid. In commerical fertilizer manufacturing, it is used to designate orthophosphoric acid, H_3PO_4. In fertilizer labeling, it is the common term used to represent the phosphate content in terms of available phosphorus, expressed as percent P_2O_5.

P_2O_5. Phosphorus pentoxide; fertilizer label designation that denotes the percentage of available phosphate.

Placic horizon. A black to dark-reddish mineral soil horizon that is usually thin, is commonly cemented with iron, and is slowly permeable or impenetrable to water and roots.

Plastic soil. A soil capable of being molded or deformed continuously and permanently, by relatively moderate pressure, into various shapes. See Consistency.

Plinthite. A nonindurated mixture of iron and aluminum oxides, clay, quartz, and other diluents that commonly occurs as red soil mottles usually arranged in platy, polygonal, or reticulate patterns. Plinthite changes irreversibly to ironstone hardpans, or irregular aggregates on exposure to repeated wetting and drying.

Plow pan. An induced subsurface soil horizon or layer having a higher bulk density and lower total porosity than the soil material directly above and below, but similar in particle size analysis and chemical properties. The pan is usually found just below the maximum depth of primary tillage and frequently restricts root development and water movement. Also called a pressure pan or plow sole.

Polypedon. A group of contiguous similar pedons.

Potash. Term used to refer to potassium or potassium fertilizers and usually designated as K_2O.

Potassium fixation. The process of converting exchangeable or water-soluble potassium to that occupying the position of K^+ in the micas.

Primary mineral. A mineral that has not been altered chemically since deposition and crystallization from molten lava.

Profile, soil. A vertical section of soil through all its horizons and extending into the C horizon.

R layer. Hard bedrock including granite, basalt, quartzite, and indurated limestone or sandstone that is sufficiently coherent to make hand digging impractical.

Rainfall erosion index. A measure of the erosive potential of a specific rainfall event.

Reaction, soil. The degree of acidity or alkalinity of a soil, usually expressed as a pH value.

Regolith. The unconsolidated mantle of weathered rock and soil material on the earth's surface; loose earth materials above solid rock.

Residual fertility. The available nutrient content of a soil carried to subsequent crops.

Reticulate mottling. A network of streaks of different color, most commonly found in the deeper profiles of soil containing plinthite.

Rhizobia. Bacteria capable of living symbiotically in roots of leguminous plants, from which they receive energy and often utilize molecular nitrogen.

Rhizosphere. The zone immediately adjacent to plant roots in which the kinds, numbers, or activities of microorganisms differ from that of the bulk soil.

Rill. A small, intermittent watercourse with steep sides; usually only several centimeters deep and, thus, no obstacle to tillage operations.

Runoff. That portion of the precipitation on an area which is discharged from the area through stream channels.

Saline soil. A nonsodic soil containing sufficient soluble salt to adversely affect the growth of most crop plants.

Salic horizon. A mineral soil horizon of enrichment with secondary salts more soluble in cold water than gypsum.

Saline seep. Intermittent or continuous saline water discharge at or near the soil surface under dry-land conditions, which reduces or eliminates crop growth.

Saline-sodic soil. A soil containing both sufficient soluble salt and exchangeable sodium to adversely affect crop production under most soil and crop conditions.

Salt balance. The quantity of soluble salt removed from an irrigated area in the drainage water minus that delivered in the irrigation water.

Sand. (1) A soil particle between 0.05 and 2.0 millimeters in diameter. (2) A soil textural class.

Sapric material. One of the components of organic soils with highly decomposed plant remains. Material is not recognizable and bulk density is low.

Saprolite. Weathered rock materials that may be soil parent material.

Saturation extract. The solution extracted from a soil at its saturation water content.

Secondary mineral. A mineral resulting from the decomposition of a primary mineral or from the reprecipitation of the products of decomposition of a primary mineral.

Self-mulching soil. A soil in which the surface layer becomes so well aggregated that it does not crust and seal under the impact of rain, but instead serves as a surface mulch upon drying.

Series, soil. See Soil series.

Sesquioxides. A general term for oxides and hydroxides of iron and aluminum.

Siderophore. A nonporphyrin metabolite secreted by certain microorganisms and plant roots that forms a highly stable coordination compound with iron.

Silica-alumina ratio. The molecules of silicon dioxide per molecule of aluminum oxide in clay minerals or in soils.

Silt. (1) A soil separate consisting of particles between 0.05 and 0.002 millimeters in equivalent diameter. (2) A soil textural class.

Site index. A quantitative evaluation of the productivity of a soil for forest growth under the existing or specified environment. Commonly, the height in meters of the dominant forest vegetation taken or calculated to an age index such as 25, 50, or 100 years.

Slickensides. Polished and grooved surfaces produced by one mass sliding past another; common in Vertisols.

Smectite. A group of 2:1 layer structured silicates with high cation exchange capacity and variable interlayer spacing.

Sodic soil. A nonsaline soil containing sufficient exchangeable sodium to adversely affect crop production and soil structure under most conditions of soil and plant type.

Sodium adsorption ratio (SAR). A relation between soluble sodium and soluble divalent cations that can be used to predict the exchangeable sodium percentage of soil equilibrated with a given solution. Defined as:

$$SAR = \frac{(sodium)}{(calcium + magnesium)^{1/2}}$$

where concentrations, denoted by parentheses, are expressed in moles per liter.

Soil. (1) The unconsolidated mineral or material on the immediate surface of the earth that serves as a natural medium for the growth of land plants. (2) The unconsolidated mineral or organic matter on the surface of the earth, which has been subjected to and influenced by genetic and environmental factors of parent material, climate, macro- and microorganisms, and topography, all acting over a period of time and producing a product—soil—that differs from the material from which it is derived in many physical, chemical, biological, and morphological properties and characteristics.

Soil association. A kind of map unit used in soil surveys comprised of delineations, each of which shows the size, shape, and location of a landscape unit composed of two or more kinds of component soils, or component soils and miscellaneous areas, plus allowable inclusions in either case.

Soil conservation. A combination of all management and land-use methods that safeguard the soil against depletion or deterioration by natural or human-induced factors.

Soil genesis. The mode of origin of the soil with special reference to the processes or soil-forming factors responsible for development of the solum, or true soil, from unconsolidated parent material.

Soil horizon. A layer of soil or soil material approximately parallel to the land surface and differing from adjacent genetically related layers in physical, chemical, and biological properties or characteristics such as color, structure, texture, consistency, kinds and number of organisms present, degree of acidity or alkalinity, and so on.

Soil loss tolerance. (1) The maximum average annual soil loss that will allow continuous cropping and maintain soil productivity without requiring additional management imputs. (2) The maximum soil erosion loss that is offset by the theoretical maximum rate of soil development, which will maintain an equilibrium between soil losses and gains.

Soil management groups. Groups of taxonomic soil units with similar adaptations or management requirements for one or more specific purposes, such as adapted crops or crop rotations, drainage practices, fertilization, forestry, and highway engineering.

Soil monolith. A vertical section of a soil profile removed from the soil and mounted for display or study.

Soil productivity. The capacity of a soil to produce a certain yield of crops, or other plants, with optimum management.

Soil science. That science dealing with soils as a natural resource on the surface of the earth, including soil formation, classification and mapping, geography and use, and physical, chemical, biological, and fertil-

ity properties of soils per se: and those properties in relation to their use and management.

Soil separates. Mineral particles, less than 2.0 millimeters in equivalent diameter, ranging between specified size limits.

Soil series. The lowest category in the U.S. system of soil taxonomy: a conceptualized class of soil bodies (polypedons) that have limits and ranges more restrictive than all higher taxa. The soil series serves as a major vehicle to transfer soil information and research knowledge from one soil area to another.

Soil solution. The aqueous liquid phase of the soil and its solutes.

Soil structure. The combination or arrangement of primary soil particles into secondary particles, units, or peds.

Soil survey. The systematic examination, description, classification, and mapping of soils in an area.

Soil texture. The relative proportions of the various soil separates in a soil.

Soil water potential (total). The amount of work that must be done per unit quantity of pure water in order to transport reversibly and isothermally an infinitesimal quantity of water from a pool of pure water, at a specified elevation and at atmospheric pressure, to the soil water (at the point under consideration).

Solum. The upper and most weathered part of the soil profile; the A, E, and B horizons.

Spodic horizon. A mineral soil horizon that is characterized by the illuvial accumulation of amorphous materials composed of aluminum and organic carbon with or without iron.

Spodosols. Mineral soils that have a spodic or a placic horizon that overlies a fragipan. A soil order.

Strip cropping. The practice of growing crops that require different types of tillage, such as row and sod, in alternate strips along contours or across the prevailing direction of the wind.

Structural charge. The charge (usually negative) on a mineral caused by isomorphous substitution within the mineral layer.

Subsoiling. Any treatment to loosen soil below the tillage zone without inversion and with a minimum of mixing with the tilled zone.

Surface charge density. The excess of negative or positive charge per unit area of surface area of soil or soil mineral.

Thermic. A soil temperature regime that has mean annual soil temperatures of 15° C or more, but less than 22° C and more than 5° C difference between mean summer and mean winter soil temperatures at a 50-centimeter depth below the surface.

Thermophile. An organism that grows readily at temperatures above 45° C.

Till. (1) Unstratified glacial drift deposited by ice and consisting of clay, silt, sand, gravel, and boulders, intermingled in any proportion. (2) To prepare the soil for seeding; to seed or cultivate the soil.

Tilth. The physical condition of soil as related to its ease of tillage, fitness as a seedbed, and its impedance to seedling emergence and root penetration.

Top dressing. An application of fertilizer to a soil surface, without incorporation, after the crop stand has been established.

Toposequence. A sequence of related soils. The soils differ, one from the other, primarily because of topography as a soil-formation factor.

Torric. A soil-moisture regime defined like aridic moisture regime but used in a different category of the U.S. soil taxonomy.

Truncated. Having lost all or part of the upper soil horizon or horizons.

Tuff. Volcanic ash usually more or less stratified and in various states of consolidation.

Udic. A soil moisture regime that is neither dry for as long as 90 cumulative days nor for as long as 60 consecutive days in the 90 days following the summer solstice at periods when soil temperature at 50 centimeters below the surface is above 5° C.

Ultimate particles. Individual soil particles after a standard dispersing treatment.

Ultisols. Mineral soils that have an argillic or kandic horizon with a basic cation saturation of less than 35 percent when measured at pH 8.2. Ultisols have a mean annual soil temperature of 8° C or higher. A soil order.

Umbric epipedon. A surface layer of mineral soil that has the same requirements as the mollic epipedon with respect to color, thickness, organic carbon content, consistence, structure, and phosphorus content, but that has a basic cation saturation less than 50 percent when measured at pH 7.

Ustic. A soil moisture regime that is intermediate between the aridic and udic regimes and common in temperate subhumid or semiarid regions, or in tropical and subtropical regions with a monsoon climate. A limited amount of water is available for plants but occurs at times when the soil temperature is optimum for plant growth.

Value, color. The relative lightness or intensity of color and approximately a function of the square root of the total amount of light. One of the three variables of color.

Vermiculite. A highly charged layer-structured silicate of the 2:1 type that is formed from mica.

Vertisols. Mineral soils that have 30 percent or more clay, deep wide cracks when dry, and either gilgai microrelief, intersecting slickensides, or wedge-shaped structural aggregates tilted at an angle from the horizon. A soil order.

Vesicular arbuscular. A common endomycorrhizal association produced by phycomycetous fungi of the family Endogonaceae. Host range includes most agricultural and horticultural crops.

Waterlogged. Saturated or nearly saturated with water.

Water potential. See Soil water potential.

Water table. The upper surface of groundwater or that level in the ground where the water is at atmospheric pressure.

Water table, perched. The water table of a saturated layer of soil that is separated from an underlying saturated layer by an unsaturated layer (vadose water).

Wilting point. See Permanent wilting point.

Xeric. A soil moisture regime common to Mediterranean climates having moist, cool winters and warm, dry summers. A limited amount of water is present but does not occur at optimum periods for plant growth. Irrigation or summer fallow is commonly necessary for crop production.

Xerophytes. Plants that grow in or on extremely dry soils.

Yield. The amount of a specified substance produced (e.g., grain, straw, total dry matter) per unit area.

Zero point of charge. The pH value of a solution in equilibrium with a particle whose net charge from all sources is zero.

Zero tillage. See No-tillage system.

INDEX

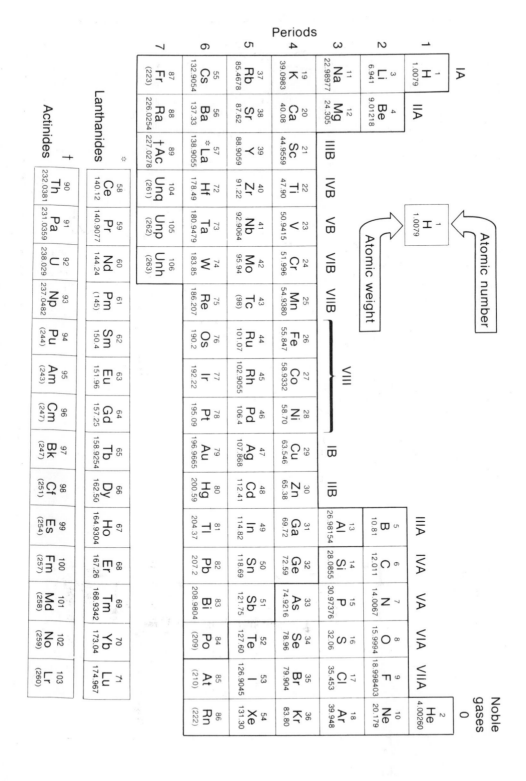